An Introduction to the
Physics and Chemistry
of Petroleum

JE PENSE LA VERITÉ

An Introduction to the Physics and Chemistry of Petroleum

ROBERT RICHARD FRANCIS KINGHORN

Department of Geology, Royal School of Mines,
Imperial College of Science and Technology

JOHN WILEY & SONS

Chichester · New York · Brisbane · Toronto · Singapore

Library of Congress Cataloging in Publication Data:

Kinghorn, Robert Richard Francis.
  An introduction to the physics and chemistry
of petroleum.

  Includes index.
  1. Petroleum—Analysis.   I Title.
TP691.K56 1983   553.2'8   82-13623
ISBN 0 471 90054 0

British Library Cataloguing in Publication Data:

Kinghorn, Robert Richard Francis
  An introduction to the physics and chemistry
  of petroleum.
  1. Petroleum
  I. Title
  553.2'82   TN863

  ISBN 0 471 90054 0

Typeset by Activity, Salisbury, Wilts
and printed by Page Bros., (Norwich) Ltd.,

Atque Inter Silvas Academi Quaerere Verum

HORACE 65–8 BC

Contents

Preface

This book is an attempt to put together in one volume the basic essentials of the chemical and physical behaviour and properties of petroleum, associated oilfield waters, and the rocks they inhabit. The book is primarily intended for recent graduates and students, both under- and post-graduate, whose major subject is geology and whose formal knowledge of chemistry, especially organic chemistry, and physics is limited. It is intended to be used in three ways. Its first objective is to teach to the geologist the language of organic geochemistry. Secondly, it should be used as a textbook to instruct the reader in the basic principles of oilfield fluid origin, behaviour, alteration, and application. Lastly, this volume should be a reference book where the reader may find the sources of the studies upon which this work is based. Besides being an aid to students who are being taught the fundamentals of petroleum, the book is intended to be of value to recent graduates who did not receive such instruction whilst students, but who need to know and understand the principles of the subject in order to perform satisfactorily within the petroleum industry.

The author considers that a full understanding of the history of petroleum, organic geochemistry, and its application cannot be obtained without a wider knowledge, which is not normally included in textbooks on the chemistry of petroleum. For this reason a significant part of the book is devoted not only to some of the background chemistry, but also to the thermodynamic properties of hydrocarbons, physical properties of the host rocks, and the nature of the associated waters. The author has used his experience gained during the teaching and research with which he has been concerned in the Department of Geology, The Imperial College of Science and Technology of the University of London. This book is not intended to be a research text, but is an introduction to the story of petroleum and the use of organic geochemistry in exploration. It is an attempt to cover in one volume the essentials of the physics and chemistry of subsurface fluids that are required for an adequate understanding of the formation and habitat of petroleum, together with the use of organic geochemistry in exploration. It is not intended for the petroleum geochemist but for the petroleum geologist who wants a deeper insight into the object of his quests, a knowledge of the great assistance organic geochemistry can be to him in that quest, and to be able to understand the language of petroleum geochemists. Petroleum geologists and geochemists are akin to the British and American peoples who have been described as two races separated by a common language.

I am grateful to all those authors whose work I have quoted. The amount of literature on petroleum, organic geochemistry, and related studies is growing at what is nearly an exponential rate, and it is impossible but to quote only a fraction of the published and relevant work. I have tried to quote a representative sample of the most important papers. The omission of reference to any particular work probably reflects more on this author than on those omitted. I would like to express my very special gratitude to all those who have assisted me in this work, especially Jenny Curtis for the excellent typing, Tony Brown, Haraldo Cantanhede and Ella Ng Chieng Hin for the drafting of the diagrams, Mokhles Rahman for the photographs of the organic matter, and for the many long and helpful discussions, Gill Davies; and last, but by no means least, my two best friends, one for the present of the pen with which this book was written and without whose encouragement and support it would probably be unfinished and the other, Caesar, for giving up many walks and sitting patiently at my feet whilst I was working.

Chapter 1

Introduction and Basic Concepts

INTRODUCTION

This book is primarily concerned with petroleum, liquid, gaseous, and solid concentrated accumulations of organic matter which can be extracted from the earth and used as sources of fuel and as raw materials. Hydrocarbon occurrences have been known for centuries and they have been used in various ways by the inhabitants local to these occurrences. At first there was no conscious effort made to discover the source of these hydrocarbons, which were normally present on the surface as seeps, to discover more seeps, or even to trade in these materials. Later reports indicate that some trade developed and that some production was from hand-dug pits and wells, but nothing was recorded as to the necessity or means of discovering new, especially underground, sources.

The relatively small world population had a need for oil products, mainly for illumination and lubrication, which could be met from animals and plants. It was realization that the distillation of mineral oils could provide products which were superior to those obtained from vegetable and animal sources that was the encouragement for prospectors to look for hitherto unknown sources of petroleum. The demand for hydrocarbon products has increased as the population of the world has increased and as the means of discovering and producing oil have improved. Oil has progressed from principally being a source of illumination to being a major contributor to the world's energy supplies and a source of raw materials for the chemical industry. Thus the changing and increasing uses of petroleum products, together with an ever-growing world population which expects continually increasing standards of living, means that the search for oil and gas is more urgent than ever.

Petroleum exploration is a very expensive business and it is a high-risk venture with very large amounts of money having to be expended with no guarantee of a return. Although a large number of preliminary tests and examinations can be made, the only way to ascertain whether or not there is oil in a particular region is to drill a well or wells. The preliminary work will serve to locate the prospect and minimize the risks as far as possible, but only the

1

drilling of wells will provide the answer as to whether or not oil is present and if so in what quantity and of what quality. Drilling on shore is expensive but many of the world's on-land prospects have already been evaluated and explored, so that the search is being extended to continental shelves. The cost of operating off shore, especially where the weather can be bad for long periods, is greater by several orders of magnitude than drilling on land. On-shore exploration is still taking place, but is confined either to locating small accumulations in already known petroleum provinces or to exploration in the more remote regions of the world. The former offers small returns and the latter requires very large capital investment, and thus every technique which can improve the chances of successful exploration is a worthwhile investment.

In the early days of petroleum exploration the preferred drilling location was on top of an anticline. Experience has shewn that anticlines are not the only geological formation in which petroleum can occur. Geologists nowadays have to have a much wider understanding of the geological environment which is conducive to the formation, migration, accumulation, preservation, and production of oil and gas. The methods which have been used by geologist and his associates to obtain this understanding have increased. For instance, geophysics is a widely used exploration tool, but its benefits are maximized when used in conjunction with other methods. Geochemistry is now becoming one of the very important exploration facilities because it enables petroleum geologists to assess the chances that hydrocarbons have been generated and to indicate of what type and in what quantity those hydrocarbons will be. The location of a porous and permeable reservoir is essential for the discovery of oil but for that reservoir to be of any value it must contain hydrocarbons. Organic geochemistry can indicate whether or not a suitable source of hydrocarbons exists and can thus enable the petroleum geologist to fit another piece in to the exploration puzzle. Geochemistry is an extra source of information which, together with all the others, allows the petroleum geologist to make a better and more valuable judgement of the petroleum prospects of an area. This can result in the saving of large sums of money. There will be no point in drilling into a reservoir sand if there is no source of organic matter with which to fill that reservoir. Organic geochemical investigations will be continued until the completion of the exploration and development. Initial analyses may involve surface samples from the edge of the basin, but later investigations involve material obtained from wells. Each well will provide more material which will allow a greater insight in the organic geochemistry of that basin.

Unfortunately, very few geologists have been taught the organic chemistry or the physical properties of the organic components of sedimentary rocks. Most geology courses include instruction in the chemistry and physics of the inorganic mineral constituents of rocks. Whilst those are of importance to petroleum geologists, petroleum is an organic substance, normally fluid, and it is a knowledge of organic chemistry and the physical properties of organic compounds which are of greater value to the petroleum geologist. Thus many geologists enter the petroleum industry to find their colleagues talking in a different language about subjects of which they are ignorant. In spite of this, many important recommendations and decisions have to be made based upon organic geochemical reports by geologists whose chemical knowledge is very limited.

'Petroleum' is a name derived from the Latin words *petra* (rock) and *oleum* (oil) and it is a general term used to describe mixtures of organic compounds, whether liquid, gaseous, or solid, which occur within the earth and which can be extracted. One of the principal components of petroleum is the hydrocarbon fraction which contains compounds composed solely of hydrogen and carbon. These compounds have great commercial value as they can be used as various types of fuel and as feedstock for the petrochemical industry. The word petroleum can be used to describe all three phases of extractable organic compounds found in the earth although the three phases can each have separate names. Gaseous petroleum is normally called natural gas; strictly speaking it should be called natural hydrocarbon gas as inorganic gases also occur in the ground. Natural gas is 'associated' if it occurs with liquid petroleum whilst 'non-associated' gas is that which does not overlie oil. Liquid petroleum, as extracted, is known as crude oil so as to distinguish it from the refined oil which is derived from crude oil. The semi-solid and solid forms of petroleum are called asphalts, tars, bitumens, pitches, or localized names such as Albertite or Gilsonite. The name petroleum is only applied to secondary organic matter, that is matter which has been produced by the thermal breakdown of kerogen. Thus oil shale would never be referred to as petroleum, but an oil or tar sand would be so described. The oil shale contains unmatured organic matter in an insoluble solid form and the action of heat is required to convert it to fluid products. An oil or tar sand contains material which has been produced by the thermal breakdown of organic matter similar to that found in an oil shale. It is relatively soluble in organic solvents and can be a fluid, even if a viscous fluid, when warmed.

In most cases these occurrences of petroleum are associated with aqueous solutions of inorganic salts. These brines are the fluids which filled the pore spaces when the sediments were deposited, and although much water is lost during compaction and at the same time the composition of the brine alters, the pores will still be filled with water. This water has to be displaced by the oil when the oil fills the reservoir. Because of the association of oil and water, a description of the character of oilfield brines is included in this book. In addition, the application of oilfield brine analysis, especially in exploration, is discussed.

Because very little organic chemistry has been taught in geology courses, the remainder of this chapter will be devoted to chemistry. There will be a description of the bonding in carbon compounds and this will be followed by an outline description of the types of compounds to be discussed in this book, the optical activity of organic compounds and the biosynthesis of organic carbon compounds.

The majority of this chemistry is organic chemistry, i.e. the chemistry of carbon, because carbon is unique among the earth's elements in that it can form compounds in which there are long chains of atoms. No other element can form chains of the length of those formed by carbon.

STRUCTURE OF THE ATOM

Matter is composed of molecules and molecules are composed of atoms, which are the basic building units of matter. Every atom consists of protons, neutrons, and electrons and the number of the protons and electrons determines the chemical properties of that atom. Protons and neutrons are of almost identical mass, whereas electrons are 1/1850 the mass of a proton. Protons have a unit positive charge and electrons carry a unit negative charge (Table 1.1).

Table 1.1

Particle	Mass (atomic mass units)*	Charge†
Proton	1.00732	+1
Electron	0.00055	−1
Neutron	1.00866	0

*1 a.m.u. is $^1/_{16}$ of the mass of an oxygen atom. The absolute weight of an electron is 9.1×10^{-28} g and that of a proton is 1.67×10^{-24} g.
†unit electron charge -1.60×10^{-19} coulomb

An atom is made up of a nucleus which contains protons and neutrons, in approximately equal numbers for the light elements, but with an excess of up to 50% of neutrons for the heavy elements, and the nucleus is surrounded by sufficient electrons to make the whole atom electrically neutral. The atomic number of an element is the number of protons or electrons in one atom of that element. All atoms of the same element have the same number of protons and electrons and they thus have the same chemical properties, as these properties are determined by the arrangement of the electrons. If two atoms have the same atomic number but different atomic weights it is because the number of neutrons in the nucleus varies. For instance, chlorine has atomic number 17 (i.e. it contains 17 protons and 17 electrons) but a chlorine atom normally contains either 18 or 20 neutrons. The natural occurrences of chlorine atoms with atomic weights 35 and 37 are such that the average atomic weight of chlorine is 35.46.

Experiments have shewn that atoms have diameters about 10^5 as great as their nuclei. Thus an atom has a compact, dense nucleus surrounded at some distance by electrons. These electrons are believed to be in motion around their nucleus and were once considered to circle their nucleus like planets around a sun. As they have the opposite electrical charge to the nucleus they should, according to classical physics, be attracted to the nucleus, but because they do not collapse into the nucleus some agency must prevent this. In 1913 Niels Bohr, a Danish physicist, suggested that the total energy of an electron is quantized, that is restricted to certain values, and thus an electron cannot have any energy but only particular energy levels. Thus there is a minimum energy level for electrons around atoms and this keeps an electron in the lowest orbit around an atom.

The electrons in an atom are not all of the same energy and the energy levels in an atom are discrete, limited, and only able to contain a specified number of electrons. The number of electrons a level or shell can hold depends upon the particular level. The maximum electron population of any energy level is $2n^2$, where n is the number of the level. (n has whole-number values: 1, 2, 3, etc., and electrons in the lowest energy level of $n = 1$ are referred to as being in the K shell or orbit. These electrons are the most tightly bound. The higher shells are lettered L, M, N, etc., corresponding to $n = 2,3,4$, etc.) Thus the first shell can contain only two electrons, the second shell can hold up to eight, and so on. Not all the electrons within a shell may have the same energy, as energy sub-levels or sub-shells within a main shell is from zero to n (where n is the principal quantum number). The lowest sub-shell within a main shell is designated s, the second sub-level is p, etc. An s sub-shell can contain two electrons and a p sub-shells six electrons. These are the only sub-levels that concern organic chemistry

From the above discussion it will be seen that in the K shell there can be a maximum of two electrons and these will be in the s sub-level, while in the L shell there can be up to eight electrons and these can be in the s and p sub-levels. These are known as the $2s$ and $2p$ electrons, respectively. Table 1.2 shews the electronic configuration of the common elements referred to within this book.

Table 1.2   Electronic configuration of common elements

Element		H	C	N	O	S	Na	Mg	Cl
K		1	2	2	2	2	2	2	2
L	s		2	2	2	2	2	2	2
	p		2	3	4	6	6	6	6
M	s					2	1	2	2
	p					4			5

The electrons are considered to be in pairs in orbitals within a sub-level. Thus, the two $1s$ electrons will be in one orbital and the two $2s$ electrons in another similar orbit but with higher energy. The six $2p$ electrons will be in the three orbitals all of the same energy but higher than that of the $2s$ orbitals. The Pauli exclusion principle states that only two electrons can be in one orbital and that these electrons will have opposite spins. When this occurs the electrons are said to be paired. Where electrons occupy equivalent orbitals singly, they will have parallel spins and such elements will have a resultant magnetic field. Such elements are weakly attracted to magnets and are called paramagnetic. This phenomenon should be compared with those solids, such as iron, which are strongly attracted to magnets and are

6

ferromagnetic, or some substances which are weakly repelled by magnets are diamagnetic.

The orbitals in which the electrons surrounding the nucleus of an atom occur have been described as shells or energy levels, and it would be very convenient if one could allocate to each electron a type of motion associated with the energy level of that electon. The Heisenberg uncertainty principle (1927) states that it is impossible simultaneously to know the position and momentum of an electron. Hence the position of an electron around a nucleus is best described as the probability or relative chance of finding that electron at a particular location within the atom. It is not within the scope of this book to discuss the calculations involved, but the likelihood of finding an electron does depend upon the shell and sub-shell the electron occupies. Figure 1.1 shews the probability of finding 1s, 2s, and 2p electrons at distances from the nucleus. Nowhere is the probability of finding an electron zero, but there is a distance at which the probability of finding an electron is the greatest.

Figure 1.1 shews only a segment of the space around a nucleus and for an s electron the maximum location probability is a sphere around the nucleus whose centre is that of the nucleus. The radius of the sphere will increase from 1s to 2s to 3s, etc. However, for p electrons the lines of maximum probability of locating an electron are dumb-bell shaped with three dumb-bells, mutually perpendicular one to another (Figure 1.2). These volumes where there is a reasonable chance to find an electron are known as orbitals.

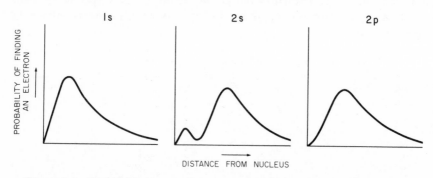

Figure 1.1 Probability curves for the location of electrons around nuclei

An electron in a p orbital has an equal probability of being found in either half of any one of the orbitals. Thus in the L level, which can contain eight electrons, two electrons are in the 2s orbital and two electrons in each of the three dumb-bell shaped orbitals which are mutually perpendicular one to another. The orbitals of atoms are often called atomic orbitals.

BONDING AND VALENCY

Bonding involves the sharing and interchange of electrons in the outer shells of reacting atoms. The inert gases, helium, neon, argon, krypton, xenon, and radon, have outer shells which are completely full of electrons and these full

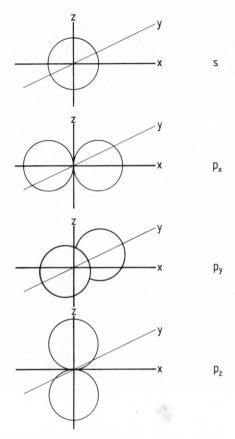

Figure 1.2 Configuration of *s* and *p* orbitals

outer shells are believed to account for their chemical stability. There is no vacancy for an incoming electron and it would be energetically unfavourable to use higher shells, so these gases are chemically unreactive. These elements, irrespective of their size, are monoatomic gases and never form diatomic molecules such as the other common elemental gases oxygen, nitrogen, and hydrogen.

Bonding is believed to occur by the sharing or transference of electrons within the outer shells of combining atoms so that the outer shells of those atoms are, at least for part of the time, filled with the maximum number of electrons. The simplest example is the hydrogen molecule which is composed of two hydrogen atoms. Each hydrogen atom contains one electron in its outer shell which, being an *S* shell, can contain a maximum of two electrons. Two hydrogen atoms form a hydrogen molecule by sharing the two available electrons so that for part of the time the *S* shell in each hydrogen atom has two electrons. This system of filling the outer shells allows two primary types of bonding to take place. One is the ionic bond which involves the transfer of electrons, and the other is the covalent bond in which electrons are shared.

Ionic bonding is achieved by the transfer of electrons from the outer shell of one atom to the outer shell of another. An example of this is the salt, sodium chloride. Table 1.2 shews us that sodium (Na) contains one electron in its outer shell (M) and eight electrons in its penultimate (L) shell, whereas chlorine (Cl) has seven electrons in its outermost (M) shell. The transfer of an electron from the M shell of the sodium ($3s$) electron to the outer shell of the chlorine will allow each atom to obtain the ideal arrangement of eight electrons in the outer shell. The outermost shell of sodium in the ionized state will be the L shell, and not the M shell of neutral sodium.

Ionic bonding requires that the bonded species are charged, thus in the above example the sodium, which loses an electron, becomes positively charged, whilst the chlorine, which gains an electron, becomes negatively charged. The overall electrical neutrality of the system remains despite the movement of electrons. Ionic compounds do not exist as discrete aggregates of atoms (molecules) in the way that covalently bonded molecules do. A molecule of hydrogen always contains two hydrogen atoms (H_2) but one cannot define a molecule of sodium chloride. The formula NaCl refers to the relative number of sodium and chlorine molecules required to obtain electrical neutrality. Many ionic compounds are crystals with each positive ion or cation surrounded by a number of negative ions or anions within the crystal lattice. Ionic compounds when dissolved in water dissociate completely.

Covalent bonding involves the sharing of electrons between the bonding atoms so that each atom has a full outer electron shell. As an example, let us consider carbon (C) because it is the compounds of carbon which are of primary interest to us. As shewn in Table 1.2, carbon has only four electrons in its outer (L) shell and thus needs four more electrons to complete the octet. In the compound methane, four hydrogen atoms each contribute one electron to the outer (L) shell of the carbon atom and thus the carbon atom has a share of eight electrons. The hydrogen atoms also 'borrow' one electron each from the carbon atom and thus have a share in two electrons, the maximum that the outer shell of hydrogen can hold. Hence each bond is composed of two electrons, one from each component being bonded, antiparallel in spin and spending part of their time with each atom. Similar covalent bonds can be made between carbon atoms to form the long chains which are a feature of organic compounds. In this case, the carbon atom in the middle of a long chain will share two of its electrons with two other carbon atoms and two electrons with two hydrogen atoms; and at the same time it shares one electron from each of these atoms with which it is bonding.

This type of bonding may be also explained in terms of combinations of the atomic orbitals described earlier. These atomic orbitals denote the probability of finding an electron in the vicinity of its nucleus. The shape of these orbitals is related to their energy and Figure 1.2 shewed the shapes of the s and p orbitals. The description of the bonding of carbon atoms using atomic orbitals also relates the shapes of carbon compounds to their bonding. When the atomic orbitals of individual atoms combine in the formation of molecules, the new combined orbitals are called molecular orbitals. An electron is not tied to one or other atom but can move within the fields of more than one nucleus.

It has already been observed that a carbon atom has in its outer (L) shell two $2s$ electrons, two $2p$ electrons, and vacancies in its $2p$ sub-shells for four more electrons. In saturated carbon compounds all the four valencies are equivalent irrespective of the fact that they involve both s and p electrons. This is due to hybridization of the L shell orbitals to form new orbitals of equal energy. The L orbitals of a carbon normally involve a $2s$ spherical and three $2p$ dumb-bell shaped, mutually perpendicular orbitals. There are two electrons in the lower energy $2s$ sub-shell and two electrons in one of the three $2p$ orbitals, which are of higher energy. These orbitals are hybridized in three different ways to form single, double, and triple bonds.

In tetragonal hybridization all the orbitals are hybridized to form four new orbitals of equal energy which are directed to the corners of a regular tetrahedron. Bonding occurs because the orbitals of another atom overlap with one of those of the carbon. Because the atomic orbitals overlap, or form a new molecular orbital, the electrons of each constituent are in the orbitals of both atoms. Molecular orbitals form the strongest bonds and the bond energy is greatest when there is the maximum overlap of atomic orbitals. In methane, each of the hybrid $2sp^3$ orbitals overlaps with the spherical $1s$ orbital of a hydrogen atom. In a carbon–carbon bond one $2sp^3$ orbital of each carbon must overlap and the bond will be strongest if the atomic orbitals and the two nuclei are in line (Figure 1.3).

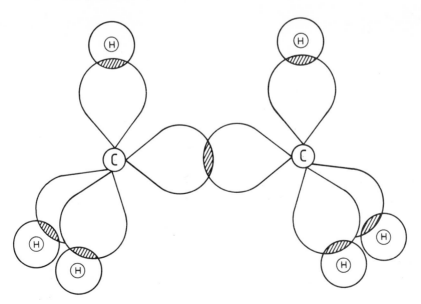

Figure 1.3 Molecular orbital diagram of ethane

These bonds involving *sp* hybridized orbitals are known as sigma σ bonds and the groups at either end of a sigma bond have the potential to rotate. This cannot occur in cyclic compounds, where the atoms are joined in a ring, but only in acyclic compounds. This is important where elements attached to the carbon atoms are not identical and the molecule will be more stable if the larger elements are as far away from each other as possible. Figure 1.4 shews how in dichloroethane the carbon–carbon bond can rotate and the molecule is more stable when the two chlorine atoms are diagonally opposite each other.

Dichloroethane

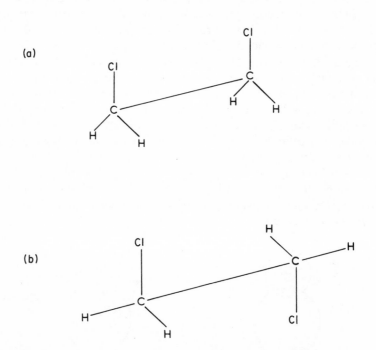

Figure 1.4 Rotation around a carbon–carbon single bond (dichloroethane)

In trigonal hybridization only the $2s$, $2p_x$ and $2p_y$ orbitals are hybridized, resulting in three equivalent orbitals at 120° to each other in the xy plane. The remaining $(2p_z)$ orbital stays undisturbed in the z plane. The three $2sp^2$ orbitals can form sigma bonds with other atoms such as carbon and hydrogen. The electrons in the $2p_z$ orbital form bonds by the overlapping of the $2p_z$ atomic orbitals of two adjacent carbon atoms. Ethylene has a double carbon–carbon bond formed of one sigma and one pi bond. The two carbon atoms and the four hydrogen atoms are in the xy plane and joined by sigma bonds. Above and below the xy plane the $2p_z$ orbitals of the carbon atoms overlap to form a π bond. The electrons in the π bond are known as π electrons.

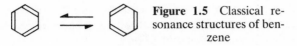

Figure 1.5 Classical resonance structures of benzene

When a π bond is involved between two carbon atoms, the atoms cannot rotate as when a sigma bond alone is involved.

The classical explanation of the structure of benzene (C_6H_6) has been a resonance hybrid between two forms (Figure 1.5). However, it has been shown that all the bonds are chemically equivalent and of the same length. The molecular orbital description of benzene satisfactorily explains this. The classical picture of benzene is akin to three ethylene molecules joined together in a ring by single bonds between adjacent ethylenes and the elimination of six hydrogen atoms, two from the position of each new bond, to give the alternate double and single carbon–carbon bonds. Now consider the molecule described in terms of molecular orbitals. The six sigma carbon–carbon single bonds will originate from two of the $2sp^2$ atomic orbitals on each carbon. (The third $2sp^2$ orbital forms a molecular orbital with the $1s$ orbital of a hydrogen atom.) All the carbon–carbon sigma bonds will be of the same length. The $2sp_z$ electrons are above and below the plane of the molecule and can overlap to form molecular orbitals with those π electrons of a neighbouring carbon atom, as occurs in ethylene. However, in the case of benzene the π orbitals on any carbon atom can overlap with the orbitals of the carbon atoms on both sides. Thus the π atomic orbitals form a molecular orbital which connects all the atoms of benzene. This gives benzene a cloud of electrons above and below it and accounts for benzene's chemical properties. All the electrons in benzene are said to be delocalized as they are able to travel anywhere around the molecule. Any system of alternate double and single bonds, such as that in B-carotene (See Figure 6.10), will have delocalized electrons.

In diagonal hybridization the $2s$ and $2p_x$ orbitals are hybridized to form two equivalent orbitals in the x plane, one either side of the carbon atom. These can form single sigma bonds with other atoms. Thus in acetylene (C_2H_2) all four atoms are in a line and joined by sigma bonds formed by the overlap of two sp orbitals from the carbon–carbon bond and one $2sp$ orbital and one $1s$ orbital for the carbon–hydrogen bond. The $2p_y$ and $2p_z$ orbitals are available to form π bonds at right angles to each other, one along the y axis and one along the z axis. Thus acetylene will have a cloud of electrons above and below as well as on each side of the carbon atoms.

These bonds are those most commonly found in organic chemistry. The π bonds are more reactive than the σ bonds and the majority of petroleum compounds involve σ bonds. When a compound has one or more π bonds it is known as an unsaturated compound. Such a compound has multiple carbon–carbon bonds and not as much hydrogen as a saturated compound with the same number of carbon atoms. Such unsaturated compounds can be reacted with hydrogen to form saturated compounds. This will involve a rehybridization of orbitals from $2sp$ and $2p_x$ and $2p_y$ or from $2sp^2$ and $2p_z$ to $2sp^3$. When the electrons are delocalized, as in benzene, the π bond is much less reactive than a normal isolated π bond.

TYPES OF CHEMICAL COMPOUNDS

Because the majority of this book concerns organic compounds, albeit in a geological context, the majority of the rest of this chapter will be spent describing the commonest organic compound types. When organic compounds are drawn there are several ways of illustrating them; however, because one is attempting to depict a three-dimensional object in two dimensions, there will always be some distortion. Firstly, all the elements can be shewn in approximately the correct places with single lines for single bonds, double lines for double bonds, etc. The next most common method of drawing organic compounds, especially cyclic compounds, is to draw only the bonds and any heteroatoms (atoms other than carbon or hydrogen). The bonds are depicted by solid lines and where two bonds meet there is assumed to be a —CH_2— group. Where three bonds (lines) meet there is a CH group and if four lines or bonds meet there will be a carbon atom. A bond sticking out from a molecule is assumed to have a methyl group (CH_3) at its end. Double bonds are represented by two parallel lines. Benzene is often represented as a hexagon with alternate double and single bonds although it is known that all the bonds are identical. The saturated cyclic compounds are sometimes drawn in perspective.

Open chain hydrocarbons

Aliphatic compounds are acyclic and composed of chains of carbon atoms with hydrogen atoms occupying the remaining valencies. The word 'aliphatic' derives from the fact that the first compounds of this class were obtained from fatty acids and *aliphos* is the Greek for 'fat'. The noun corresponding to aliphatic is alkane, i.e. an alkane is an acyclic aliphatic hydrocarbon. An alternative name for these compounds, based upon their relative unreactivity, is paraffins (Latin *parum affinis* = 'little affinity'). Alkanes have a general formula C_nH_{2n+2} and the simplest member of the series is methane. All the bonds involve sp^3 hybridization and are sigma bonds. At room temperature and pressure alkanes are gaseous up to C_4 (i.e. those with up to four carbon atoms), liquid between C_5 and C_{15}, and larger homologues are solids. The largest normal (i.e. all the carbon atoms in one chain) alkane so far prepared is heptcontane $C_{70}H_{142}$ (melting point 105° C). Above C_4 the carbon atoms can be arranged in several ways to give isomers. The boiling and melting point of the isomers decrease with the increase in the amount of branching, although the chemical formula remains the same. The largest paraffin that has been synthesized is $C_{94}H_{190}$ with a molecular weight of 1318 and it comprises a chain of 76 carbon atoms to which are attached 18 methyl groups (CH_3).

Because the four valencies of carbon point to the corners of a regular tetrahedron the so-called 'straight chain' hydrocarbons in fact have a saw-toothed shape and may be folded up, hence the preference to use the prefix 'normal' to describe these compounds.

Some hydrocarbons contain carbon atoms with sp^2 hybridization and thus a σ and one or two π bonds join those carbon atoms. These compounds are unsaturated. Compounds with a double bond are olefins or alkylenes, whereas those with a triple bond are alkynes. Olefins are found in crude oils and have the general formula C_nH_{2n}. They are more reactive than the saturated paraffins.

Cyclic hydrocarbons

Hydrocarbons may also exist in the form of rings of carbon atoms. Although these alicyclic (*ali*phatic *cyclic*) compounds or naphthenes have ring sizes from C_3 upwards, the commonest compounds in crude oil are those with five and six carbon atoms. These saturated compounds have a general formula C_nH_{2n}. Some of the larger alicyclic compounds are naturally odoriferous and are used in the perfumery trade. Alicyclic compounds can have more than one ring, with several of the rings sharing carbon atoms. Many of the compounds of this type which occur in petroleum are derived natural products and are based upon isoprene molecule $CH_2=C(CH_3)CH=CH_2$. The terpenes are compounds with formula $(C_5H_8)_n$. The $C_{10}H_{16}$ compounds are monoterpenes and the higher members have special names, e.g. the sesquiterpenes ($C_{15}H_{24}$), the diterpenes ($C_{20}H_{32}$), etc. The steroids are also naturally occurring compounds with four rings, three six-membered and one five-membered (Figure 1.6). Although these compounds are formed from the unsaturated isoprene they can be saturated or unsaturated.

In addition, aliphatic side chains may be attached to the rings in place of hydrogen. Alicyclic compounds may have some double bonds and be unsaturated.

Aromatic hydrocarbons

Aromatic hydrocarbons are based upon the benzene nucleus. As discussed in the section on bonding, these compounds have a different chemistry to saturated ring compounds. The term aromatic is derived from the Greek (*aroma* = 'fragrant smell'). Derivatives of benzene can be of two types; either with aliphatic side chains on the aromatic ring or several aromatic rings fused together. Because benzene rings must be planar, only two carbon atoms are common to each ring. The longer aliphatic side chains can be attached to the aromatic nucleus at more than one place to give compounds with a mixture of aromatic and alicycle rings (Figure 3.7).

14

Steroid nucleus

Sesquiterpene nuclei

Farnesane

Eudesmane

Guriane

Dotted lines indicate isoprene units

Figure 1.6 Natural product compounds constructed from isoprene units

Organic acids (fatty acids)

Organic acids have the general formula RCOOH where, with the exception of formic acid, R is an alkyl (C_nH_{2n+1}) or alkylene (C_nH_{2n-1}). Formic acid has the formula HCOOH and is the simplest member of the series. The name 'fatty acid' arises from the fact that two of the higher members of the

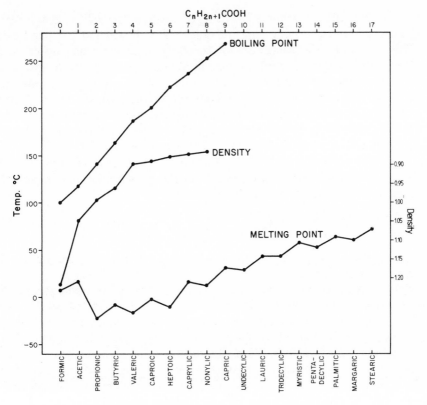

Figure 1.7 Physical properties of fatty acids

series, palmitic and stearic, which were among the first to be isolated, were prepared from animal fats. Organic acids are weakly acidic and formic acid is the strongest of them. Figure 1.7 shews the melting points, boiling points, and densities of some of these monocarboxylic acids. The first three normal fatty acids are pungent smelling liquids, the acids from C_4 to C_9 are oily liquids which smell like goat's butter, and the higher acids are odourless, colourless solids. It is interesting to note the 'saw-tooth' nature of the melting point curve (Figure 1.7) and to compare it with that for the alkanes (see Figure 3.1). Although these acids have systematic names, most of them are known by their trivial names which are generally derived from the Latin name of a source of that acid. Formic acid was obtained by the distillation of ants (*formica* = 'ant') and acetic acid is the principal component of vinegar (*acetum* = 'vinegar'). The systematic names are formed by adding the ending '-oic' to the name of the alkane corresponding to the longest carbon chain containing the carboxyl group, e.g. HCOOH = methanoic acid; CH_3COOH = ethanoic acid.

Formic, acetic, propionic acid, and *n*-butyric acids are miscible with water in all proportions, but for higher acids the solubility decreases rapidly with increasing molecular weight. Iso-butyric acid $((CH_3)_2CHCOOH)$ has solubility of 20 g per 100 g of water at 20° C; 4 g of *n*-valeric acid will dissolve in 100 g of water; and the next three acids will only dissolve in water at less than 1 g per 100 g of water. The remainder of the fatty acids are water-insoluble in molecular solution but significant quantities may dissolve by the formation of colloidal or micellar solutions. (See Chapter 8, p.301). The acidic group in a fatty acid is the hydrophilic group and is water soluble and whilst the acid group is the largest group in the molecule it will determine the molecule's water solubility; but when the alkyl group becomes larger than the acid group the acids behave like alkanes, which they then closely resemble.

Reactions of fatty acids include decarboxylation to give alkanes (see (A) below), a reaction which often occurs in the presence of alkalis; and the reaction with alcohols to form esters (B).

$$RCOOH \rightarrow RH + CO_2 \tag{A}$$
$$RCOOH + R'OH \rightarrow RCOOR' + H_2O \tag{B}$$

The formation of the ester ($RCOOR'$) is called esterification and the reverse reaction is saponification.

Some monocarboxylic acids have an unsaturated hydrocarbon group. The three most interesting of such acids are all derivatives of stearic acid ($C_{17}H_{35}COOH$) and commonly occur, together with palmitic and stearic acids, in fats and oils (see Table 1.3). They are oleic acid ($C_{17}H_{33}COOH$) which has a double bond in the 9–10 position (the carbon atom of the carboxyl group is numbered 1); linoleic acid ($C_{17}H_{31}COOH$) with double bonds in the 9–10 and 12–13 positions; and linolenic acid ($C_{17}H_{29}COOH$) with double bonds in the 9–10, 12–13, and 15–16 positions. Oils containing the latter two acids are known as drying oils because on exposure to air they change into hard solids, e.g., linseed oil. The 'drying' process is believed to involve oxidation and polymerization centred on the mutiple bonds. This process is believed to be similar to some of the reactions involved in the formation of kerogen. Dicarboxylic acids, both saturated and unsaturated, are also known and these have two acidic groups.

Alcohols

Alcohols have the general formula ROH where R is an alkyl group and OH is the alcoholic or hydric group. The general formula for alcohols is $C_nH_{2n+1}OH$ and the simplest member of the series is methyl alcohol or methanol (CH_3OH). The second member of the series, C_2H_5OH, ethanol or ethyl alcohol, is probably the best known. Alcohols can be primary, secondary, or tertiary and are RCH_2OH, R_2CHOH, and R_3COH respectively, where the alkyl groups do not have to be identical. The principal reactions are esterification (see above) and dehydration to give olefins.

When functional groups are added to alkyl chains the number of isomers increases greatly. Consider the C_4 and C_5 alkanes and alcohols (Figure 1.8). Butane can have only two isomers but there are four isomeric butyl alcohols,

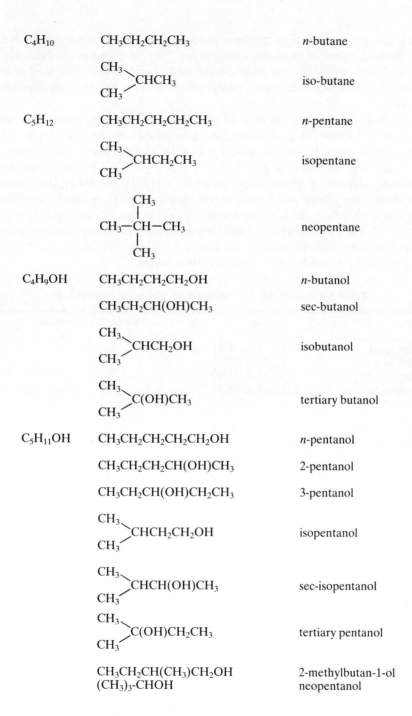

C_4H_{10}	$CH_3CH_2CH_2CH_3$	n-butane
	CH_3 $>CHCH_3$ CH_3	iso-butane
C_5H_{12}	$CH_3CH_2CH_2CH_2CH_3$	n-pentane
	CH_3 $>CHCH_2CH_3$ CH_3	isopentane
	CH_3 | $CH_3-CH-CH_3$ | CH_3	neopentane
C_4H_9OH	$CH_3CH_2CH_2CH_2OH$	n-butanol
	$CH_3CH_2CH(OH)CH_3$	sec-butanol
	CH_3 $>CHCH_2OH$ CH_3	isobutanol
	CH_3 $>C(OH)CH_3$ CH_3	tertiary butanol
$C_5H_{11}OH$	$CH_3CH_2CH_2CH_2CH_2OH$	n-pentanol
	$CH_3CH_2CH_2CH(OH)CH_3$	2-pentanol
	$CH_3CH_2CH(OH)CH_2CH_3$	3-pentanol
	CH_3 $>CHCH_2CH_2OH$ CH_3	isopentanol
	CH_3 $>CHCH(OH)CH_3$ CH_3	sec-isopentanol
	CH_3 $>C(OH)CH_2CH_3$ CH_3	tertiary pentanol
	$CH_3CH_2CH(CH_3)CH_2OH$ $(CH_3)_3$-CHOH	2-methylbutan-1-ol neopentanol

Figure 1.8 Chain branching and structural isomerization in alkanes

while the three isomers of pentane produce eight isomeric pentyl alcohols. Acetic acid esters of alcohols, especially of butyl and pentyl alcohols, have strong fruity smells and the twelve butyl and pentyl acetates are used in the artificial flavouring industry (tertiary pentyl acetate has the smell of pear drops).

Alcohols can exist with more than one alcoholic (or hydric) group. Dihydric alcohols are known as glycols and have a general formula $C_nH_{2n}(OH)_2$. The commonest glycol is ethylene glycol (CH_2OHCH_2OH) and is well known as an anti-freeze additive (m.p. -11.5 °C, b.p. 197 °C). The only important trihydric alcohol is glycerol or glycerin, $CH_2OHCHOHCH_2OH$, which occurs in many animal and vegetable fats combined with a series of fatty acids. Typical acids which are combined with glycerol in fats as esters are shewn in Table 1.3. The esters of glycerol and fatty acids are known as glycerides. These saturated fatty acids are some of the fatty acids commonly found in sediments together with the acids with 14 and 20 carbon atoms.

Table 1.3 Acids which combine with glycerol in fats and oils

Name	Formula	Double bond position
Lauric acid	$C_{11}H_{23}COOH$	—
Myristic acid	$C_{13}H_{29}COOH$	—
Palmitic acid	$C_{15}H_{31}COOH$	—
Stearic acid	$C_{17}H_{35}COOH$	—
Oleic acid	$C_{17}H_{33}COOH$	9–10
Linoleic acid	$C_{17}H_{31}COOH$	9–10, 12–13
Linolenic acid	$C_{17}H_{31}COOH$	9–10, 12–13, 15–16

Table 1.4 Melting points of simple triglycerides

Triglyceride	m.p. (°C)
Tripalmitin	60
Tristearin	71
Triolein	17
Trilinolein	below 0

The glycerides are important constituents in animal fats and vegetable oils. The triglycerides which contain unsaturated acids are liquids at room temperature and are unsuitable as edible fats, and are known as oils (Table 1.4). However, by converting the unsaturated acids into saturated acids by hydrogenation, a process known as hardening, the oils may be turned into fats. Butter is unusual among natural fats, because it contains the glycerides of a variety of acids, one of which is butyric acid. It is the liberation of butyric acid from the glyceride which makes butter rancid. The butyric acid can be removed either by evaporation with steam or treatment with sodium carbonate solution.

Margarines are vegetable oils with some animal fat which are sufficiently hydrogenated to be solid so they can be used as edible fats. The softer margarines have greater amounts of unsaturated material present within them.

Mineral waxes have been shewn to be the high molecular weight alkanes and these mixtures were called waxes because of their similarity to animal waxes. Animal waxes are the fatty acid esters of alcohols other than glycerol (the alcohols are normally monohydric). It has already been seen how the high molecular weight alcohols and acids have similar physical properties to alkanes because they consist of long alkane chains, but with a small functional group at one end. It is not surprising that the esters of such long chain alcohols and acids should also be similar to alkanes.

Beeswax is myricyl palmitate ($C_{15}H_{31}COOC_{30}H_{61}$), while cholesteryl palmitate is a wax found in blood plasma. Other esters of chloresterol occur in wool wax and some of the waxes of the skin contain hydroxylated fatty acids (i.e. acids with an alcohol group elsewhere on the alkane chain) esterified with higher aliphatic alcohols. The biological significance of these waxes is uncertain but it is possible that the waxes of the skin help keep it pliable and waterproof.

Aldehydes and ketones

These compounds contain a carbonyl group as found in organic acids. Ketones have two alkyl groups attached to the carbonyl group whereas aldehydes have a single alkyl group and one hydrogen on the carbonyl group.

$$\begin{array}{cc} \underset{R}{\overset{R}{>}}C=O & \underset{R}{\overset{R}{>}}C=O \\ \text{Ketone} & \text{Aldehyde} \end{array}$$

Aldehydes and ketones are fairly reactive compounds and are often involved in polymerization and condensation reactions to give larger molecules. Aldehydes can react with alcohols to give acetals:

$$RCHO + 2R'OH \rightarrow RCH(OR')_2 \quad \text{(acetal)}$$

Carbohydrates

These are polyhydric alcohols with a carbonyl group and general formula $C_nH_{2n}O_n$. Aldohexoses are sugars with six carbon atoms of formula:

$$CH_2(OH)CH(OH)CH(OH)CH(OH)CH(OH)CHO$$

There are sixteen possible isomers with this formula. It is because they contain an aldehyde group that they are known as aldohexoses. These compounds are optically active (see below), existing in *D* and *L* forms. The *D* and *L* represent the configuration of the carbon atom next to the CH$_2$OH. (+) and (−) indicate the direction of rotation of plane-polarized light (Figure 1.9).

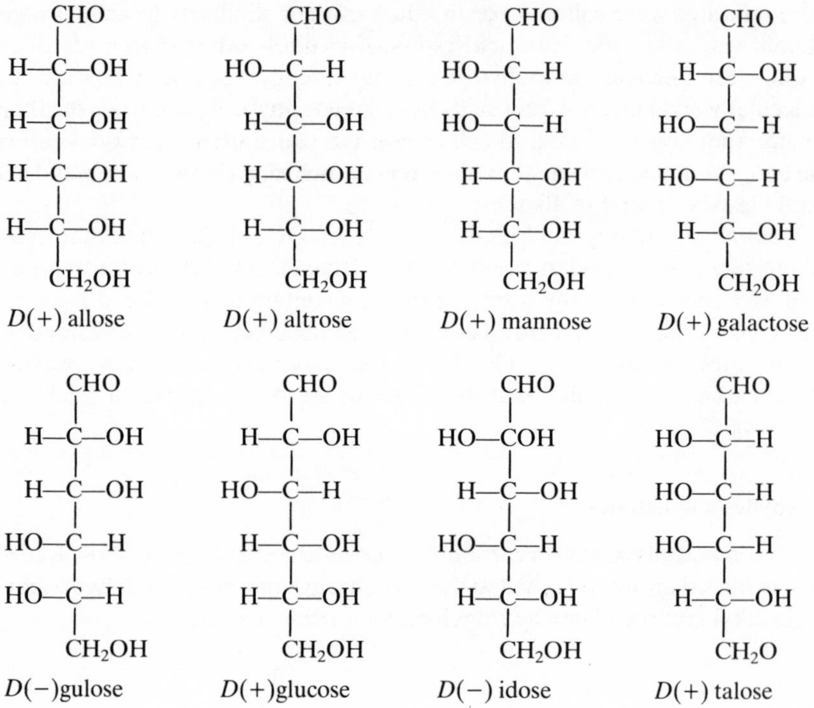

CHO	CHO	CHO	CHO
H—C—OH	HO—C—H	HO—C—H	H—C—OH
H—C—OH	H—C—OH	HO—C—H	HO—C—H
H—C—OH	H—C—OH	H—C—OH	HO—C—H
H—C—OH	H—C—OH	H—C—OH	H—C—OH
CH$_2$OH	CH$_2$OH	CH$_2$OH	CH$_2$OH
D(+) allose	*D*(+) altrose	*D*(+) mannose	*D*(+) galactose

CHO	CHO	CHO	CHO
H—C—OH	H—C—OH	HO—COH	HO—C—H
H—C—OH	HO—C—H	H—C—OH	HO—C—H
HO—C—H	H—C—OH	HO—C—H	HO—C—H
HO—C—H	H—C—OH	H—C—OH	H—C—OH
CH$_2$OH	CH$_2$OH	CH$_2$OH	CH$_2$O
D(−)gulose	*D*(+)glucose	*D*(−) idose	*D*(+) talose

Figure 1.9 Hexoses

Ketohexoses also have the formula C$_6$H$_{12}$O$_6$ but have a ketonic carbonyl group instead of an aldehydic group. The only important ketose is *D* fructose:

CH$_2$(OH)CH(OH)CH(OH)CH(OH)COCH$_2$OH

Pentoses are sugars with five carbon atoms of which *L*(+) arabinose, *D*(+) xylose, and *D*(+) ribose are the most important.

It has already been noted that alcohols and aldehydes react together. Because of the zig-zag nature of carbon chains the aldehyde group on a sugar can be close enough to an alcoholic group for a reaction to occur to give the cyclic form of the sugar (Figure 1.10).

When a sugar forms a six-membered ring (including the oxygen), it is known as a pyranose. Sugars in the five-membered ring form are known as furanose. Furanose forms are formed when the ketone group reacts with the alcohol group on the third carbon from it, instead of the fourth as in pyranoses. Thus the cyclic sugar in Figure 1.10 should be called α -*D*-glucopyranose.

Sugars commonly occur in combination rather than individually. Sucrose (or cane sugar) contains one molecule of glucose and one molecule of fructose,

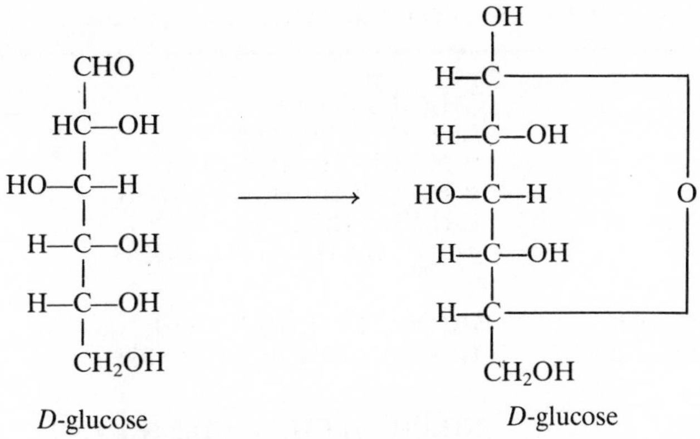

$$
\begin{array}{c}
\text{CHO} \\
| \\
\text{HC—OH} \\
| \\
\text{HO—C—H} \\
| \\
\text{H—C—OH} \\
| \\
\text{H—C—OH} \\
| \\
\text{CH}_2\text{OH}
\end{array}
\qquad\longrightarrow\qquad
\begin{array}{c}
\text{OH} \\
| \\
\text{H—C} \\
| \\
\text{H—C—OH} \\
| \\
\text{HO—C—H} \qquad\quad \text{O}\\
| \\
\text{H—C—OH} \\
| \\
\text{H—C} \\
| \\
\text{CH}_2\text{OH}
\end{array}
$$

D-glucose *D*-glucose

Figure 1.10 Formation of pyranose sugars

which can be obtained by the hydrolysis of sucrose:

$$C_{12}H_{22}O_{11} + H_2O \rightarrow C_6H_{12}O_6 + C_6H_{12}O_6$$

$D(+)$glucose $D(-)$fructose

Maltose, a product of the enzymatic hydrolysis of starch, contains two molecules of glucose whereas lactose, which occurs in milk, contains one molecule of glucose and one of galactose. Many plant components, e.g. starch and cellulose, are polymers of sugars, that is, they are made up of many sugar molecules. In addition many polymers are made from aminosugars where one of the hydroxyl (OH) groups is replaced by an amino ($-NH_2$) group. Chitin is a polymer of such aminosugars.

Amines and aminoacids

Amines are organic nitrogen compounds with a general formula R_3N where R can be either alkyl groups or hydrogen atoms. The common amines are the primary amines, RNH_2.

The aminoacids are compounds where the alkyl chain has both amino ($-NH_2$) and carboxyl ($-COOH$) groups. Table 1.5 shows some of the aminoacids derived from proteins. Some of these aminoacids are neutral (equal numbers of acid and amino groups), while others are acidic or basic. Aminoacids are the molecules from which the protein polymers are made. Many different aminoacids are incorporated into proteins, normally in identical repeating units. Proteins occur in the cells of animal bodies and in plants. A deficiency in some of these proteins will inhibit growth and may even cause death. With the exception of glycine, all aminoacids contain at least one asymmetric carbon atom and therefore exhibit optical activity (see below).

Other heteroatomic compounds

Besides the compounds above and the thiols (or mercaptans) (RSH) and diakyl

Table 1.5 Aminoacids derived from proteins

A	Neutral	
	glycine	$CH_2(NH_2)COOH$
	alanine	$CH_3CH(NH_2)COOH$
	valine	$(CH_3)_2CHCH(NH_2)COOH$
	leucine	$(CH_3)_2CHCH_2CH(NH_2)COOH$
	phenylalanine	$C_6H_5CH_2CH(NH_2)COOH$
	tyrosine	$p\text{-}HOC_6H_4CH_2CH(NH_2)COOH$
B	Acidic	
	aspartic acid	$HOOCCH_2CH(NH_2)COOH$
	glutamic acid	$HOOCCH_2CH_2CH(NH_2)COOH$
C	Basic	
	lysine	$NH_2CH_2CH_2CH_2CH_2CH(NH_2)COOH$
	ornithine	$NH_2CH_2CH_2CH_2CH(NH_2)COOH$

sulphides (R_2S), which are the sulphur analogues of the alcohols and the ethers respectively, the majority of heteroatomic compounds in oils are cyclic in nature. They are generally five- or six-membered rings with oxygen, nitrogen, or sulphur replacing a carbon atom. These compounds may be saturated or unsaturated. Figure 1.11 shews some of the more common heterocyclic compounds. Thiophen occurs in coal tar and shale oils, while acridine, pyrolle, and pyridine are also found in coal tar. Indole occurs in jasmine flowers, in orange blossom, and coal tar, and some indole derivatives are plant growth promoters. Many of these heterocyclic compounds are to be found in dyes.

OPTICAL ACTIVITY

Light is a transverse wave motion consisting of electric and magnetic vibrations occurring in all possible planes containing the ray of light and at right angles to the direction of motion. The electric and magnetic vibrations are mutually at right angles. If all the electric vibrations are in one plane, and thus all the magnetic vibrations are in the plane at right angles, the light is said to be plane polarized. When light is passed through a Nicol prism (made of iceland spar) it becomes polarized. Polarized light will only pass through a second Nicol prism if it is parallel to the first, or polarizing, prism.

Certain compounds have the ability to rotate the plane of polarized light and these compounds are said to be optically active. Compounds which rotate the light to the right (or sun-wise) are dextrorotatory. The opposite rotation is called laevoratation. These are indicated by $(+)$ and $(-)$ respectively before a compound. The prefixes D and L refer to stereochemical relationships, the way the groups in a molecule are arranged in relation to one another. The amount by which these compounds rotate the light may be measured in a polarimeter. Monochromatic light is passed through a Nicol prism (the polarizer), then through the sample, and finally through a second Nicol prism.

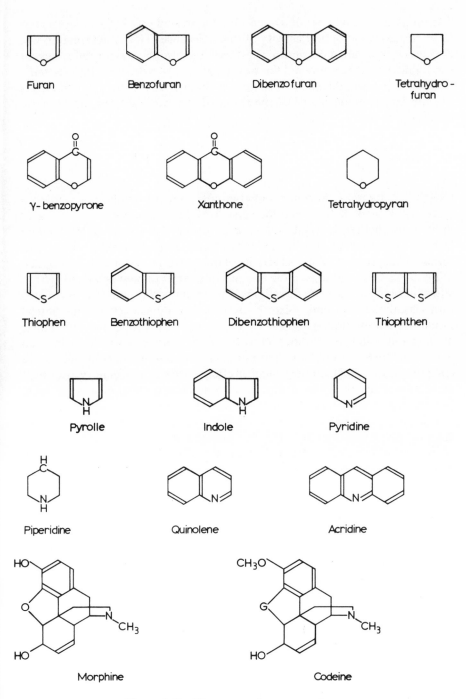

Furan   Benzofuran   Dibenzofuran   Tetrahydro - furan

γ- benzopyrone   Xanthone   Tetrahydropyran

Thiophen   Benzothiophen   Dibenzothiophen   Thiophthen

Pyrolle   Indole   Pyridine

Piperidine   Quinolene   Acridine

Morphine   Codeine

Figure 1.11 Heteroatomic compounds

The direction and number of degrees that the second prism (the analyser) has to be turned is the optical activity of that substance. The actual rotation will depend not only upon the optical substance, but also the temperature, solvent used, thickness traversed, and the wavelength of the light used. If the specific rotation is $[\alpha]$, the thickness of sample in decimetres is l, the temperature is $t°C$, the density of the liquid is d, the D-line of sodium is used, and the observed rotation is α, then

$$[\alpha]_D^t = \frac{\alpha_D^t}{l \times d}$$

(NB for solutions d is the number of grams of solute per millilitre of solution.)

In 1848 Pasteur noticed that the crystals of the sodium salt of tartaric acid differed in whether the crystals faces were arranged in a right- or left-handed fashion with respect to the crystal axis. Pasteur carefully picked out the two forms and measured the optical activity of the two forms. Although the original salt was inactive, solutions of the two sets of crystals rotated polarized light to equal amounts in opposite directions. The equimolecular mixture of the two forms is called the racemic mixture. Besides their optical activity, the optical isomers (enantiomorphs) also shew other variations in properties. Often their melting and boiling points are slightly different, a property which can be used to separate them. In addition, their rates of reaction with other optically active compounds vary and their physiological effect varies. Laevorotatory nicotine is more poisonous than the dextro form and D-laevorotatory ascorbic acid (vitamin C) is more efficient than the L-dextro compound.

This property is due to the molecules involved being asymmetric, i.e. they have no plane of symmetry. The two enantiomorphs of an optically active compound will be mirror images of each other. The four valencies of a carbon atom are directed to the four corners of a regular tetrahedron, and if four different groups are attached to these valencies they can be arranged in two ways. These two arrangements are the two enantiomorphs and are non-super-imposible mirror images of each other. Figure 1.12(a) illustrates this. The number of enantiomorphs is 2^n where n is the number of asymmetric carbon atoms in the molecule. Thus for the molecule shewn in Figure 1.12(a) there are two enantiomorphs. For a compound with two asymmetric carbon atoms there will be four enantiomorphs. If, however, the groups around each carbon atom are the same then two of these forms will be the same and they will be an internally compensated form. It is possible to obtain asymmetric molecules which do not contain an asymmetric carbon atom, e.g. hexahelicene. In this molecule one end ring is raised up to avoid the other and thus it is the first spiral of a helix. Thus the molecule has no plane of symmetry and it is optically active. Figure 1.12(b). Figure 1.12(c) shews the various forms of tartaric acid. Equimolecular mixtures of the laevo and dextro forms are optically inactive (the racemic mixture) and are externally compensated, while the mesotartaric acid is internally compensated. Many compounds produced by living

(a)

(b) Hexahelicene

(c) Tartaric acid

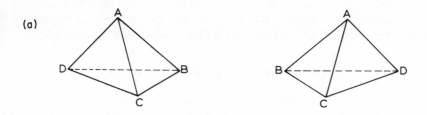

COOH
HO — C — H
H — C — OH
COOH

Laevotartaric acid

COOH
H — C — OH
HO — C — H
COOH

Dextrotartaric acid

COOH
H — C — OH
H — C — OH
COOH

Mesotartaric acid

RACEMIC ACID

Figure 1.12 Enantiomorphs (optical isomers)

organisms are optically active, but no compounds produced abiochemically are optically active. In biological reactions one enantiomorph is preferentially produced but in abiochemical reactions the racemic mixture is produced.

BIOSYNTHESIS AND THE CARBON CYCLE

Photosynthesis is the means whereby inorganic carbon is converted into organic carbon, and it is believed to have become a worldwide occurrence on a large scale some 3×10^9 years ago. Photosynthesis utilizes the energy of sunlight to convert carbon dioxide into organic matter. The primary organic products are sugars, with oxygen as in important by-product. The reaction can be summarized as follows:

$$6CO_2 + 6H_2O \xrightarrow{\text{677 kcal}} C_6H_{12}O_6 + 6O_2$$

The 677 kcal of energy are required to produce 1 g molecule of sugar.

This reaction occurs in both land and aquatic plants using the carbon dioxide which is available in the atmosphere or the sea respectively. Because atmospheric carbon dioxide is isotopically lighter than the CO_2 dissolved in the sea (i.e. it contains less of the isotope ^{13}C), terrestrial plants are isotopically lighter than marine plants. The reactions require the presence of chlorophyll which in the higher plants is concentrated in their leaves and gives them a green colour. Plants actually emit carbon dioxide as a product of respiration in addition to the oxygen expired as a result of photosynthesis. For every 35 g of carbon dioxide assimilated and converted to organic matter, about 10 g is produced by respiration and emitted. When the plants die, the carbon is trapped in the sediment where bacteria convert it to carbon dioxide which is released to the atmosphere. The average time that organic carbon stays in a sediment before returning to the atmosphere is 100–200 years. The plants use the sugars and other chemicals derived from carbon dioxide to build cell walls, etc. Some plants are eaten by animals who use oxygen to burn this organic carbon as a fuel. The waste product of this reaction is carbon dioxide:

$$C_6H_{12}O_6 + 6O_2 \rightarrow 6CO_2 + 6H_2O + \text{energy}$$

The carbon cycle can be split into land and sea subcycles with the atmosphere involved in both (Figure 1.13). The atmosphere holds about 7×10^{11} tonnes of organic carbon, carbon stored as dead organic matter in the soil. Of this total, about 3.5×10^{10} tonnes is the mobile part of the cycle with plants assimilating 3.5×10^{10} tonnes and respiring 1×10^{10} tonnes. The balance is carbon dioxide generated by bacterial action on the dead organic matter in the soil.

The sea cycle is more complex (Figure 1.13), but the same in essentials. The sea plants (phytoplankton) absorb carbon dioxide from the water and use the sun's energy to convert this carbon dioxide into organic compounds and oxygen (which is returned to the sea). Thus the majority of life in the oceans occurs in the top few metres where sunlight penetrates. This is the photic or euphotic zone. As on land, sea animals (zooplankton) eat the plants and use oxygen to

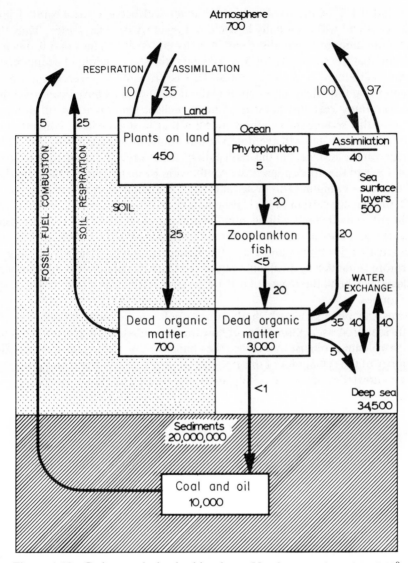

Figure 1.13 Carbon cycle in the biosphere. Numbers are in units of 10^9 tonnes of organic carbon (Reproduced with permission from Open University course T100, units 26–27, © The Open University Press)

convert the carbon in the plants to carbon dioxide and energy. Not all the organisms are eaten by the next predator in the food chain and those which die a 'natural death' can be consumed by bacteria or stored as dead organic material. The relative capacities of phytoplankton, zooplankton and fish, and dead organic matter are 5×10^9, 5×10^9, and 3×10^{12} tonnes respectively.

The carbon cycle has an annual turnover of about 7×10^{10} tonnes. The carbon cycle is not completely secure and small amounts leak out. Between

0.01 and 0.1% of the total annual turnover will become incorporated into sediments and will eventually become sedimentary organic matter. Thus the maximum amount being absorbed annually is 7×10^7 tonnes and if this has continued at the same rate for 3×10^9 years, the total amount of sedimentary organic carbon is 21×10^{16} tonnes. This is probably an overestimate as it assumes that the rate of production of the flora and fauna have been constant, which is almost certainly not true. A more realistic figure is about 10^{13} tonnes. Some of this sedimentary carbon (i.e. not that in soils) will be converted to coal, oil, and gas and will eventually return to the biosphere. Oil seeps will be oxidized at the surface, but the main path by which organic carbon returns from fossil fuels to the atmosphere is by combustion by man. This accounts for 5×10^9 tonnes per annum. This cycle takes several million years to complete.

Of the organic carbon which leaves the soils and becomes incorporated in the sediment, only 0.006% will become oil and gas and 0.17% will become coal. Thus the amounts of organic matter entering the sediment to become oil and coal are 42×10^3 tonnes and 1.2×10^5 tonnes per annum respectively. Man is therefore squandering a valuable resource. The total amounts of oil (plus gas) and coal will be in the region of 6×10^8 to 1.26×10^{13}, and 1.7×10^{10} to 3.6×10^{14} tonnes.

Thus the process by which the fossil fuels are generated seems a very inefficient process. A knowledge of how this conversion takes place will enable the petroleum geologist to increase the chance of discovering oil and gas. The majority of the remainder of this book attempts to explain these processes and how a knowledge of them can help an exploration petroleum geologist.

Chapter 2

Oilfield Brines

INTRODUCTION

Petroleum was first discovered, as oil and asphalt seepages, many centuries ago and the first production, from hand dug pits and wells, was very limited. Until the middle of the nineteenth century the small population of the world required only limited amounts of oil for illumination and lubrication. Such a small demand was easily met by the use of vegetable and animal products. This limited market postponed the introduction of modern techniques in the search for oil and gas. However, in 1855 it was discovered that the distillation of petroleum gave light oils which could be used as a superior alternative to the animal oils then in use. This encouraged the search for oil and the expansion of methods to discover oil accumulations.

Salt has always had a great commercial value, and at times a great political significance, and much effort was put into discovering sources of salt water which could be evaporated to obtain their dissolved salts. Peoples inhabiting coastal regions were at an advantage because of a steady supply from the sea, or if in cooler latitudes where evaporation is slow, by trading with sea-merchants. Thus the search for salt water was concentrated in the centre of the land masses. It was during efforts to increase salt water production that petroleum was encountered as an unwanted byproduct. Once it was realized that light fuel, illuminating oils, as well as the heavy lubricants could be obtained from these petroleums, the salt was used to find the oil and the salt water was the unwanted by-product.

The actual significance of the concurrence of oil and water was somewhat slow in being realized (Schiltuis, 1938). Munn (1920) advanced theories as to the role that moving underground water might have as a primary cause of migration and accumulation of oil. Petroleum engineers and geologists now accept that waters are associated with petroleum and that all sediments in the upper part of the earth's crust, with the possible exception of a thin layer near the surface, are saturated with water. These waters are believed to be the alteration products of the original sea water which was present when the sediments were deposited.

The use of analyses to determine the composition of oilfield waters has many applications. It has been shewn that the chemical composition of the waters in the different porous strata will, in most cases, vary significantly and knowledge of these compositions allows the differentiation of the strata. If petroleum is associated with the brine, the analysis of the brine is a quicker means of identifying the horizon whence it came. The various strata can be correlated

29

over reasonable differences by use of the detailed knowledge of the composition of their waters. If large volumes of water are produced with oil, the analysis of that water can indicate whence it came and action can be taken to reduce the volume. If the water is bottom water, the bottom of the well can be plugged. Edge water is more difficult to deal with except by the abandonment of the well. If water is coming from a higher horizon, perhaps through a leaking casing, this is readily recognized by chemical analysis or even by a concentration measurement. If the type of water, as well as the volume of water, has changed then the difference must be due to another source and this can often be identified by analysis. Once the source is known the appropriate corrective action can be taken.

The disposal of oilfield brines can also cause problems by interference with animal and plant life, both land and/or aquatic, as well as by interaction with host waters or sediments. If the produced water is simply dumped, the high salt concentrations may cause irreparable damage to the flora and fauna of the area. If, however, the water is pumped into a producing sand to aid production or into a deep aquifer purely for disposal, care has to be taken that there will not be an unwelcome interaction with the water already present in that formation or with the minerals that constitute that formation. The chemical nature of both host and introduced waters must be known so as not to cause reduction of the porosity of a sand by the precipitation of insoluble salts. An example of this would be the addition of a brine containing large amounts of calcium to a strong solution of carbonate which could, in certain circumstances, e.g. lack of dissolved carbon dioxide, cause the precipitation of calcium carbonate. There are other combinations of solution of radicals which will produce the precipitation of insoluble solids and reduce porosity and permeability. Because the careful injection of water into producing wells can increase the production of oil, the sensible use of unwanted oilfield waters can have economic advantages by the use of a potential pollutant.

Quantitative borehole measurements such as the electrical resistivity logs require a knowledge of the pore waters, which will be related to the concentration of the salts dissolved in the water. The value of the resistivity of the connate waters can be estimated from the self-potential log but this is done on the assumption that the brine is a sodium chloride solution of high concentration and hence the chloride ion is the resistivity-determining ion. In brines of low chloride salinity the other ions present, especially the calcium and the magnesium, will make material differences from the estimated chloride resistivity and the only way of ensuring that the correct resistivity of the formation water is known is by direct measurement.

The production engineer can also benefit from a knowledge of the salt content of the waters coproduced with the oil because the oil is more valuable without the salt. Much of the water coproduced with the oil is in the form of an oil-in-water emulsion which can be very difficult to break. The stability of such emulsions increases with time and thus the removal of the salt-containing water is easier and cheaper if it is done at the wellhead. Salt water is denser than oil and if it is not removed from the oil at the wellhead it will be occupying space in the transport used to take the oil to the refinery which could be used to transport more oil. If the salt does reach the refinery it will attack the refinery equipment and attract a lower price. Thus oil containing salt has a lower value, costs more to transport, and is more readily desalted when fresh. All these are sound reasons for wellhead removal of coproduced water. The methods used to remove the salt water will depend, in some degree, upon the composition of the brine.

Care has to be exercised when waters are mixed or changes in conditions, such as temperature or pressure, occur as these can cause the precipitation of solids, especially the alkaline earth metal sulphates and carbonates. The precipitation can occur in the most inconvenient places and can block both the pores of a reservoir rock and well fittings. Carbon dioxide will increase the solubility of calcium carbonate due to the formation of the more soluble bicarbonate:

$$CaCO_3 + CO_2 + H_2O \rightleftharpoons Ca(HCO_3)_2$$

Because this reaction is reversible, any adverse change in conditions which favours the liberation of carbon dioxide will drive the equilibrium to the left with the result that calcium carbonate precipitates. Most compounds have their solubility determined by temperature and normally the higher the temperature, the more soluble the compound. The above reaction shews that the reverse case is true for calcium carbonate, assuming that no external pressure is applied to the system in order to keep the carbon dioxide in solution. Thus a careful balance often has to be maintained, as the following example shews.

If the temperature of the water leaving a well drops from 88 to 77 °F, then 0.02 g of calcium sulphate per 100 g of water will precipitate. This is equivalent to 3.2 g/bbl (barrels; 1 barrel = 42 U.S. gallons) 0.167 lb per 1000 gallons. The addition of steam will help keep the brine warm and dilute it, both of which will help prevent the precipitation of calcium sulphate. However, the steam may cause the liberation of carbon dioxide from the brine, thus causing the precipitation of calcium carbonate. Both calcium carbonate and calcium sulphate are relatively insoluble compounds which can cause serious problems if they precipitate out of solution in awkward places.

Drilling engineers are also interested in water composition as brines can upset the physical properties of the drilling mud to such an extent that the hole can be in danger of collapse. The pressure gradient in the well will be affected because fresh water provides a considerably smaller pressure than does the equivalent column of a strong brine. In addition to this, concentrated brines are corrosive and can attack drilling equipment.

NAMING OF WATERS

Oilfield brines can be named either by their relative position to the associated oil accumulations or in an attempt at a generic classification.

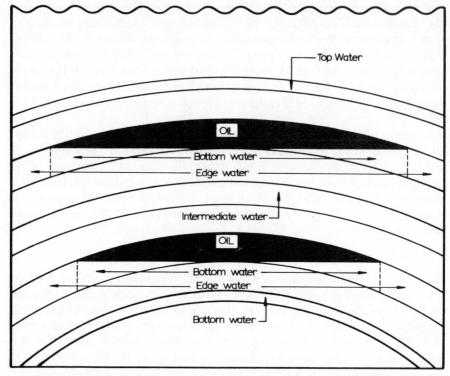

Figure 2.1 Naming of oilfield waters by position relative to oil

The first group of names includes 'bottom' waters, 'edge' waters, 'top' waters, and 'intermediate' waters. These are illustrated in Figure 2.1. The free water which is below the crude oil is known as the bottom water or the edge water. These two terms are derived from the position of the water relative to the oil as seen in cross-section and in plan respectively. As the oil and gas production declines and the reservoir is depleted, the production of free edge or bottom water will increase. If the water is bottom water the well may be plugged above the oil–water contact so that only oil is produced, but if the well is on the flank of a structure and edge water reaches the well there is little that can be done except abandon the well or use it for water injection. Top waters are the salt solutions which are present in porous strata, overlying and completely separate from the topmost oil-bearing horizon. Any brine found between two oil or gas accumulations is known as intermediate water.

The second classification is an approximate generic classification based upon the supposed history of the waters. 'Meteoric' waters are those which recently, in geological terms, have been in atmospheric circulation and they thus contain considerable amounts of oxygen and carbon dioxide. Such waters are derived from ones similar to those now in the seas, rain, snow, rivers, etc. These meteoric waters are characterized by relatively low concentrations of ions, often less than 10,000 mg/l, and they are carried into the ground where the oxygen they contain is able to react with sulphide through the rocks to produce sulphate ions, whilst the carbon dioxide is readily converted to carbonate and bicarbonate ions. The presence of these radicals in an oilfield water suggests that it is totally, or in part, meteoric water which has recently been in atmospheric circulation; or it may have incorporated when a now buried unconformity was exposed and atmospheric waters were allowed to enter sediment.

'Connate' waters are those in which the brines are believed to be original sea water in which the marine sediments were deposited and which once filled all of the pore spaces. However, it is very doubtful if such connate waters are the original sea water as most of these waters are quite different from modern sea waters and they are derived from the ancient sea water by a series of reactions. The more accepted term is 'interstitial' water which is the water in place at the time of drilling, and it really gives no clue as to the origin of the water. Most connate or interstitial waters are strong chloride brines and contain little sulphate, carbonate and bicarbonate. The main cation is often sodium and these brines are normally many times more concentrated than modern sea waters. Hence if ancient sea water was similar to modern sea water, significant changes to the brines must have occurred during their time in the sediment. Sometimes the interstitial water is classed as syngenetic (formed at the same time as the enclosing rocks) or as epigenetic (due to subsequent infiltration). Diagenetic water is a term used to denote those waters which are believed to have undergone significant chemical alteration before, during, or after sediment consolidation.

'Mixed' waters have multiple origin containing elements derived from the chloride-rich connate or interstitial waters as well as from the sulphate, carbonate, or bicarbonate bearing meteoric waters. The presence of sulphate in oilfield wates indicates that meteoric waters are involved.

Oilfield brines can also be described as 'formation' waters, i.e. those which occur naturally in rocks and present in them immediately before drilling. As has already been seen, oilfield waters can be classified according to origin or position and it is possible to unite some of these different classifications.

OILFIELD BRINE COMPOSITION

Typical compositions of rivers, seas, and oilfield brines are given in Table 2.1 and it is possible to see dramatic differences in composition, both absolute and relative, between the three. The concentration of oilfield waters can range from almost fresh to upwards of 300,000 mg/l of dissolved solids. The commonest ions are calcium, sodium, magnesium, chloride, sulphate, and bicarbonate. Of these all but the sodium are readily determined by classical wet chemical methods of analysis.

The cations are the positively charged ions within a solution and are mostly metals and ammonium.

The alkali metals are a series of reactive monovalent metals. Sodium is the most abundant element, not only of this group of metals, but also all metal ions found in oilfield brines and the other alkali metals in brines are present in much lower concentrations (Table 2.1).

Table 2.1 Comparison of composition of sea water, river water, and oilfield brine

	River water (mg/l)	Sea water (mg/l)	Oilfield brine (mg/L)	
			average	maximum
Li		0.8	6	500
Na	42	11,000	43,450	120,000
K	14	350	166	1,200
Rb		0.1	0.36	4
Ce		0.5	0.05	1
Ca	140	500	3,000	30,000
Mg	21	1,300	1,600	30,000
Sr		7	250	4,500
Ba		0.03	35	670
B		5	22	450
F	1	1.3	2	5
Cl	42	19,000	90,0000	270,000
Br		65	475	6,000
I		0.05	50	1,410
HCO_3			233	3,000
CO_3	245		50	300
SO_4	84		394	8,400
Organic acids			188	2,300
Ammonium			144	3,300
Silica	84		30	500
Fe	21	0.01	10	1,000
NO_3	7			

Sodium is abundant in the earth's crust, of which it accounts for 2.8% by weight, and the vast majority of sodium compounds are very soluble and not easily removed from solution once dissolved. The oceans and the evaporitic sediments are the major reservoirs of sodium on the surface of the planet. Igneous rocks are the next major source of sodium with very little in non-evaporitic sediments. Classical wet analytical methods for the quantitative determination of sodium are long and far from easy and thus the sodium content of a brine is often calculated from other analytical results, either by simple subtraction of the determined elements from the total dissolved solids or by the more accurate method of using reacting values (see later). More modern analytical techniques such as atomic absorption can give a direct measurement of the sodium content of a brine. Unfortunately, the method used to obtain values for the concentration of sodium in brines is rarely stated and thus the degree of reliance that can be placed upon such unqualified results is low.

Of the other alkali metals, the second most abundant in the earth's crust is potassium (2.55% by weight) however, potassium readily forms stable minerals, such as feldspar and mica, which are resistant to leaching by water; a fact which may account for the relatively low potassium concentrations to be found in natural waters, both above and below ground. Potassium is more readily retained in soils, clays, and sediments than is sodium. Thus sodium tends to be the major metallic constituent of sea water and oilfield brines.

The other alkali metals are lithium, rubidium, and caesium, all of which can be present in oilfield brines, but always in very small amounts. Quite often the alkali metal content of a brine is assumed to be its sodium content, either as calculated or occasionally measured.

The alkaline earth metals are the second most abundant group of metals which occur in oilfield brines. Calcium is the most predominant alkaline earth metal, but magnesium is a close second, unlike the situation in the alkali metal group. (Table 2.1). In fresh waters the alkaline earth metals are more common than the alkali metals, but in sea water the position is reversed. Magnesium salts are generally more soluble than the corresponding calcium salts and this situation is reflected in the relative magnitudes of the two ions in sea water. However, in oilfield brines calcium is the more common ion of the two. There are various explanations of this phenomenon, one of which is dolomitization. In this process calcium ions in the rock are replaced by magnesium ions in the pore water and, because the ionic volume of magnesium is less than that of calcium there will be an increase in the pore sizes of that rock with a consequent increase in porosity and permeability. Thus it is common to ascertain the concentrations of both magnesium and calcium in oilfield waters.

Magnesium comprises about 2.1% by weight of the earth's crust and is dissolved out of the crust materials by chemical weathering especially in the presence of waters containing carbon dioxide. Calcium, at 3.6% by weight, is even more abundant in the crust but less so overall the whole earth. The presence of carbon dioxide facilitates the solution of the calcium from the crustal rocks. The solubility of calcium is related to temperature and the amount of dissolved carbon dioxide. Calcium is soluble in the presence of dissolved carbon dioxide because the more soluble bicarbonate is formed:

$$CaCO_3 + H_2O + CO_3 \rightleftharpoons Ca(HCO_3)_2$$

Thus an increase in temperature will cause the liberation of carbon dioxide and the precipitation of calcium carbonate.

Of the other alkaline earth metals only barium and to a lesser extent strontium are important because of the sparingly soluble nature of some of their salts, notably the carbonates and sulphates. Mixing of waters containing sulphates or carbonates with those with barium or strontium will cause plugging of rock pores or well equipment by the precipitation of these insoluble salts.

Many other metals have been identified in oilfield waters but, in general, they occur in relatively low concentrations, often being the minor dissolution products of the rock cements, clays, and even casings and well equipment with which the brine has been in contact. Common metals in this category are iron and aluminium and to a lesser degree manganese, zinc, copper, lead, mercury, and boron. These are very rarely separately determined.

Ammonia is also observed in oilfield brines, but as it is a dissolved gas the quantity in any brine will depend upon the temperature and pressure of the brine. These are normally different at the sampling point to within the reservoir and thus accurate measurements of the concentration of ammonia in brines can be difficult.

Anions are the negatively charged species which together with cations form salts. The halogens are the most predominant anions in both true connate waters and sea water, with chlorine as the major representative of the halogens in these waters. The chlorides of all common metals are soluble, which accounts for the frequency of their occurrence in most waters. The other halogens in oilfield brines occur in much lower concentrations than chloride and are very rarely separately determined.

Fluorine occurs in several minerals including fluorspar (CaF_2). However, calcium fluoride is sparingly soluble and waters will only be able to contain significant fluoride ions if there is no calcium present. Bromine occurs in mixed salts by the replacing of chloride ions. Iodine is accumulated by marine plants and thus the concentration of iodine in subsurface waters may depend upon the presence of nearby organic rich argillaceous deposits. The content of iodine and bromine in oilfield waters is rarely greater than a few hundred milligrams per litre litre although in some Trinidadian waters the iodine content can be high and of significance (Parker & Southwell, 1929).

Sulphate is often present in oilfield brines, although it is considered that true connate waters should not contain sulphate and its presence is due to the invasion of surface waters where the dissolved oxygen has oxidized pyrite to sulphate. Alternatively, the brine may have dissolved sulphate-containing minerals from the host rocks. The very low solubility of some metal sulphates means that it is important to know whether sulphate is present in case mixing with other waters, containing these metals ions, is likely to occur.

Sulphur can also occur, especially above the oil zone, in the form of sulphide, and knowledge of its presence is important for corrosion and water-compatability studies. Much of the sulphide in oilfield brines is in the form of hydrogen sulphide which originated from the action of anaerobic bacteria. These bacteria obtain the oxygen they require for life from sulphate ions which are reduced to sulphide ions in the process. Because hydrogen sulphide is normally a gas and the amount dissolved in a brine will depend upon temperature and pressure, analysis is difficult and will be related to the concentration at the time of analysis and not necessarily to subsurface conditions.

Carbonates and bicarbonates are found in many oilfield waters and the concentrations of these two ions are related. The carbonates of the alkaline earth metals are sparingly soluble but in the presence of dissolved carbon dioxide the more soluble bicarbonates are formed:

$$H_2O + CO_2 + CaCO_3 \rightleftharpoons Ca(HCO_3)_2$$

The two ions, carbonate and bicarbonate, are each determined and the results of these analyses are reported. The actual figures of the relative concentration of carbonate and bicarbonate will depend upon the conditions of temperature and pressure prevailing at the time of analysis. These conditions probably will be different from those at which the brine existed in the subsurface and thus the relative amounts of carbonate and bicarbonate reported may be different from the proportions of those ions in the subsurface.

Silica is a very abundant component of sediments and it is therefore to be expected that silica will be present in oilfield waters. In such waters silica normally occurs as an un-ionized colloidal solution and because of this it is of no great importance and is very rarely determined.

Petroleum originates from the organic matter which is deposited in aqueous sediments and as these sediments compact the organic matter is converted via humic acids and kerogen into petroleum and pyrobitumen. At the same time the compaction causes the expulsion of water and it is not surprising that formation waters contain dissolved organic compounds in significant amounts.

The solubilities of petroleum hydrocarbons in water increase with temperature and decrease as the dissolved inorganic components increase. This is known as salting-out, where a more soluble component expels a less soluble one from solution. There are certain exceptions to this rule and the exceptions involve the aromatic hydrocarbons such as bezene, toluene, the xylenes, and napthalene, which increase in solubility in water as the concentration of silver ions increase. This is believed to be due to the formation of a simple 1 : 1 complex (Andrews & Keefer, 1949). This effect of silver is a minority effect and an increase in the concentration of the majority of metal salts will cause the removal from solution of increasing amounts of organic matter. The effectiveness of the different metals in keeping organic matter out of solution varies greatly (McDevit & Long, 1952).

Organic compounds containing nitrogen are often considerably more soluble in water than are pure hydrocarbons. Concentrations of aminoacids found in petroleum-related waters can be as high as 230 µg/l (Degens *et al.*, 1964). Typical aminoacids found in waters are serine, glycine, alanine, and arginine.

The fatty acids which contain a hydrophilic group, the acidic group, tend to be reasonably soluble in water by the formation of colloids or micells and their solubility increases with temperature. Fatty acids which have been identified in sea water include lauric, myristic, palmitic, stearic, hyristoleic, palmitoleic, oleic, linoleic, and linolenic (Slowey *et al.*, 1962). Straight-chain fatty acids have been isolated from petroleum-associated waters. The acids had between 14 and 30 carbon atoms with the even-numbered acids predominating (Cooper, 1962).

Naphthenic acids are also found in natural waters and oilfield brines. In general, such naphthenic acids are frequently found in waters associated with naphthenic oils and the more naphthenic the crude, the more naphthenic acids will be found in the associated waters. Many of the naphthenic and fatty acids could play an important part in the migration of the generated oil from the source to the reservoir. This will be discussed more fully in Chapter 8.

The concentrations of the organic compounds which are found in oilfield brines are used in petroleum will be discussed more fully later in this chapter.

In general the concentration of subsurface brines gradually increases with depth; however, this increase in concentration is greater in sands than in shales. This difference may be because the shales allow the passage of water, but not of the dissolved ions. There are exceptions to this general trend of increasing concentrations with depth. The two commonest cases are under unconformities, where meteoric water may have been trapped, and in zones of overpressure. An examination of subsurface brines across a basin shews that the salinity increases towards the centre. This is probably because those brines at the edges of the basin can be in contact with meteoric waters and because the deeper brines in the basin centre are more liable to alteration.

ANALYTICAL METHODS

The units used to describe the concentrations of ions in oilfield brines include milligrams per litre (mg/l), grams per gallon (US or imperial), parts per million (ppm), or percentage. The most commonly used are milligrams per litre and parts per million which are related by the specific gravity, thus:

ppm = (mg/l)/specific gravity

Sometimes the ions are described in terms of their reacting values which are obtained from the concentration of that ion (in mg/l) multiplied by the reciprocal of the equivalent weight of the ion or radical or ratio of the valency of the ion to its molecular weight, is called the reaction coefficient and its units are milliequivalents per milligram (me/mg). A solution of ions must be neutral in terms of electric charge and thus the sum of the reacting values of the anions and the cations must be the same. This allows for the calculation of the concentration of an ion which has not been determined analytically, as is often the case with sodium. An example is given in Table 2.2 (Reistle & Lane, 1928).

The specific gravity of a brine must be measured and it is defined as the ratio

Table 2.2

Ion	Concentration (mg/l)	Reaction coefficients	(me/mg)	[Reacting value] (me/l)		(%)	
Na	207	1/23	(0.0435)	9		31.09	
K	20	1/39	(0.0256)	0.5	} 14.5	1.7	} 50
Ca	100	2/40	(0.0499)	5		17.3	
SO₄	504	2/96	(0.0208)	10.5		36.0	
Cl	106.51	1/35.5	(0.0282)	3	} 14.5	10.6	} 50
CO₃	30	2/60	(0.0333)	1.5		3.4	

of the weight of a volume of the brine to the weight of an equal volume of water, normally measured at 60 °F. The specific gravity is required to convert milligrams per litre to parts per million and vice versa. The specific gravity will give indications of the amount of total dissolved solids in a sample as shewn in Table 2.3 (Reistle & Lane, 1928).

Table 2.3 Relationship between specific gravity and concentration of dissolved solids

Specific gravity	Total dissolved solids (ppm)
1.000	0
1.005	7,000
1.010	13,700
1.015	20,500
1.020	27,500
1.030	41,400
1.040	55,400
1.050	69,400
1.060	83,700
1.070	98,400
1.080	113,200
1.096	128,300
1.100	143,500
1.110	159,500
1.120	175,800
1.130	192,400
1.140	210,000

The total dissolved solids value can also be obtained by the evaporation to dryness of a weighed volume of the brine. (the volume taken is often 100 ml). The sample is heated gently until precipitation of the solids is just seen to be beginning. The sample is not allowed to boil so that it will not spit and thus eject small quantities of concentrated brine out of the flask. Once precipitation has started the flask is transferred to an oven at 105 °C where the evaporation to dryness is completed. The total dissolved solids, in ppm, can be calculated from the weights of solid residue and original sample. Care has to be taken when any of the potential products of the evaporation are hydroscopic. Thus brines which contain chloride and calcium can give solid calcium chloride which can absorb up to twice its own weight of water. Hence once such a dry precipitate is obtained it will be absorbing water from the atmosphere. Storage in a dessicator is required, but even so, dubious results can be obtained.

The resistivity of a brine may also need to be measured, especially for accurate interpretations of resistivity logs where the resistivity is affected by the type and concentration of the ions in the pore waters. Resistivity is measured using a resistivity cell which contains platinum electrodes and resistance-measuring equipment. Commercial equipment is readily available, using high-frequency currents to avoid polarization effects. Resistivity is the specific resistance of a material or the resistance offered by a cube of material at 0 °C, and is given by $\rho = RA/l$, where ρ is the resistivity, R is the resistance of a uniform conductor of length l and cross-sectional area A. The units of resistivity are ohm-metres, whilst conductivity is measured in reciprocal ohms (or mhos) per metre. Either resistivity or conductivity of pore waters can be measured.

Resistivity is greatly affected by temperature (Figure 2.2) and it often is not convenient to measure the resistivity of a brine either at 0 °C or at the temperature at which it exists in the formation, and thus corrections have to be made.

The acidity of oilfield brines is often measured by its pH value, which is the concentration of the hydrogen ions in the brine. The pH scale runs from 0 to 14 with low numbers being acidic, high numbers alkaline, whilst pure water and other neutral solutions have pH values of 7.

Commercially available pH meters are readily available which utilize a hydrogen-ion sensitive lithium glass electrode, with a full range from 0 to 14, and a mercury calomel rod immersed in saturated potassium chloride as the reference electrode. Because pH is temperature sensitive these instruments have built-in temperature compensation. Normally a polythene sheath, which contains both electrodes and a resistance thermometer, is placed into the solution whose pH is to be ascertained.

These pH meters need to be calibrated before measurements are made and this is achieved by the use of standard buffer solutions, normally of pH values, 4, 7, and 9.5. Wherever possible it is best to use two buffer solutions, one either side of the value of the sample to be measured. The electrodes should be kept in distilled water, washed with a little of the sample to be analysed before the test and washed with water after the test. The pH of a brine should be measured as soon as possible at the wellhead because exposure to the atmosphere can cause oxidation of some dissolved ions which would cause a change in the pH of the solution.

The pH of a solution will determine the solubility of many of the dissolved salts. Especially important is the solubility of calcium carbonate. Acids (low pH) are known to dissolve calcium carbonate with the liberation of carbon dioxide:

$$CaCO_3 + 2H^+ \rightarrow Ca^{++} + H_2O + CO_2$$

Conversely, calcium carbonate is precipitated from water if that water is strongly alkaline (high pH). Thus the pH of an oilfield brine will determine the solubility of the less soluble salts. Whether on the surface or in the subsurface, calcium carbonate is precipitated from alkaline solutions and in these circumstances the orifice through which the brine is passing will be reduced in size. The carrying capacity of surface piping can be greatly reduced by such precipitation. In the subsurface the result will be a reduction in the porosity and the permeability of the strata. Acid solutions will dissolve metal salts, especially those such as calcium carbonate, which will increase the porosity and permeability of the rocks. Solutions of extreme pH can be very corrosive and

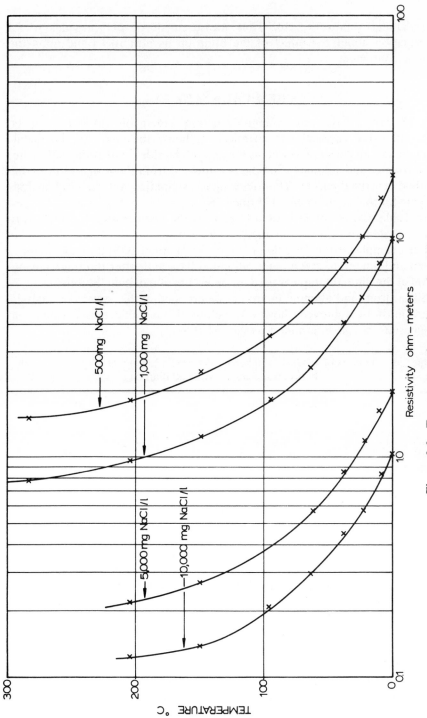

Figure 2.2 Temperature dependence of resistivity

can damage pipework, etc. Thus measurement of pH provides useful information about the effect of the brine on its host rocks and upon the drilling equipment.

CHEMICAL ANALYSES

The majority of the ions commonly determined in oilfield brines can be analysed using classical wet chemical methods. Included are the metals calcium and magnesium as well as the anions chloride, hydroxide, carbonate, bicarbonate, and sulphate. Sodium is estimated from these figures using the method described earlier. The following is a simplified wet chemical analysis scheme for the common oilfield brine ions.

In Table 2.4, factor D is used to convert the volume used for analysis to the equivalent of 1 litre of the original sample.

For example, if the specific gravity is 1.071, then 100 ml of the water is taken and diluted to 500 ml in a calibrated flask and, after thorough mixing, 50 ml of the diluted sample is used for determination of metallic ions.

Silica, iron, and aluminium are not determined in this system of analysis. They must be removed, however, before the metallic ions calcium and magnesium can be determined, as shewn below.

Table 2.4 Quantity of water to be taken for determination of metallic ions: relation between specific gravity and dilution

Sp.gr. at 15.6 °C.	Original sample Take	Dilute to	Sample for analysis Use	Factor D
1.000–1.003	500 ml	—	500 ml	2
1.004–1.015	100	—	100	10
1.016–1.030	50	—	50	20
1.031–1.070	100	500 ml	100	50
1.071–1.120	100	500	50	100
1.121–1.150	50	500	50	200
Above 1.150	50	1000	50	400

The appropriate amount of sample for analysis (see Table 2.4) is placed in a porcelain evaporating dish, made slightly acid with HCl, and evaporated to dryness; it is then baked in an oven at 105 °C for at least 6 hours to render the silica insoluble. To the contents of the dish 5 ml of concentrated HCl and 50 ml of distilled water are added, boiled for 15–30 seconds, transferred to a 9 cm filter paper, and washed thoroughly with hot water. The filter paper and its contents are rejected; the filtrate and washings are treated to remove iron and aluminium. To oxidize all the iron present a few drops of nitric acid are added and the solution boiled. It is evaporated if necessary, to a volume of about 100 ml, then 10 ml of a 10% solution of NH_4Cl are added, made slightly alkaline by adding dilute NH_4OH, and boiled for about 10 minutes. The precipitated iron and aluminium, if present, are removed by filtering through a 9 cm filter paper and washing with hot water. The precipitate is rejected.

Determination of calcium

The combined filtrate and washings from the iron and aluminium precipitation are concentrated, if necessary, to approximately 200 ml, made distinctly alkaline with ammonia, and heated to boiling. A saturated solution of ammonium oxalate is added drop by drop with constant stirring until no further precipitation occurs; then 10 ml more of the ammonium oxalate solution is added rapidly and the whole is boiled for 2 minutes, stirring constantly, if necessary, to prevent loss by bumping. The solution is kept warm for 3 hours, filtered through a 9 cm ashless filter paper, and washed thoroughly with hot water. The filtrate and washings are reserved for the magnesium determination.

The filter paper containing the calcium oxalate is punctured and the precipitate washed with hot water into the beaker in which it was precipitated. It is well to place 10 ml of hot 10% sulphuric acid in the beaker before washing the oxalate. The filter paper is washed alternately with hot 10% sulphuric acid and hot water until free from the precipitate, using care that only negligible portions of the filter paper are washed into the solution. After the calcium oxalate is dissolved the solution is brought to 70 °C and titrated to a faint pink with decinormal potassium permanganate solution; when this point is reached the punctured filter paper is dropped into the solution and gently agitated, care being taken not to disintegrate it. The pink colour will remain unless the washing of the paper was incomplete and even then only a few more drops of permanganate should be required to bring back the pink colour. This quantity should be noted and added to the amount originally used. Since 1 ml of decinormal potassium permanganate solution is equivalent to 2.0035 mg of calcium, the number of millilitres of permanganate used (m), multiplied by 2.0035 times the appropriate factor D from Table 2.4 is the concentration of calcium (Ca) i.e. $m \times 2.0035 \times D = $ mg Ca per litre.

Determination of magnesium

The filtrate and washings from the calcium determination are concentrated to about 150 ml and 20 ml of 10% di-ammonium acid phosphate are added to the boiling solution, boiled 3–5 minutes, and allowed to cool. When cold, it is agitated thoroughly with a stirring rod until all the precipitate has apparently formed and then, slowly from a burette, with constant stirring, 5 ml of concentrated ammonia are added. The precipitate is allowed to stand overnight, or at least 6 hours, then filtered through an ashless filter paper, washed free from chlorides with 3% ammonia, and given a final wash with magnesia wash solution (dissolve 200 g ammonium nitrate in 400 ml concentrated ammonia and make up to 1 litre with distilled water). The precipitate and filter paper, while still moist with the magnesia wash solution, are transferred to a weighed porcelain crucible, placed in a cold muffle furnace, and brought to a full red heat which leaves the pyrophosphate in a grey condition rather than pure white, as it should be for accurate results. By this procedure the ammonium salts are volatilized and the paper burned completely at a low temperature. The final result is a snow-white mass of magnesium pyrophosphate, $Mg_2P_2O_7$. This is cooled in a dessicator and weighed. Since 1 g of magnesium pyrophosphate contains 0.2184 g of magnesium, the weight (w) in mg multiplied by 0.2184 times the appropriate factor D from Table 2.4 is the concentration of magnesium (Mg) in milligrams per litre, i.e. $w \times 0.2184 \times D$ = mg Mg per litre.

Determination of total alkaline earths

Alternatively, direct titration of a brine with the disodium salt of ethylene-diaminetetra-acetic acid (trade names Sequestrene, Versene) gives the combined calcium and magnesium content. This method is used for the determination of hardness of a water and the result is usually expressed in terms of the equivalent amount $CaCO_3$ in ppm. The indicator is Erichrome Black T and for the proper colour change to be obtained it is essential that the pH of the solution be kept at a constant level. For this reason it is necessary to add a buffer solution as well as indicator. The reagent solution is usually supplied at a strength such that 1 ml of reagent solution is equivalent to 1 mg of calcium carbonate.

Procedure:
1. Pipette 1 ml of the brine and 50 ml of distilled water into a conical flask.
2. Add 0.5 ml of the buffer solution and 5 drops of the indicator solution. If neither calcium nor magnesium is present in the brine, the solution will be blue; if either or both are present, it will be coloured red.
3. Assuming calcium or magnesium to be present, titrate the solution with the reagent from a 10 ml burette until the last traces of red disappear and the solution in the conical flask is a clear blue colour.
4. The number of millilitres (m) of reagent used is equal to the number of milligrams of $CaCO_3$ equivalent to the combined Ca and Mg content of 1 ml of the brine. This number, multiplied by 1,000, is recorded as the total hardness of the brine in milligrams of $CaCO_3$ per litre of brine. This is multiplied by 0.40 to obtain the concentration of calcium in mg/l.
5. If appreciably more than 10 ml of the reagent are used for 1 ml of brine, it is preferable, when possible, to dilute the brine and repeat the titration. The dilution should be such that between 1 and 10 ml of reagent are required for 10 or 20 ml of the diluted brine. The amount of solution in the conical flask should be made up to 50 ml with distilled water before starting titration.
6. If less than 1 ml of reagent is required for 1 ml of the original brine, the titration should be repeated with a larger volume of brine such that between

1 and 10 ml of reagent are required. The amount of solution in the conical flask should be made up to 50 ml with distilled water before starting titration.

Soluble sulphate content of waters

The following method gives the order of magnitude of the sulphate content. It therefore gives more information than the standard API test on mud filtrates, which merely records the $BaSO_4$ precipitate as light, medium, or heavy. If, on the other hand, an exact figure is required, the separation and determination of the sulphate content must be carried out by standard methods of quantitative chemical analysis.

1. Make up a solution of sodium sulphate such that 1 ml of solution contains 0.5 mg of sulphate ion.
2. Add the sulphate solution to six test-tubes, each of ½-inch diameter, so that the first tube has 1 ml of solution, the second tube has 2 ml, and so on. Then add 1 ml of acidified $BaCl_2$ solution to each tube. Finally, add distilled water so that the total volume in each tube is 10 ml. Now stopper with a rubber bung and shake.
3. The six tubes prepared in 2 above form a set of standards and should be labelled '0.5 mg SO_4 in 10 ml', '1.0 mg SO_4 in 10 ml', and so on.
4. To determine the approximate sulphate content of a water sample, take a clean, ½-inch test-tube and add 1 ml of the sample. Then add 1 ml of acidified $BaCl_2$ solution and make up to a total volume of 10 ml with distilled water. Shake and compare the degree of turbidity with that of the freshly shaken standards
   (a) If the turbidity falls between that of standard 2 and that of standard 3, for example, then the sulphate content of the sample is reported as being between 1000 and 1500 mg/l.
   (b) If the turbidity is less than that of the first standard, repeat, using now 2 ml of the test sample. Continue in this manner, increasing the amount of test sample by 1 ml each time, until the turbidity comes within the range of the standards. Suppose 4 ml of the test sample were required and the turbidity then was between that of standards 3 and 4. Then the sulphate content of the sample is reported as being between

$$\frac{1.5 \times 1000}{4} \quad \text{and} \quad \frac{2.0 \times 1000}{4} \quad \text{mg}$$

i.e., between 375 and 500 mg/l. Suppose, on the other hand, that 9 ml of the test sample were used and the turbidity still did not reach that of standard 1. Then the sulphate content of the sample is reported as being less than 55

$$\text{i.e.,} \quad \frac{0.5 \times 1000}{9} \quad \text{mg}$$

(c) If the turbidity is greater than that of standard 6, dilute 20 ml of the test sample to 100 ml and repeat the tests with the diluted sample. For example, suppose the test water has been diluted 5 times in this manner and its degree of turbidity then falls between standards 3 and 4. Then the sulphate content of the sample is reported as being between 7,500 and 10,000 mg/l.

The following method gives a quantitative analysis of sulphate. Of the original filtered sample, 100 ml are measured into a 250 ml beaker, evaporated to dryness on a steam bath or hot plate, and then baked overnight at 105 °C. The residue is moistened with 10 ml of concentrated hydrochloric acid, dissolved in 100 ml of water, and the solution boiled and then filtered to remove silica and insoluble material. The filter is washed thoroughly and to the boiling filtrate a hot 10% solution of barium chloride is added, drop by drop, with constant stirring, until no further precipitation occurs; then 10 ml in excess are added rapidly and the solution allowed to digest for half an hour at the boiling point. The solution is covered and allowed to stand at room temperature for at least 12 hours.

The precipitate of barium sulphate is filtered and thoroughly washed with warm water, using a 9 cm ashless, washed filter paper of dense, firm texture. The precipitate and filter paper are placed in a weighed No. 00 porcelain crucible, ignited in an electric muffle furnace, cooled in a desiccator, and weighed. Since 1 mg of barium sulphate represents 0.4115 mg of sulphate and 100 ml of sample are used, the weight of barium sulphate in milligrams multiplied by 4.115 is the concentration of sulphate (SO_4) in parts per million (milligrams per litre).

Determination of chloride

For this determination, 10 ml of the original filtered water sample is titrated with 0.100N silver nitrate, using 1 ml of a potassium chromate solution as indicator, until the reddish colour of silver chromate is permanent. Now 1 ml of 0.1N silver nitrate is equivalent to 3.5457 mg of chlorine; therefore, since 10 ml of water is taken for titration, the number of millilitres of 0.1N silver nitrate used, multiplied by 354.57, gives the concentration of chloride (Cl) in milligrams per litre.

This is because a 'normal' solution contains one equivalent weight of the reacting species per litre and this reacts with one equivalent weight of reactant. Thus a normal solution of silver nitrate contains 107.87 g of silver per litre and this would react with 35.457 g of chlorine to give silver chloride. Because 'normal' solutions are too strong, decinormal solutions are used.

The equivalent weight of an ion or radical is its weight divided by its valency, thus the equivalent weight of chloride ion is 35.5457, i.e. 35.5457 divided by 1. The equivalent weights of the other elements commonly determined in oilfield waters are: carbonate (CO_3) = 30 (60/2; bicarbonate (HCO_3) = 61 (61/1;

hydroxide (OH) = 17 (17/1); calcium (Ca) = 20 (40/2); sulphate (SO_4) = 48 (96/2); and sodium (Na) = 23 (23/1).

Hence in this case the concentration of chloride ions (in mg/l) is given by:

ml of N/10 acid (titre) × dilution factor × 3.5457 × 100
(only 10 ml brine were used, not 1000 ml).

A satisfactory end-point cannot be obtained when more than 8 to 10 ml of 0.1N silver nitrate is required. If the sample is acid, it is neutralized with sodium carbonate; if hydroxide is present, dilute acetic acid is added until the cold solution will just discharge the colour of phenolphthalein. Acidity due to chlorides having an acid reaction, such as aluminium chloride, is treated with an excess of a neutral solution of sodium acetate and titrated as usual. If the solution is too highly coloured to titrate, those ions which give the colour are precipitated by sodium hydroxide or sodium carbonate and the filtrate is neutralized with acetic acid before titration. To obtain trustworthy results, sulphide waters should be boiled with a few drops of nitric acid and then neutralized.

Many oilfield waters or brines contain large quantities of the chlorides of sodium, calcium, and magnesium. If the chloride content of the water is high, a small amount is diluted with distilled water free from chlorides and an aliquot portion taken for analysis; if very low in chlorides, a portion of the sample is concentrated for the analysis.

When a quantity of water other than 10 ml is taken, the appropriate factor must be used. For example, 50 ml of a water sample is diluted to 1 litre and 10 ml, corresponding to 0.5 ml of the original water, is taken for titration. The result must therefore be multiplied by 2,000. If 9.3 ml of 0.1N silver nitrate were used in titration, 9.3 times 3.5457 and multiplied by 2,000 gives 65,950 milligrams per litre of chlorine as chloride (Cl).

Determination of hydroxide, carbonate, and bicarbonate

For this, 100 ml of the original filtered sample are placed in a 250 ml beaker and three or four drops of phenolphthalein indicator solution added. A red coloration indicates the presence of normal carbonate. The solution is titrated with decinormal hydrochloric acid until the coloration just disappears. The number of millilitres used corresponds to the value P of Table 2.5. Then, to the same solution, two drops of methyl orange indicator are added and the titration carried to the neutral point with this indicator. The total number of millilitres of decinormal hydrochloric acid used in both titrations corresponds to T of Table 2.5, in which are shown the relations between alkalinity to phenolphthalein and to methyl orange in the presence of hydroxide, carbonate, and bicarbonate.

One millilitre of decinormal hydrochloride acid is equivalent to 1.7 mg of hydroxide, 3.0 mg of carbonate, or 6.1 mg of bicarbonate. Since 100 ml of the original sample are used for the titration, these values multiplied by 10 give the

Table 2.5 Relations between alkalinity to phenolphthalein and alkalinity to methyl orange in presence of hydroxide, carbonate, and bicarbonate

Result of titration	Value of radical expressed in terms of ml of 0.1 N sulphuric acid		
	Hydroxide	Carbonate	Bicarbonate
$P = 0$	0	0	T
$P < T/2$	0	$2P$	$(T–2P)$
$P = T/2$	0	$2P$	0
$P > T/2$	$(2P–T)$	$2(T–P)$	0
$P = T$	T	0	0

Note: T = total alkalinity in presence of methyl orange
P = alkalinity in presence of phenolphthalein

corresponding figures in terms of 1 litre of the original sample, namely, 17 mg for hydroxide, 30 mg for carbonate, and 61 mg for bicarbonate. The latter values, when multiplied by the proper figures calculated according to Table 2.5, give the concentration milligrams per litre for hydroxide (OH), carbonate (CO_3), and bicarbonate (HCO_3).

Thus concentrations are given by:

(i) value of titre from Table 2.5 $\times$ 30 $\times$ 10 mg carbonate
per litre

(ii) value of titre from Table 2.5 $\times$ 61 $\times$ 10 mg bicarbonate
per litre

(iii) value of titre from Table 2.5 $\times$ 17 $\times$ 10 mg hydroxide
per litre

Example 100 ml of the sample is titrated with 0.1N hydrochloric acid, using 1.5 ml in the presence of phenolphthalein and a total of 16 ml in both titrations. Thus $P = 1.5$ ml and $T = 16$ ml. P is less than $T/2$; therefore, from Table 2.5, no hydroxide is present, the total carbonate is $2P$, and the bicarbonate is $(T - 2P)$. These figures correspond, respectively, to 3 ml and 13 ml of 0.1N hydrochloric acid. Hence there are present $(3 \times 30) = 90$ mg/l of carbonate (CO_3) and $(13 \times 61) = 793$ mg/l of bicarbonate (HCO_3).

Sodium concentration

The concentration of the sodium in a brine is often calculated using either the value of the total dissolved solid subtracting the concentration of ions determined as above, or by using the reacting values of the ions already determined.

Wiggins & Wood (1935) and Collins (1975) give details of the methods used to determine the above and the less commonly investigated ions, such as iron, potassium, ammonium, bromine, and iodine. They also give alternative methods of analysing the ions discussed here.

Alternative methods

Many of the titrations described in the preceding sections may be carried out using automatic titrimeters. The solution which is to be analysed is constantly stirred and its pH measured. (Sometimes electrodes specific to the ion being determined are required). The reagent is added from a burette via a valve controlled by the instrument with the rate of reagent addition decreasing with closeness to the nearness of the solution's required final pH. The end-point of the titration is specific to the ion being determined. When the titration is complete the volume of reagent used is read from the burette.

There are many instrumental methods of determining the concentration of ions in solutions. These are mostly spectroscopic methods and include Atomic absorption and flame emission spectroscopy which involve the absorption or emission of wavelengths when samples are analysed as free atoms in the vapour phase. All these methods have high accuracies and repeatabilities and are quick to perform. They are thus very suitable to the analysis of large numbers of samples. However, all of these methods have one great disadvantage as far as oilfield brine analysis is concerned. The methods are only suitable for the determination of the metallic components and the non-metallic radicals will need to be determined by classical wet-chemical methods.

Graphical representation of results

The results of these analyses are often presented in diagrammatic form for ease of comparison. There are three diagrams which are the most commonly used and they will be illustrated using the data for two brines given in Table 2.6.

The Tickell diagram (Figure 2.3) was suggested in 1921, and is a six-axis plot on which the percentage reacting values of the ions are placed. The two brines detailed in Table 2.6 are greatly different in overall concentration and have minor relative differences in composition. However, the percentage reacting values will submerge small differences and completely overlook differences in concentration. Examination of Figure 2.3 will show this to be true of the Tickell diagram in its original form. The diagram can be improved by plotting concentrations (in mg/l) or reacting values (in me/l). Care then has to be taken that the scales are fully labelled and that the same scales are used when two brines are being compared.

Reistle (1927) proposed a diagram (Figure 2.4) where the ion concentrations (in mg/l) are plotted on a vertical scale. The cations are plotted above the central zero line and the anions below. This diagram, because it involves concentrations in mg/l, does shew up differences between brines. It is especially useful for regional correlations or for plotting variations in the waters along a formation. Its narrowness allows many analyses to be plotted on one piece of paper.

The Stiff diagram (Figure 2.5) is a diagram where ion concentrations, normally in me/l, are plotted and differences can be identified. Cations are plotted on the left of the central zero, anions to the right, and the intercepts are

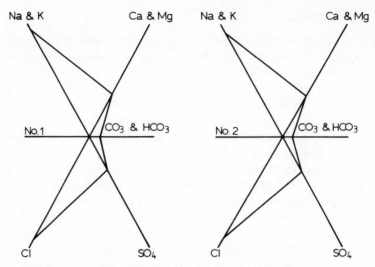

Figure 2.3 Tickell diagram of water analysis. (Reproduced by permission of California Division of Oil and Gas)

Table 2.6 Brine analysis reporting using reacting values

Reaction coefficients = reciprocal of equivalent weight of the ion.

Reaction value (me/l) = concentration (mg/l) × reaction coefficient (me/mg)

Equivalent weight of ion = weight of ion divided by its valency

Ion	Reaction coefficient	Concentration		Reacting values			
		Brine I	Brine II	Brine I (me/l)	(%)	Brine II (me/l)	(%)
Ca^+	0.0499	4,322	2,247	215.67	10.56	112.13	9.96
Mg^+	0.0823	960	553	79.01	3.87	45.51	4.04
Na^+	0.0435	17,412	9,319	725.42	35.52	405.38	36.00
Cl^-	0.0282	26,889	14,059	728.27	35.66	396.46	35.21
SO_4^{2-}	0.0208	11,188	6,437	232.71	11.46	133.89	11.89
CO_3^{2-}	0.0333	1,832	456	61.01	2.96	15.18	1.35
HCO_3^-	0.0164	—	1,064	—	—	17.45	1.55

Reaction properties (Palmer method)

	Brine I	Brine II
primary salinity	71.04	72.00
secondary salinity	23.20	22.20
primary alkalinity	0.00	0.00
secondary alkalinity	5.66	5.80

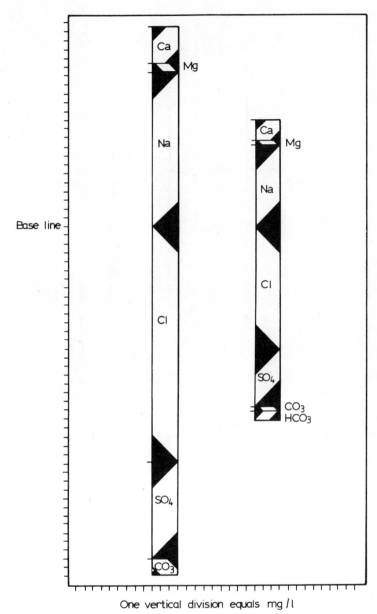

Figure 2.4 Reistle diagram of water analysis (Reproduced by permission of Bureau of Mines, US. Dept. of Interior)

often connected. The scales need to be clearly labelled and the same scales used when brines are being compared. Quite often the sodium and chlorine scales are ten times the scales for the other ions. This diagram is generally considered to be the most useful and most easily used and constructed, especially by non-technical staff.

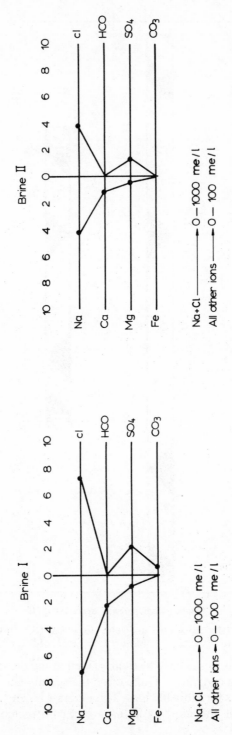

Figure 2.5 Stiff diagram of water analysis (Reproduced with permission from *J. Petroleum Technology*, 1951, **3**, 3, p. 15. © 1951 SPE-AIME)

CLASSIFICATION OF OILFIELD BRINES

The object of classification systems is to provide a basis for grouping waters which are closely chemically related and these classifications systems are based upon the dissolved constituents of waters. Many of the classification systems consider only the major dissolved inorganic constituents and ignore the matter and minor and trace inorganic species. The ions which are most commonly used in classification systems are those normally determined, i.e. magnesium, calcium, carbonate, bicarbonate, sulphate, and chloride, together with sodium which is estimated by difference as described earlier.

Palmer

Palmer (1911) observed that salinity (salts of strong acids) and alkalinity (salts of weak acids) determined the properties of natural waters. He also used the terms 'primary' and 'secondary' to qualify the general water properties on the principle that the soluble decomposition products of the oldest rocks are the alkali metals (primary) whilst the decomposition products of more recent rock formations are sources of the alkaline earth metals (secondary). However, these assumptions are not necessarily true; but using them Palmer described five properties of waters:

1. Primary salinity (alkali salinity): the strong acid radicals (sulphate, nitrate, chloride) combined with the primary bases (sodium and potassium), i.e. salinity not to exceed twice the sum of the reacting values of the radicals of the alkalis.
2. Secondary salinity (permanent hardness): the strong acids combined with secondary bases (the alkaline earths calcium, magnesium, barium), i.e. the excess, if any, of salinity over primary salinity not to exceed twice the sum of the reacting values of the radicals of the alkaline earths group.
3. Tertiary salinity (acidity): this is any excess of salinity over primary and secondary salinity.
4. Primary alkalinity (permanent alkalinity): this is the weak acids carbonate, bicarbonate, and sulphide, combined with the primary bases. These are soft waters and are the excess, if any, of twice the sum of the reacting values of the alkalis over salinity.
5. Secondary alkalinity (temporary alkalinity or temporary hardness): the weak acids combined with secondary bases, or the excess of twice the sum of the reacting values of the radicals of the alkaline earths over secondary salinity.

The brines shown in Table 2.6 illustrate the use of the Palmer method. For brine II the primary salinity is the alkali metals (405.38 me/l) with the radicals of the strong acids; chloride (396.46 me/l) and sulphate (133.89 me/l).

The 405.38 me/l of sodium will be available to react with 405.38 me/l of chloride and sulphate, the total of which is 530.35 me/l (chloride = 396.46 me/l and sulphate = 133.89 me/l). This will leave 124.97 me/l of those acid radicals and give a primary salinity equal to the concentration (in me/l) of the primary bases and an equivalent amount of strong acid radicals. Thus the primary salinity of this brine will be 810.76 me/l (2 × 405.38) or 72% of the total of 1126 me/l. The remaining 124.97 me/l of the strong acid radicals will be available for combination with an equivalent amount of the alkaline earth metals, i.e. 124.97 me/l out of a total of 157.64 me/l (Calcium = 112.13 me/l and magnesium = 45.51 me/l). This gives a secondary salinity of 249.94 me/l or 22.20%. There will be 32.67 me/l of the alkaline earth metals available for combination with the 15.18 me/l of carbonate and the 17.46 me/l of bicarbonate to give a secondary salinity of 65.34 me/l or 5.80% The salinities and alkalinities observed in waters are related to the class of water and may be calculated using either the concentration in me/l (as above) or the percentage values (following).

Primary salinity and secondary alkalinity are normally present in oilfield waters and if the strong acids exceed the primary bases then the third property will be secondary salinity, otherwise the third property will be primary alkalinity.

This system can use the reacting values of the ions expressed as a percentage and thus allows the data to be displayed graphically (Figure 2.6). Because of the use of percentage values these diagrams suffer the same drawbacks as the

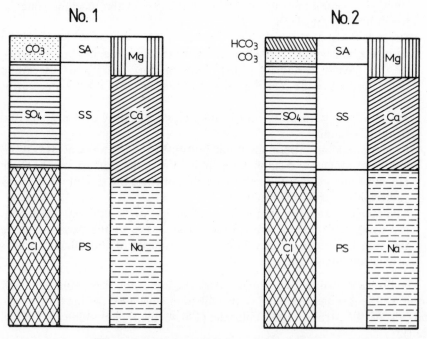

Figure 2.6 Palmer diagram of water analysis

Tickell diagram, i.e. absolute concentration is ignored and minor differences are lost.

Five classes of water can be observed based upon the relative concentrations of their ions. If a is the percentage reacting value of the alkali metal cations, b is the percentage of reacting value of the alkaline earth metal cations, and d is the reacting value of the strong acid radicals, the five classes and the salinities and alkalinities are defined as follows:

1. $d < a$
   $2d$ = primary salinity
   $2(a - d)$ = primary alkalinity
   $2b$ = secondary alkalinity
2. $d = a$
   $2a = 2d$ = primary salinity
   $2d$ = secondary alkalinity
3. $d > a; d < (a + b)$
   $2a$ = primary salinity
   $2(d - a)$ = secondary salinity
   $2(a + b - d)$ = secondary alkalinity
4. $d = a + b$
   $2a$ = primary salinity
   $2d$ = secondary salinity
5. $d > (a + b)$
   $2a$ = primary salinity
   $2d$ = secondary salinity
   $2(d - a - b)$ = tertiary salinity

All these classes of water were shewn by Palmer to exist in nature.

Examination of the two brines in Table 2.6 will shew that they are both class 3 brines. E.g. for brine I the percentage of strong acid anions, d (35.66% + 11.46% = 47.12%), is greater than the value of the alkali metals, a (35.52%), and the strong acid anions, d (47.12%), is less than the sum of the alkali metals, a (35.52%), and the alkaline earth metals, (10.56% + 3.87% = 14.43%). Thus the primary salinity = $2a$ = 2×35.52 = 71.04%; the secondary salinity = $2(d - a)$ = $2[(Cl + SO_4^=) - Na^+]$ = $2[(35.66 + 11.46) - 35.52]$ = 23.20% and the secondary salinity = $2(a + b - d)$ = $2[Na^+ + (Ca^{++} + Mg^{++}) - (Cl^- + SO_4^{2-})]$ = $2[35.52 + (10.56 + 3.87) - (35.66 + 11.46)]$ = 5.66%

Ostroff (1967) has noted that the Palmer system groups together constituents which are not closely chemically related. Absolute ionic concentration is ignored by the use of percentage reacting values, small differences are smoothed out, and the saturation conditions relating to sulphate and bicarbonate solubility and precipitation are not considered.

Sulin

Sulin (1946) describes a method of classifying waters according to chemical type, subdivided into groups, subgroups, and classes. He found four basic environments for the generation of natural waters, two of which produce meteoric waters while the other two yield connate waters. The continental conditions promote the formation of sulphate waters by allowing soluble sulphate components to be added to the water (sulphate–sodium type waters). Alternatively, continental conditions can aid the formation of waters rich in sodium bicarbonate (bicarbonate–sodium type waters). Marine conditions give rise to the formation of a chloride–magnesium type of water whilst deep sub-surface conditions yield chloride–calcium waters. The continental derived waters are meteoric whilst the other two environments yield typical connate waters either by incorporation of sea water in to the sediment at the time of deposition or by water formation under stagnant conditions in the earth's crust. Table 2.7 shews the Sulin classification of major water classes. Sodium concentrations include all alkali metals and chloride includes all halogens.

Table 2.7   Major water classes—Sulin

Water type	ratios of % reacting values		
	Na/Cl	(Na–Cl)/SO_4	(Cl–Na)/Mg
sulphate–sodium	> 1	< 1	< 0
bicarbonate–sodium	> 1	> 1	< 0
chloride–magnesium	< 1	< 0	< 1
chloride-calcium	< 1	< 0	> 1

Sulin then split these water types into groups similar to and related to those described by Palmer. The alkalinities are due to a predominance of carbonate and bicarbonate ions while salinities are due to chloride and sulphate ions. Primary waters have high concentrations of the alkali metals, secondary waters contain predominant amounts of alkaline earth metals, whilst iron and aluminium are the major metals in tertiary waters.

Class A_1 is primary alkalinity in which the carbonates and bicarbonates of the alkali metals are the main contributors.

Class A_2 is secondary alkalinity where the carbonates and bicarbonates of the alkaline earth metals are predominant.

Class A_3 is tertiary alkalinity and is due to the preponderance of the iron and aluminium carbonates and bicarbonates.

Class S_1 is primary salinity and is caused by the sulphates and chlorides of the alkali metals.

Class S_2 is secondary salinity where the alkaline earth sulphates and chlorides are predominant.

Class S_3 is tertiary salinity and is due to the chloride and sulphates of iron and aluminium.

Waters are classified by decreasing values of these Palmer characteristics, e.g. $S_1A_1A_2$ represents Palmer's class 1 with primary salinity dominant and secondary alkalinity the least important.

Because the Palmer system does not take account of the relationships between the sulphate and chloride ions and between calcium and magnesium, Sulin introduced the ratios SO_4/Cl and Ca/Mg which establish additional subgroups.

Hence the Na/Cl ration will separate meteoric and connate waters. Meteoric waters are split by the $(Na-Cl)/SO_4$ ratio in to sulphate–sodium and bicarbonate–sodium types, whilst the $(Cl-Na)/Mg$ ratio splits the connate waters into chloride–magnesium and chloride–calcium types. The Palmer system is used to define the water group and the subgroups are obtained from the SO_4/Cl and Ca/Mg ratios.

Chebotarev

Chebotarev (1955) classified waters on the basis of their dissolved chloride, sulphate, and bicarbonate concentrations. He then subdivided these groups into genetic types determined by the absolute concentrations of the dissolved constituents, as expressed in percentage reacting values. The bicarbonate group contains three genetic subtypes: bicarbonate, carbonate–chloride and chloride–bicarbonate. The sulphate group contains two genetic types, sulphate and sulphate–chloride; whereas the chloride group contains the chloride–bicarbonate, chloride–sulphate, and chloride genetic subtypes.

Schoeller

Schoeller (1955) classified waters on the basis of dissolved components and obtained six primary types based upon the amount of dissolved chloride and four subgroups based upon concentrations of sulphate. The amounts of carbonate and bicarbonate were also used for an additional differentiation of the water types. The index of base exchange was noted to indicate the relative likelihood of the exchange of ions in the water with those in the host clays.

ORIGIN AND ALTERATION OF OILFIELD BRINES

It is accepted that petroleum source rocks and the majority of petroleum reservoir rocks are deposited in aquatic, normally marine, conditions. The water that was incorporated into these sediments will therefore normally be a salt solution of the same composition as the ancient sea water. There is no reason to believe that ancient sea waters were significantly different from modern sea waters.

It will be noted (Table 2.1) that the vast majority of oilfield brines are considerably different from river and sea waters. These differences involve absolute concentration (oilfield brines are more concentrated) and relative concentrations (e.g. calcium–magnesium ratio).

Very low salinities can be explained by dilution with meteoric waters where infiltrating fresh water has mixed with subsurface water. Brines in formations which have been exposed at an unconformity may well have been concentrated by evaporation or diluted by the influx of fresh meteoric water. The composition of such brines can be significantly changed in these circumstances. Very concentrated waters can be, in some cases, explained where the waters are in direct contact with salt, e.g., the Permian basin of the West Texas Gulf coast or in some of the Romanian fields.

In basins with restricted circulation and high evaporation the sea water, before it is incorporated into the sediment, will be more concentrated than normal and thus the original pore water contains more dissolved salts. This could, in part, account for some of the more concentrated brines.

As sea water is evaporated the carbonates and then the sulphates are the first to precipitate and this would normally also remove the alkaline earth metals such as barium, strontium, calcium, and magnesium. This is the precipitation of limestone, dolomite, and gypsum. Sodium chloride, or halite, begins to precipitate out at a chloride concentration of 275,000 mg of chloride per litre and with it a little bromide is removed from solution. Bromine does not form salts by precipitation but replaces chlorine in the sodium chloride lattice with the result that the residual brine becomes enriched in bromine. Sylvite (potassium chloride) begins to precipitate at a chloride concentration of 360,000 mg/l.

During this process both the bromide and potassium ions are concentrated with respect to chlorine and sodium respectively. The incorporation into a sediment of a more concentrated sea water from an evaporitic basin will help to explain some, but not all, of the differences between modern sea water and connate waters. The increase in relative concentration of bromine and potassium with respect to chlorine and sodium is explainable in such a situation, but not the calcium–magnesium and alkali metal–alkaline earth differences.

In addition, most oilfield brines are more concentrated than sea water but not every connate water could have originated from sea water which had been enriched by evaporation. Therefore other processes must have been involved in the alteration of the trapped sea water and these processes involve interactions between the brine and the host rock.

Certain elements, especially bromine, boron, and iodine, together with ammonium ions, are concentrated by corals and seaweeds. These organisms absorb these ions from the sea water and over their life build up great concentrations of these ions within their bodies. The bromine, iodine, boron, and ammonia can be leached out of the debris of dead corals and seaweeds and will enrich the pore water of the sediment into which the debris was incorporated.

Many of the clays which are hosts of the pore waters are chemically very similar to the materials used in ion exchange reactions. Such reactions are reversible and stop when an equilibrium has been established. The reactions are also governed by pressure, temperature, concentration, the bonding strength of the ions, the number of sites available, and the law of mass action. Grim (1952) has shown that the replacing power of some ions to vary with the clay used. Thus for a kaolinite in ammonia the order is:

$$Cs > Rb > K > Ba > Sr > Ca > Mg > H > Na > Li$$

whilst on montmorillonite in ammonia the order becomes:

$$Cs > Rb > K > H > Sr > Ba > Mg > Ca > Na > Li$$

These results can be interpreted to indicate that caesium and rubidium are most likely to be removed from solution whereas sodium and lithium are more likely to be left in solution or added to the solution from the clay in exchange for other ions. The relative positions of calcium and magnesium change depending on the clay used and the enrichment of potassium in brines is not directly explained by ion exchange mechanism. Experiments have shewn that the potassium concentration of many oilfield brines is less than that obtained by the evaporation of sea water to a similar total salt concentration. It has been suggested that ion exchange reactions could remove potassium from a brine which originated from a partially evaporated sea water.

Calcium is enriched in oilfield brines and hypotheses involving cation exchanges with clays have been advanced to explain the calcium enrichment. Such reactions could be illustrated thus:

$$2Na^+(soln) + Ca(clay) = Ca^+(soln) + 2Na(clay)$$

In some cases the sodium to calcium and magnesium ratio decreased as the total of dissolved solid contents rises and this phenomenon has been related to the base exchange index and shews that the waters have exchanged their alkali metals for the alkaline earth metals of the clays. (Collins, 1975; Schoeller, 1955). There have also been suggestions that some non-metallic elements can be involved in the exchange reactions of the clays. Bromide or iodide can be removed from the clay and added to the water in exchange for chloride ions. Collins (1969) suggests that this would explain the enrichment of iodine in many oilfield brines.

In many basins mixtures of dolomite and anhydrite occur which indicate that sulphate may be removed from brines by dolomitization as well as by bacterial reduction. These dolomitization reactions principally affect the quantity of alkaline earth metals present in a brine. Sea water contains much more magnesium than calcium which, on the basis of solubilities, is to be expected because magnesium salts are more soluble than their calcium equivalents. In contrast to this, oilfield brines contain more calcium than magnesium. The following discussion shews how dolomitization could account for this reversal in the relative amounts of calcium and magnesium.

Magnesium in solution could react with calcite minerals ($CaCO_3$) in the host rock to form dolomite, thus reducing the magnesium concentration of the brine.

$$Mg^{++}(soln) + 2CaCO_3(rock) \rightleftharpoons Ca^{++}(soln) + MgCa(CO_3)_2(rock)$$

If little or no sulphate is present in the brine, the calcium concentration of the brine will increase while the magnesium content decreases, and thus the total alkaline earth metal concentration will remain constant, ignoring any overall concentration of the brine. If sulphate is present the calcium will precipitate as anhydrite and the total alkaline earth metal concentration falls.

$$Ca^{++}(soln) + SO_4^{--}(soln) \rightarrow CaSO_4(rock)$$

The overall effect of these two reactions is:

$$Mg^{++}(soln) + SO_4^{--}(soln) + 2CaCO_3(rock) \rightarrow$$
$$CaSO_4(rock) + CaMg(CO_3)_2(rock)$$

In this case the magnesium is removed from solution and the calcium stays in solution. Thus the concentration of magnesium in the brine decreases. Reactions such as these explain not only the decrease in magnesium when its salts are more soluble than calcium's, but also the generation of dolomite and anhydrite and the removal of sulphate from solution without the aid of bacteria. The dolomitization reaction is especially welcome because magnesium has a smaller ionic radium than calcium and hence when it replaces calcium in the rock the porosity of the rock is increased.

Strontium can replace calcium in this process and this could be one of the reasons that oilfield brines are enriched in strontium as compared with sea water. Certain reactions can concentrate or dilute oilfield brines. A typical reaction of this sort is the hydration of anhydrite to form gypsum.

$$CaSO_4 + 2H_2O = CaSO_4 \cdot 2H_2O$$

Thus the absorption of water or its release from the rock minerals can affect brine concentrations.

Collins (1975) suggests that the reactions between brines and rock minerals can also account for the depletion of dissolved alkali metals in brines by the formation of silicates and typical reactions could be

$$3Al_2Si_2O_5(OH)_4 + 2K^+ \rightarrow 3KAl_3SiO_{10}(OH)_2 + 3H_2O + 2H^+$$
$$KAl_3SiO_{10}(OH)_2 + 6SiO_2 + 2K^+ \rightarrow 3KAlSi_2O_8 + 2H^+$$
$$Al_2Si_2O_5(OH)_4 + 4SiO_2 + 2Na^+ \rightarrow 2NaAlSi_2O_8 + H_2O + 2H^+$$

Besides accounting for a depletion in alkali metal concentrations, such reactions will release hydrogen ions into the brine which will decrease the pH of the brine and allow it to dissolve other metallic ions as well as converting bicarbonate to carbon dioxide and bisulphide to sulphide.

Compaction of the sediment will expel the freshest water containing the fewest ions and the residual water becomes more concentrated. This fresh water will be expelled at shallow depths and analyses shew that pore waters do become more saline and concentrated with depth. Hence water which is expelled at depth will be much more concentrated than the original sea water which was incorporated into the sediment. As the sediments are more deeply buried, both temperature and pressure increase and higher temperatures and pressures favour the solution of minerals in the pore water. (Most salts are more soluble in hot water.) Thus when water is expelled from a clay at depth it is more likely to be more concentrated and able to dissolve some of the host minerals.

IMPORTANCE OF OILFIELD WATER ANALYSIS

The composition of oilfield waters can be of great importance because the organic matter and the oil derived from it are always in contact with water. The original organic matter is deposited in water, there is water present during the conversion of that organic matter into oil, and the generated oil has to migrate through water-filled pores to a reservoir where the oil is in contact with water. Thus the composition of those waters could be related to the genesis, migration, accumulation, and alteration of the oil. Uses of brine analyses include the use of oilfield waters to identify formations and to assist the interpretation of borehole measurements such as electrical resistivity logs. Variations in water composition between formations may be related to the geological history of the area and may provide a clue to oil distribution. Alterations in the water coproduced with oil could indicate encroachment of edge water or physical damage to casings, etc.

Philippi (1965) suggested that petroleum and natural gas can be generated by low-temperature reactions, and the catalytic effect of the surrounding sediments, especially shales, will influence the rate of reaction and the type of hydrocarbon formed. Since water is present in all the sediments which contain organic matter destined to become oil and gas, it is not unreasonable to assume that the water plays a part in the transformation process. Water is certainly important as a factor in determining the depositional environment of petroleum precursors and the salinity of the water may affect the oxygenation of the water, which will affect the preservation of the organic matter. Similarly, the pore water may affect the composition of the clays and thus their effectiveness as catalysts.

The compaction of sediments results in the expulsion of water and some solubilized organic matter. This organic matter can be true petroleum or the precursor of petroleum, a 'protopetroleum' which is converted to petroleum in the reservoir. The solubility of hydrocarbons in water depends upon temperature and the quantity of other dissolved species. The more saline is the brine, the less hydrocarbons are able to be carried in solution by that brine. However, elevated temperatures increase the solubility of petroleum hydrocarbons in water. Hence a hot, dilute brine is best for transporting oil from source to reservoir by molecular solution of the hydrocarbons, although there are other forms of solution to be considered.

Peake and Hodgson (1966, 1967) have shewn that hydrocarbons in the range C_{12}–C_{36} can dissolve in water in concentrations up to 100 ppm. The carbon preference index (CPI) measures the relative number of alkanes with odd-numbered carbon chains to those with even-numbered chains. For ancient sediments and crude oils there are approximately equal numbers, but recent sediments have a preponderance of odd-numbered chains (high CPI). It has been suggested that this high CPI disappears, in part at least, because the water tends to dissolve equal amounts of odd and even length alkanes. Welte (1965) has shewn that by extracting bituminous rocks with a hydrocarbon-free oilfield brine, equal amounts of odd and even paraffins were obtained.

There are many non-hydrocarbon compounds such as ester ketones, and acids, which are water-soluble and can be transported to the reservoir and there converted to petroleum hydrocarbons. However, they can also form micelles (Baker, 1960), and Cordell (1973) has proposed that these soap micelles could be the means by which petroleum is transported from the source to the reservoir, where a suitable mechanism to release the petroleum is required. Kartzev et al., (1959) have shewn that some oilfield waters can contain up to 8,000 mg/l of dissolved organic acid salts and they have suggested that as little as 50 mg/l of such salts could solubilize and transport significant amounts of hydrocarbons. These organic acid salts are similar to soaps and can form micelles in which the oil is transported. The different types of compounds comprising the micelles will determine the hydrocarbons carried in the water and thus the type of hydrocarbon accumulation.

The accumulation of petroleum requires a mechanism whereby the petroleum hydrocarbons can be released from solution in a place where they can be trapped, often in structural traps such as anticlines or in stratigraph traps such as permeability barriers.

One such mechanism to release the oil from the water is a change in coarseness of the sand grains. Experiments have shewn that if an oil-in-water emulsion is passed through sands, then as the sand gets finer the oil the released, (Cartmell & Dickey, 1970). If the fining of the sand continues it may prove to be an effective permeability barrier to the further movement of the oil.

The concentration of salt in a brine will determine the amount of hydrocarbons it can dissolve and the greater the amount of inorganic salts, the less organic matter can be dissolved. Thus weak brines are more effective at transporting hydrocarbons by molecular solution. Any factor which increases the salt content of a brine will favour the release of the oil. Thus contact with salt formations will tend to release oil which will be able to accumulate if a suitable reservoir is nearby.

If the oil is transported by surfactants in micellar solution a mechanism is required to decompose and render ineffective these surfactants so that the oil can be released from the water and, assuming the correct geological conditions existed, would accumulate.

Brines are able to dissolve greater amounts of hydrocarbons if they are at high temperatures and pressures. Most source rocks are deeper than the reservoirs they source and the natural movement of oil-carrying waters, because of buoyancy, is upwards. This will take the brine to cooler regions with lower pressures which favour the release of dissolved hydrocarbons.

Hydrocarbons are susceptible to alteration, and changes in the water with which they are associated could affect this alteration. Long-chain alkanes can be split into two or more shorter compounds by chemicals in the water. Oxidation of organic matter will be effected by changes in redox potential while changes in pH will alter the solubility of organic acids, thus affecting the amount of hydrocarbon material that can be solubilized by these acids.

Water is a very important factor in the alteration of reservoired crude and is discussed more fully in Chapter 8. Water can dissolve organic compounds and this process is especially important where fresh water infiltrates through an outcrop and flows through the zone containing a structurally trapped hydrocarbon occurrence. Because water is the host for bacteria and all that they require for life, water is essential for the bacteriological degradation of oil. Bacteria require oxygen, nutrients, and food, all of which can be supplied in the water. Bacteria are only able to function at low temperatures and in the absence of poisons such as hydrogen sulphide. Bacteria are very selective in the organic compounds that they consume, with the n-alkanes being the first eaten and aromatic compounds, due to their toxicity, being untouched (Winters & Williams, 1971). Besides removing paraffinic material, bacteria add multi-ring heteroatomic compounds to the oils and altered oils will have higher correlation indices, lower API (American Petroleum Institute) gravities (see Chapter 3, p.94 for definition), high asphaltene contents, and higher viscosities. Because many of the compounds added to oil are optically active, the optical rotation of the oil changes. Whereas the bacterial alteration of paraffinic crudes will leave them more viscous but probably still liquid, the alteration of asphaltic oils can produce solid, immobile tars.

Oilfield water analyses are used in a variety of ways. The knowledge of the electrical resistivity properties of oilfield waters is essential in the detailed interpretation of electric well logs. In fields where there are many reservoirs the different producing sands can often be distinguished and correlated by comparison of the composition of the water within each sand. Such correlations are normally only possible over relatively small areas.

Changes in the volume and concentration of the water coproduced with the oil can give indications as to the advance of edge or bottom water or of mechanical failure, e.g. leaks in the casing. A knowledge of the composition of all the formation waters in the area will allow the operator to locate the source of the new water and to take the appropriate action.

Oilfield waters are often analysed during the exploration for petroleum and the analyses fall into two categories. Firstly, there are the analyses of the organic content of the water and, secondly, the changes in the inorganic constituents of the brine as oil accumulations are neared. Many claims have been made about the success of purely geochemical methods of discovering oil and gas accumulations. Davis & Yarborough (1969) have patented a geochemical method of hydrocarbon prospecting which involves the analysis of formation waters for saturated hydrocarbons. Buckley *et al.* (1958) have studied the dissolved hydrocarbon gases in the waters of petroleum-bearing strata and concluded that methane was the commonest dissolved gas and it generally was to be found in edge waters.

Zarella *et al.* (1967) believed that hydrocarbons generally diffuse vertically and not laterally and that the quantity of dissolved benzene in waters was related near to hydrocarbon accumulations. Because fatty and naphthenic acids are possibly involved in the migration of the generated oil from its source to the reservoir, many geochemical exploration methods involve determining the type and quantity of such acids (Cooper & Kvenvolden, 1967).

Kolodii (1969) has shewn that significant variations in the dissolved solids contents of waters, by as much as 300 g/l, occur when those waters are near oil accumulations as compared with background values. In particular, the concentrations of methane, carbon dioxide, and hydrogen sulphide rose whereas the amount of dissolved sulphate decreased near hydrocarbon accumulations. Serebriako & Tronko (1969) have shewn the ammonium ion concentrations to be around 800 mg/l near oil and gas but below that value in unproductive areas. Sudo (1969) noted that the sulphate contents of formation waters decreased but the iodide and bicarbonate concentrations increased as hydrocarbon pools were approached.

Collins (1975) reports that other workers have noted increases in potassium, boron, bromine, and iodine as oil pools are neared and that alkaline waters with higher than normal concentrations of sulphates, carbonates, and naphthenic acids are considered to be associated with naphthenic oils, whereas the alkaline earth chloride waters with low-numbered fatty acids are related to paraffinic oils (Krejci-Graf, 1962).

Billings (1969) has patented an exploration method using the principle that iron, cobalt, manganese, nickel, vanadium, titanium, chromium, and scandum are soluble in their reduced state but not in their oxidized state. This method involves plotting concentration trends to indicate the location of petroleum. All of these methods described here can be treated graphically. Maps with iso-concentration lines for any or all of the dissolved components can be plotted and used, with other information, to find the likely location of hydrocarbon occurrences. Hitcham & Horn (1974) considered that the concentration of magnesium and iodine are indicators of oil accumulations.

They based this on the assumption that the more magnesium is present in solution, the less dolomite has been formed and the lower is the porosity of that rock.

A summary of the uses of oilfield brine analysis in oil and gas exploration is given in Figure 2.7 (Collins, 1975).

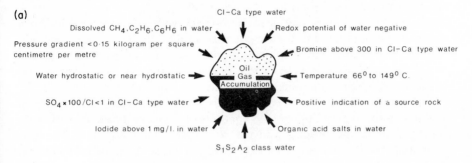

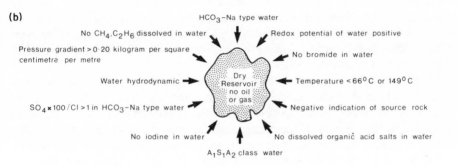

Figure 2.7 Petroleum occurrence indicators from brine analyses, (Collins, 1975. Reproduced by permission of Elsevier Scientific Publishing Co.)

An important use of oilfield brine analysis is in the making of decisions about the disposal of oilfield waters and the compatability of water used for injection. Careless disposal of brines on the surface can cause serious damage to the flora and fauna of an area. If the waste brines are injected into porous strata, either as a disposal method or to increase oil production, incompatability between the injected and the native waters can cause precipitation of solids. If this precipitation occurs on a large enough scale, the porosity and permeability of the strata will be greatly reduced and this may impede the continuing production of oil and gas from that area. The particular problems involve the insoluble carbonates and sulphates of calcium, barium, and strontium.

The surface handling of oilfield brines requires a knowledge of the constituents of the water and their solubility characteristics. Mixing of different brines can cause precipitation of the less soluble salts. In addition, the cooling of brines may cause the precipitation of part of their dissolved solid content. Precipitation of these insoluble salts can cause restrictions to flow of fluids, or in extreme cases total blockage of the pipes. The precipitation of salts not only reduces the carrying capacity of a pipe but can reduce its effectiveness as a heat exchanger. These salts and their concentrated solutions are often corrosive and they may eat away valuable wellhead fittings. This can also have safety implications. The cleaning of such pipework often requires physical descaling on the dismounted fittings as well as the applications of chemical solutions to dissolve the unwanted precipitates.

Many of the brines that are coproduced with oils contain significant quantities of minerals which have potential value. These include magnesium, calcium, potassium, boron, lithium, bromine, and iodine, and many of these are already recovered from sea water. However, the feasibility and economics of recovering these from oilfield brines varies.

The halogens have been recovered from subsurface brines in the USA but the economics depend upoon the concentration of the brine as well as the selling price of the product and the cost of the plant required. The most commonly recovered halogens are bromine and iodine. Collins (1975) quotes a case in California where brines containing 60 mg/l of iodine were used to obtain iodine. However, the increasing costs eventually made that process uneconomic. In another case quoted by Collins, 36 mg/l of iodine in the brine was sufficient because the iodine was recovered with other minerals including bromine and magnesium salts.

The magnesium in brines has the most potential for recovery as it can command a high market price; however, it is very rarely recovered because of the expense of processing and because of the requirements of other chemicals needed in the process. If these chemicals are readily available in the locality, the economics of magnesium production can be favourable for its recovery. Calcium is more readily recovered from the mining of mineral deposits even though it can occur in considerable quantities in brines.

Sodium chloride is the most abundant mineral in most oilfield brines, but its value is so low that it is rarely recovered. Lithium and potassium are, to a limited extent, recovered from oilfield brines, but rubidium and caesium are more cheaply obtained as by-products of normal lithium production.

In general, it is only commercially viable to produce minerals from oilfield brines where alternative supplies are limited and sources of other chemicals required are available in the location, and where multiproduct production is possible.

REFERENCES

Andrews, L. J., & Keefer, R. M. (1949). Cation complexes of compounds containing carbon-carbon double bonds II. The solubility of cuprous chloride in aqueous maleic acid solutions. *J. Am. Chem. Soc.,* **71,** 2394.

Baker, E. G. (1960) A hypothesis concerning the accumulation of sediment hydrocarbons to form crude oil. *Geochem. & Cosmochim. Acta,* **29,** 309.

Billings, G. K. (1969) Geochemical petroleum exploration method. US Patent No. 3,428,431.

Buckley, S. E., Hocutt, C. R., & Teggart, M. S. (1958) Distribution of dissolved hydrocarbons in subsurface waters. In *Habitat of Oil,* Weeks, L. G. (ed.), AAPG, Tulsa.

Cartmell, J. C., & Dickey, P. A. (1970) Flow of a disperse emulsion of crude oil in water through porous media. *BAAPG,* **54,** 2438.

Chebotarev, I. I. (1955) Metamorphism of natural waters in the crust of weathering. *Geochim. & Cosmochim. Acta,* **8,** 198.

Collins, A. G. (1969) Chemistry of some Anadarko basin brines containing high concentrations of iodide. *Chem. Geol.,* **4,** 169.

Collins, A. G. (1975) *Geochemistry of Oilfield Waters,* Elsevier.

Cooper, J. E. (1962) Fatty acids in recent and ancient sediments and petroleum reservoir waters. *Nature,* **193,** 744.

Cooper, J. E., & Kvenvolden, K. A. (1967) Method for prospecting for petroleum. US Patent No. 3,305,317.

Cordell, R. J., (1973) Colloidal soap as a proposed primary migration medium for hydrocarbons. *BAAPG,* **51,** 1618.

Davis, J. B., & Yarborough, H. R. (1969) Geochemical exploration. US Patent No. 3,305,317.

Degens, E. T., Hunt, J. M., Rutter, J. H., & Reed, W. E. (1964) Data on the distribution of amino acids and oxygen isotopes in petroleum brines of various geologic ages. *Sedimentology,* **3,** 199.

Grim, R. E. (1952) *Clay Mineralogy,* McGraw-Hill, New York.

Hitcham, B., & Horn, M. K. (1974) Petroleum indicators in formation waters of Alberta, Canada. *BAAPG,* **58,** 464.

Kartzev *et al.* (1959) Geochemical methods of prospecting and exploration for petroleum and natural gas (English edition by Witherspoon, P. A., & Romey, W. D.). University of California Press.

Kolodii, V. V. (1969) Origin of some hydrochemical anomalies in petroleum provinces. *Nauch. Tekh. Sb. Ser. Neftegazov. Geol. Geofiz.,* **8,** 37 (in Russian).

Krejci-Graf, K. (1962) Oilfield waters. *Erdöl, Kohle Petrochem.,* **15, 102.**

Lane, A. C. (1927) Calcium chloride waters, connate and diagenetic. *BAAPG,* **11,** 1283.

McDevit, W. F., & Long, F. A. (1952) The activity coefficient of benzene in aqueous salt solutions. *J. Am. Chem. Soc.,* **74,** 1773.

Munn, M. J. (1920) The anticlinal and hydraulic theories of oil and gas accumulation. *Econ. Geol.,* **15,** 398.

Ostroff, A. G. (1967) Comparison of some formation water classifications systems. *BAAPG,* **51,** 404.

Palmer, C. (1911) The geochemical interpretation of water analyses. *US Geol. Surv. Bull.,* **479,** 5–31.

Parker, J. S., & Southwell, C. A. P. (1929) Trinidad oilfield waters. *J. Inst. Pet. Tech.,* **15,** 138.

Peake, E., & Hodgson, G. W. (1966) Alkanes in aqueous systems I. *J. Am. Oil Chem. Soc.,* **43,** 215.

Peake, E., & Hodgson, G. W. (1967) Alkanes in aqueous systems II. *J. Am. Oil Chem. Soc.,* **44,** 696.

Philippi, G. T. (1965) Depth, time and the mechanics of petroleum generation. *Geochim. & Cosmochim. Acta,* **29,** 1021.

Reistle, C. E., (1927) Identification of oilfield waters by analysis. *USBM Technical paper,* 404.

Reistle, C. E., & Lane, E. G. (1928) *US Bureau of Mines Tech. Paper,* 432.

Rittenhouse, G., *et al.* (1969) Minor elements in oilfield waters. *J. Chem. Geol.,* **4,** 189.

Russel, W. L. (1933) Subsurface concentration of chloride brines. *BAAPG,* **17,** 1213.

Schoeller, H. (1955) Geochemie des eaux souterraines. *Rev. Inst. Francais des petroles,* **10,** 181.

Schilthuis, R. J. (1938) Connate water in oil and gas sands. *Petroleum Development and Technology,* AIME, pp. 199–214.

Serebriako, O. I., & Tronko, I. V. (1969) Ammonium content in groundwaters of the N.W. Caspian region as indicators of oil and gas. *Geol. Nefti. Gaza.,* **9,** 57.

Slowey, J. F., Jeffery, L. M., & Hood, D. W. (1962) Fatty acid content of ocean waters. *Geochim. Cosmochim. Acta,* **26,** 607.

Stiff, H. A., (1951) The interpretation of chemical water analysis by means of patterns. *J. Pet. Technol.,* **3**(10), 15.

Stiff, H. A., (1952) A method for predicting the tendency of oilfield waters to deposit calcium sulphate. *AIME Pet. Trans.,* **195,** 25.

Sudo, Y. (1969) Geochemical study of brine from oil and gas fields in Japan. *J. Japan. Assoc. Pet. Technol.,* **32,** 286.

Sulin, V. A. (1946) Waters of petroleum formations in the system of natural waters, *Gastoptekhizdat,* Moscow (in Russian).

Tickell, F. G. (1921) A method for graphical interpretation of water analyses. *Calif. State Oil Gas Superv.,* **6,** 5.

Welte, D. H. (1965) Relation between petroleum and source rock. *BAAPG,* **49,** 2246.

Wiggins, W. R., & Wood, C. E. (1935) Examination of oilfield waters. Essential protective measure. *Oil and Gas J.,* 27/June.

Winters, J. C. & Williams, J. A. (1971) Microbiological alteration of oil in muddy sandstone and other reservoirs. *BAAPG,* **59,** 369.

Zarrella, W. M., Mousseau, R. J., Coggeshall, N. D., Norris, M. S., & Schrayrt, G. J. (1967) Analysis and significance of hydrocarbons in subsurface brines. *Geochim Cosmochim. Acta,* **31,** 1155.

Chapter 3

The Nature of Petroleum

INTRODUCTION

Petroleum is a term derived from the Latin words *petra* (rock) and *oleum* (oil), which is used to describe a wide range of naturally occurring hydrocarbon mixtures. Sometimes the word 'petroleum' is limited to those mixtures which are fluid enough to be extracted via a drill pipe. The majority of petroleums are fluid, hence the qualification to the definition of the word petroleum which is mentioned above. Petroleums include components that are gaseous, liquid, and solid. Commonly all three forms are produced together with gases and solids within a liquid. However, both the solid and the gaseous forms can occur separately. Examples are natural gas, ozokerite (paraffin wax plus mineral matter), and natural asphalts.

The chemistry of petroleum is the chemistry of organic carbon, i.e. organic chemistry, which is a wide and complex subject. Somewhere in the region of half a million different carbon compounds have been identified or made and doubtless many more await discovery or synthesis. The potential is almost endless, but in the chemistry of petroleum one is interested in a narrow band of compounds. This makes the task slightly easier, but nonetheless it is still a complex one.

Petroleum deposits vary widely in their physical and, to a lesser degree, their chemical properties. In spite of this the bulk of the compounds which are in crude oils are hydrocarbons, i.e. compounds which only contain the elements of carbon and hydrogen. The remaining compounds, which are maybe as little as 2%, contain other elements, in particular nitrogen, sulphur, and oxygen, and because of this are they often called heteroatomic compounds. Having said that crude oils are predominantly mixtures of hydrocarbons, the question 'Why are various crude oils so totally different from one another?' may arise.The answer lies in the very wide range of organic compounds which can be found in crude oil. The physical and chemical properties of these crude oils reflect the different proportions of the various compounds, each with its own properties, which go to make the mixtures we call crude oil.

The range of compounds is from the simplest, methane, to the very large complex asphaltene molecules whose structure is not completely under-stood. Thus crude oil is a fluid, which is normally liquid, in which gases are dissolved and in which solids are both dissolved and suspended. The physical conditions of the reservoir, i.e. temperature and pressure, are quite different from those at the surface where the oil is produced and stored. The production of the oil will mean that the physical conditions of the oil change so that the proportion dissolved or suspended in the liquid state may alter and the stock tank oil may be considerably different to the oil in the reservoir. Hence accurate analysis of the reservoir oil is almost impossible and analysis of the stock tank oil may be difficult because of the very large numbers of compounds present. Many of these compounds are similar in chemical and physical properties and thus separate analysis of them can be difficult. API project 6 has been investigating since 1927 a Ponca City crude oil. Mair (1964) states that in 37 years, 234 compounds have been isolated and identified in this project from this oil (Table 3.1). The amounts in which the individual compounds have occurred varied from 0.0004 to 2.3% by volume (Baker, 1962).

The use to which an analysis of a crude oil is to be put will depend upon who wants to use the results of such an analysis, and thus the actual analysis may vary depending upon its final user and use. The geologist is interested in the chemical and physical properties of the oils and how these properties, and the composition of the oil, have altered during geological time. These properties, and the variations in them, are related to the origin, migration, and accumulation of the oil, and detailed knowledge of them will assist a geologist in more accurately predicting where he would expect to find oil and its quality. However, the refiner is more interested in the commercially valuable compounds that are present in the oil and which may be made from the oil by catalytic cracking and re-formation within the refinery. Many of these latter compounds have no counterparts in naturally occurring oils and gases.

The bulk of compounds which are found in crude oil are hydrocarbons and these are relatively chemically inert, and thus originally the most important chemical property of oils was that they were readily oxidized in an exothermic reaction. Their reaction with oxygen to liberate energy mostly in the form of heat meant that they could be used as fuels. Because this reaction could also produce light, the oils were used for illumination. Nowadays oil, or many fractions of it, can be used as feedstocks for the chemical industry.

Because of the great range of compounds in crude oils, their physical properties also vary. Oils can vary from greenish-brown, very fluid liquids of low viscosity and specific gravity (< 0.810) to viscous, black, heavy (> 0.985) liquids which often solidify upon cooling. In spite of all these differences in properties, the elemental analyses of crude oils shew that the same elements occur in similar quantities. Table 3.2 shows the range of elemental composition of crude oils.

If the principal difference between crude oils is the variation in the type and amount of compounds that constitute these mixtures and these variations cause great differences in physical properties, why are the elemental analyses so similar? The answer is to be found in Chapter 1, where the chemical composition of the types of compounds that constitute crude oil were described. Most of the compounds are hydrocarbons which are compounds consisting of chains of carbon atoms with hydrogen atoms attached at their

Table 3.1 Compounds identified in one crude oil

Carbon number	4	5	6	7	8	9	10	11	12	13	14	15	16	17	18	Total
branched paraffins	1	1	4	6	15	7	5				1	1				41
alkyl cyclopentanes		1	1	5	13	2										22
alkyl cyclohexanes			1	1	8	3	1									14
alkyl cycloheptanes				1												1
bicycloparaffins					3	3	5	1								12
tricycloparaffins							1									1
alkyl benzenes			1	1	4	8	22	4								40
aromatic cycloparaffins						1	4	3			1			2	1	12
fluorenes											1	2	3	1		7
dinuclear aromatics								2	12	15	5	1	1	1		37
trinuclear aromatics											1	4	1	1		7
tetranuclear aromatics													1	1	1	3
sulphur compounds										1	1		1	1		4
Normal paraffins C_1 to C_{33}																33
Grand total																234

The boiling points range −161.48 °C to 475 °C. The volume percentages, where determined, range 0.0004–2.3% (*n*-heptane), the indicated total exceeding 45.6%.

spare valencies. On average, each carbon atom uses two valencies to bond to two other carbon atoms and two valencies to bond to two hydrogen atoms. Thus the basic building-block of hydrocarbon chains is the methylene group ($-CH_2-$). Admittedly, some compounds, the unsaturated ones, will have less than two hydrogen atoms per carbon but some will have more, e.g. methyl groups ($-CH_3$), and it is reasonable to consider hydrocarbons, which are the major components of oils, as a large number of methylene groups arranged in a wide variety of ways. The elemental analysis of a methylene group shews it to contain 85.7% carbon and 14.3% hydrogen. These figures should be compared with those in Table 3.2

Table 3.2 Elemental composition of oils

Element	Weight percentage
carbon	84–87
hydrogen	11–14
oxygen	0.1–2.0
sulphur	0.06–2.0
nitrogen	0.1–2.0

Since all these hydrocarbons are so similar in their composition and chemical properties, why do the physical properties of oils vary so much? Many of the physical properties of hydrocarbons, such as specific gravity, boiling, and melting points, vary with chain length. In general, the larger the molecule the denser it becomes, and the higher its boiling and melting points. Thus the heavy, viscous oils have a higher proportion of larger molecules as compared with the lighter, less viscous oils.

TYPES OF COMPOUNDS OCCURRING IN PETROLEUM

Alkanes

The alkanes are a homologous series of saturated hydrocarbons of general formula C_nH_{2n+2}. The simplest member of this series of methane (CH_4), followed by ethane (C_2H_6), propane (C_3H_8), and butane (C_4H_{10}). In crude oils all the normal alkanes from C_1 to C_{40} (and a few beyond C_{40}) have been identified. On average a crude oil will contain somewhere between 15 and 20% of such compounds; however, this figure may rise to as high as 35% in very paraffinic crude oils, or drop to zero in the case of heavily biodegraded oils.

Once an alkane contains more than three carbon atoms in its chain, structural isomerism may occur. This means that the same number of carbon and hydrogen atoms may be put together in a variety of ways. This variety increases with the increase in the number of carbon atoms. Table 3.3 illustrates the great increase in the number of possible isomers as carbon number

Table 3.3 Number of alkane isomers for various carbon numbers

Carbon numbers	No. of possible isomers
$C_1 C_2 C_3$	1
C_4	2
C_5	3
C_6	9
C_{12}	355
C_{15}	4,347
C_{18}	60,523
C_{25}	36.8×10^6
C_{40}	62×10^{12}

increases, and Figure 3.1 shows the variation with carbon number of boiling points, melting points, and densities. Not all these compounds have been discovered or synthesized and only a very small proportion of the total number of possible isomers of C_1 to C_{40} actually occur in petroleum.

Paraffins are the most abundant hydrocarbons in both gaseous and liquid petroleum. Methane is the predominant component of natural gas and alkanes between pentane (C_5) and pentadecane (C_{15}) are the chief constituents of straight-run (uncracked) gasolene or petrol. Above C_{17} the alkanes are solid, waxlike substances and crude oils which contain high concentrations of this paraffin wax will be viscous and have high cloud and pour points (see 'Physical properties of oils' below). This paraffin wax may also contain isoalkanes, anteisoalkanes, and methylcycloalkanes.

Isoalkanes are 2-methylalkanes (Figure 3.2) and quite a number of these have been discovered in crude oils as have been the anteisoalkanes, the 3 methylalkanes (Figure 3.2). These compounds normally occur in the range up to C_{30} and they are especially prevalent with carbon numbers between C_5 and C_8, although paraffin wax contains many iso- and anteisoalkanes. In the medium molecular-weight range (C_9–C_{25}) the isoprenoid compounds, on which there is a methyl group on every fourth carbon atom of the chain, are common, although the most abundant are pristane, 2,6,10,14-tetramethyl pentadecane (C_{19}) and phytane, 2,6,10,14-tetramethylhexadecane (C_{20}) which can together account for up to 55% of all isoprenoid hydrocarbons in a crude oil (Figure 3.2). Pristane and phytane are believed to be the remnants of the phytyl side chain of chlorophyll (Figure 3.3).

Alkenes

Alkenes are reactive compounds and are thus fairly rare in crude oils; however, small quantities of *n*-hexene, *n*-heptene, and *n*-octene have been identified. Alicyclic alkenes have also been observed in crude oils (Figure 3.4). The remains of natural product species found in oils and sediments may also contain unsaturated rings.

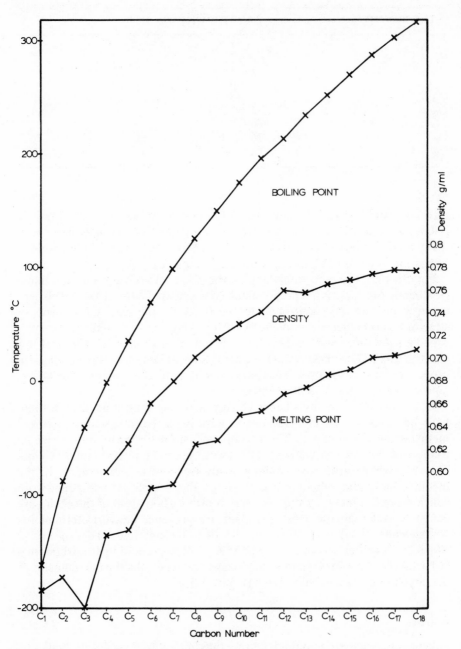

Figure 3.1 Dependence of some physical properties of *n*-alkenes upon the number of carbon atoms

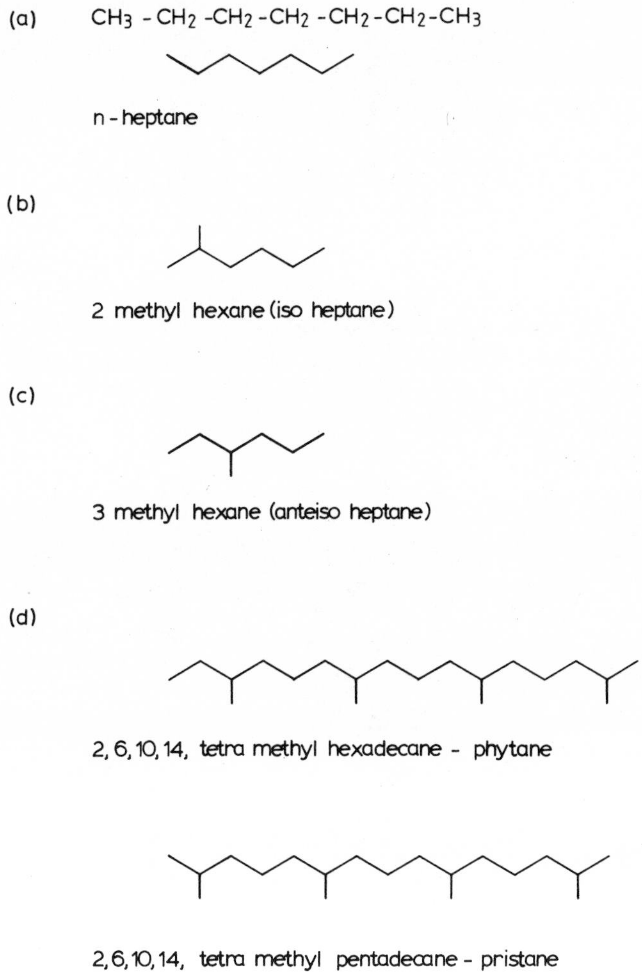

(a) $CH_3 - CH_2 - CH_2 - CH_2 - CH_2 - CH_2 - CH_3$

n - heptane

(b)

2 methyl hexane (iso heptane)

(c)

3 methyl hexane (anteiso heptane)

(d)

2, 6, 10, 14, tetra methyl hexadecane - phytane

2, 6, 10, 14, tetra methyl pentadecane - pristane

Figure 3.2   Some saturated hydrocarbons found in oils

Naphthenes (cycloparaffins, alicyclic compounds)

The naphthenes have this general formula C_nH_{2n} and are rings of methylene groups with or without substituents (Figure 3.5). The smallest of these compounds is cyclopropane, which has three carbon atoms. In crude oils the most common naphthenes are the five- and six-membered rings, and occasionally a few have rings with seven carbon atoms. The three- and four-membered rings are very reactive due to bond angle strain. Naphthenes can have substituent groups attached to the ring, e.g. methyl cyclohexane, or they can consist of two or more naphthene rings fused together, e.g. decalin (Figure 3.5). In crude oils the methyl derivative is the most abundant

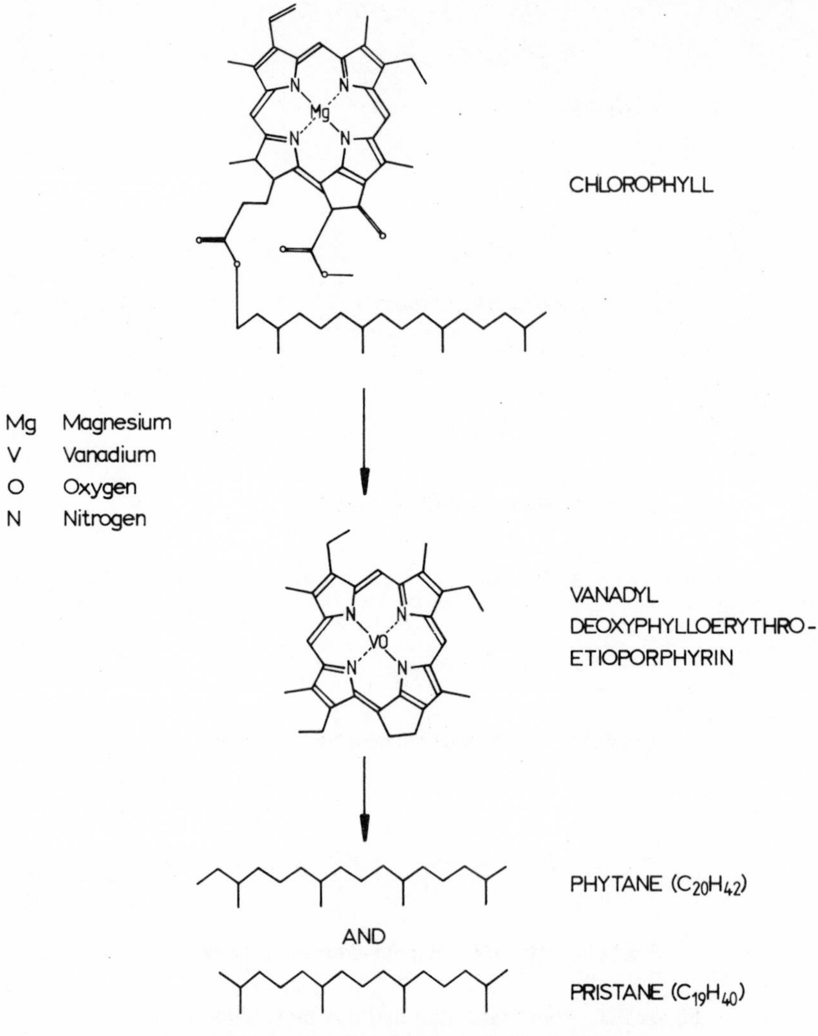

CHLOROPHYLL

Mg   Magnesium
V   Vanadium
O   Oxygen
N   Nitrogen

VANADYL
DEOXYPHYLLOERYTHRO -
ETIOPORPHYRIN

PHYTANE $(C_{20}H_{42})$

AND

PRISTANE $(C_{19}H_{40})$

Figure 3.3 Degradation of chlorophyll

compound, being present in greater quantities than the parent alicyclic compound.

Crude oil can contain up to 50% of such naphthene compounds, although a majority of these molecules may be in the higher molecular-weight fractions and are the fused-ring naphthenes rather than the simple naphthenes found in the lower-boiling fractions.

Most of the cycloparaffins below C_{10} are single-ring compounds. For alicyclics above C_{10}, about 50% are mono- or dicyclic compounds with alkyl chains as substituents. The normal combination is one long chain and several methyl or ethyl groups (Hood *et al.*, 1959). The remainder of the cycloparaffins are compounds with three to five rings of five or six members each. The

(a) Alkenes

$$CH_3-CH_2-CH_2-CH_2-CH=CH_2 \quad \text{hexene}$$

(b) Cyclo alkenes

cyclo pentene   1,3, cyclo hexadiene

Figure 3.4   Some unsaturated hydrocarbons found in oils

compounds with three rings can be less than 20% of the total cycloparaffin content of an oil. Typical compounds are perhydrophenanthrene and adamantine (Figure 3.5).

The tetra- and pentacycloalkanes can be 25% of the total naphthene content greater than C_{10} of an oil. These are normally direct descendants of the tetracyclic steroid and pentacyclic triterpene natural product compounds. The four-ring compounds normally have 27–30 carbon atoms compared with 27–32 for the five-ring compounds. These compounds have been incorporated into crude without being involved in the kerogen and with very little structural alteration, whereas the one- and two-ring compounds are often the products of the breakdown of the kerogen.

The naphthene fractions of crude oils are often analysed by mass spectroscopy to identify the percentage of compounds with one, two, three, four, and five rings. This information can be used in maturation level assessment. The tetra-and pentacycloalkanes are most abundant in young and immature crude oils.

Aromatics

The aromatic, or benzene, series of compounds are so named because many members have a strong aromatic odour (Greek, *aroma* − 'fragrant smell'), and all are derivatives of benzene. The general formula depends on the number of fused rings and is C_nH_{2n-6R}, where R is the number of rings. All aromatic hydrocarbons contain at least one benzene ring.

Aromatics rarely amount to more than 15% of the total crude oil and they are frequently concentrated in the heavy fractions, e.g. in gas oil, lubricating oil, and the residuum. The alkyl derivatives of benzene, e.g. toluene and the xylenes, are the most common aromatic compounds in petroleum and occur in greater amounts than the parent compound benzene. The other derivatives of benzene are the fused-ring compounds (Figure 3.6), the di-aromatics (naphthalenes) and tri-aromatics (phenanthrenes or, more rarely, anthracenes). Again, the compounds with alkyl substituents are more common than

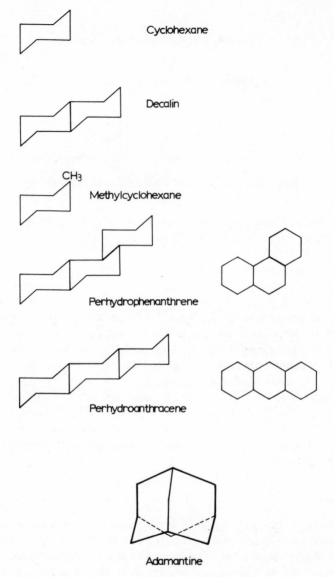

Figure 3.5 Typical cyclic hydrocarbons (naphthenes) found in oil

the parent compounds, and di- and trimethyl naphthalenes and phenanthrenes are the most frequently found multiring aromatics. This is an indication that these compounds are derived from natural product molecules.

Naphtheno-aromatics

There are in crude oils compounds which are partially aromatic and partially naphthenic. These naphtheno-aromatics and multiringed compounds, of

CH3
Benzene
Toluene
CH3
a
Naphthalene
CH3
Xylene
b
c
a. meta
b. ortho
c. para
Anthracene
Phenanthrene

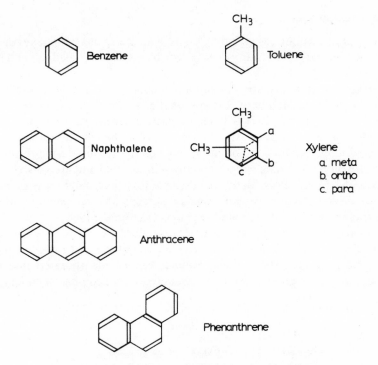

Figure 3.6 Types of aromatic hydrocarbons found in oils

which at least one ring is aromatic and at least one naphthenic. There may also be aliphatic side chains. These types of molecules are particularly abundant in shallow, immature oils and many of the compounds can be related to steroid and triterpenoid structures from which it is believed they originated (Figure 3.7).

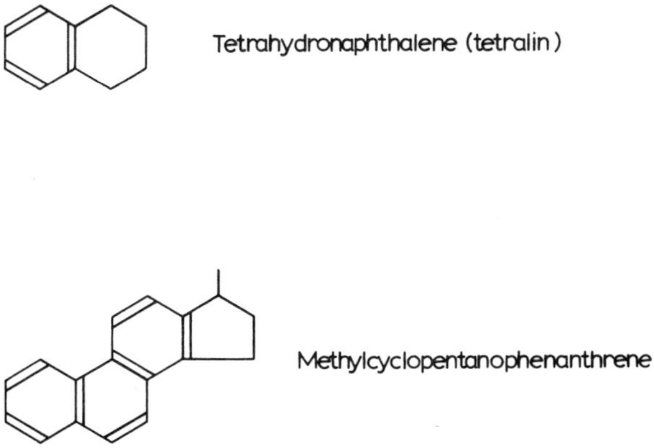

Tetrahydronaphthalene (tetralin)
Methylcyclopentanophenanthrene

Figure 3.7 Some naphtheno-aromatic compounds isolated from oils

80

Sulphur compounds

Many sulphur compounds including thiols, sulphides, and thiophenes occur in crude oils, and sulphur is fairly evenly distributed over the medium and heavy fractions of crude oils.

Mercaptans or thiols are the sulphur analogues of alcohols and have a general formula RSH, where R is any alkyl or aryl group, e.g. methanethiol, CH_3SH, or thiophenyl, C_6H_5SH. Thiols extracted from the gasolene fractions of crude oils contain up to six carbon atoms, although the oil could have contained longer-chain compounds which were destroyed during distillation. Thiols are liable to attack copper and brass in the cold and are very corrosive compounds. Most thiols found in oils have less than eight carbons atoms. Besides normal alkyl mercaptans, isoalkylmercaptans, cyclopentanethiol, and cyclohexanethiol have been found in petroleum (Smith, 1966).

The dialkyl sulphides have the general formula $R-S-R'$, where both R and R' are alkyl groups. Arylalkyl and diaryl sulphides have aryl groups in place of one or both of the alkyl groups. Sulphur can be incorporated into cyclic compounds, although only the five-membered thiophene has been identified

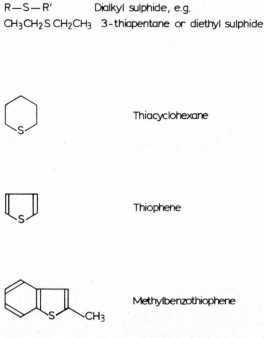

R—SH	Thiol or Mercaptan, e.g.
CH_3CH_2SH	Ethanethiol or Ethyl mercaptan
R—S—R'	Dialkyl sulphide, e.g.
$CH_3CH_2S\,CH_2CH_3$	3-thiapentane or diethyl sulphide

Thiacyclohexane

Thiophene

Methylbenzothiophene

Figure 3.8   Examples of sulphur compounds found in oils

among the aromatic compounds (Figure 3.8). Among the thiacycloalkanes, thia-adamantane has been discovered.

All sulphur compounds are foul, evil-smelling, lachrymatory, and corrosive, and lower the value of crude. Not only will the handling of the oil be more difficult, but total removal of sulphur compounds is needed; otherwise the products may be corrosive, foul-smelling, and dangerous when used.

It is worth noting that these types of compounds, as with many others in crude oil, are found in other situations to which one is more accustomed. For example, $CH_3CH_2CH_2SH$, or n-propyl mercaptan, is the chemical which gives the onion its distinctive aroma and which causes one to cry when peeling onions. $CH_3CH_2CH_2CH_2SH$, or n-butyl mercaptan, is not quite such a household chemical as it is responsible for the smell of the skunk. Dialkyl sulphides are often part of the active ingredient of spices and $CH_2=CHCH_2-S-CH_2CH=CH_2$ is the chemical which is found in garlic. Most of these sulphur compounds can be considered as derivatives of hydrogen sulphide (H_2S) which is itself corrosive, highly poisonous, and foul-smelling.

The exact quantity of sulphur compounds in crude oils is very difficult to establish. Those mentioned above are some of the compounds that have been isolated, albeit in very low abundance. Typical quantities are 0.001% by weight of the crude oil. In high-sulphur crudes a lot of sulphur is released as hydrogen sulphide during distillation, indicating that the sulphur was either present in the crude as dissolved H_2S or in compounds which decompose under the conditions of distillation. Sulphur is also known to occur in the heavy asphaltene fraction of the oils and in general the sulphur content increases as the average boiling point of the fraction increases. Gruse & Stevens (1960) quote Newton & Leach (1929) as distilling a large quantity of a West Texas crude of 2.03% sulphur. They noted that half the sulphur was released as hydrogen sulphide from all fractions, but mostly within the temperature range 300–400 ° F. The increasing sulphur content with increasing boiling point is shewn in Table 3.4 (Newton & Leach, 1929).

Table 3.4 Sulphur content of distillation fractions of a Texas oil

Fraction	Sulphur (%)
gasolene	0.34
heavy naphtha	0.57
kerosene	0.60
gas oil	0.68–1.04
lubricating distillates	1.50–2.5
coke	3.1

Nitrogen compounds

Nitrogen compounds are unwelcome in crude oils as they are responsible for the poisoning of catalysts and the formation of gums in fuel oils. The compounds that have been most investigated are those which can be easily separated and include the pyridines, quinolenes, indoles, pyrolles, and carbazoles (Seifert, 1969; Figure 3.9). However, these compounds are only a very small percentage of the total nitrogen compounds in oils. The majority of nitrogen in crude oils is in the higher-boiling fractions of the oils, and often 95% of all nitrogen compounds are found in residues with boiling points over 300° C at 30 mm Hg pressure. Porphyrins contain four indole nuclei and a metal, commonly nickel or vanadium. These compounds are remnants of the breakdown of chlorophyll moelcules (Figure 3.3). The more asphaltic crude oils have higher nitrogen contents and this confirms the idea that most of the nitrogen in oil is in the higher-boiling fractions.

TYPE   STRUCTURE

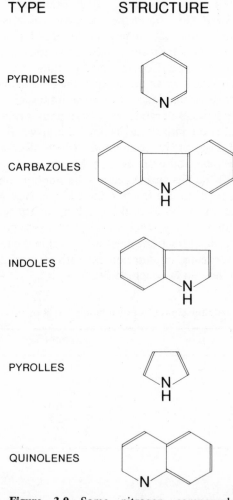

PYRIDINES

CARBAZOLES

INDOLES

PYROLLES

QUINOLENES

Figure 3.9 Some nitrogen compounds found in crude oils

Oxygen compounds

The most important oxygen compounds in oils are the organic acids and they are especially common in young and immature crude oils. These acids range in size from C_1 to C_{30} and include some isoprenoid structures. Cyclic aliphatic acids also occur, but in smaller amounts in most oils. Sometimes these are in the form of naphthenic-aromatic acids. Aliphatic acids predominate below C_8 with monocyclic acids being the major acid components between C_9 to C_{13}. Above C_{14} the bicyclic acids are dominant. There are also a few steroid carboxylic acids to be found in some crude oils (Seifert, 1973). In addition to acids, there is evidence of alcohols, phenols, ketones, esters, and anhydrides (Seifert & Howells, 1969). Figure 3.10 illustrates some of the oxygen compounds found in oils.

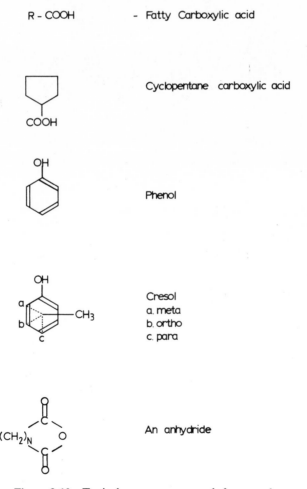

Figure 3.10   Typical oxygen compounds from crude oils

84

Oxygen, like nitrogen and sulphur, can occur in the asphaltic part of crude oil. This is the high molecular-weight component of oils and the molecules can contain oxygen, nitrogen, or sulphur, or any combination of them. As the molecular weight increases, the likelihood of finding these heteroatoms increases.

The structure of the asphalts is a series of sheets each made of varying basic units. The basic units are aromatic and naphtheno-aromatic molecules condensed into sheets. Sulphur is present in benzothiophene groups, nitrogen in quinolene groups, and oxygen in both the units, as furan and pyrans and their derivatives or as ether bridges between the sheets (Figure 3.11). As these units unite and become closer together their solubility decreases. These compounds can form complexes with metal ions.

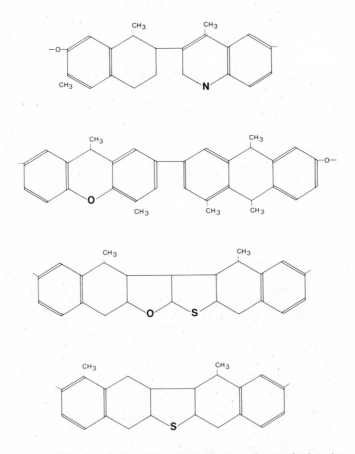

Figure 3.11 Asphalt nuclei—theoretical structures of sub-units. From *Chemical Technology of Petroleum* by Gruse & Stevens. Copyright © 1960. (Used with permission of McGraw-Hill Book Company)

Organo-metallic compounds

Most crude oils contain metals, normally in porphyrin complexes, derived from chlorophyll and haemoglobin. The commonest metals are vanadium and nickel, although copper and iron and a wide variety of other metals in lesser amounts have been reported. There is a good correlation between metal content, sulphur content, and asphaltene present in the oil.

OCCURRENCE IN PETROLEUM

In this section the products that can be obtained from crude oil will be described and discussed. The majority of these products are obtained by fractional distillation but some are removed from the oil by precipitation or solvent extraction. The fractions obtained by these latter processes include the waxes, resins, and the asphaltenes. At the end of the distillation there will remain a residuum. The distillation tower is so designed that the various products are removed at different levels, with fresh crude being continuously introduced. In general, the lighter fractions will contain more paraffins than aromatic and asphaltic compounds.

Gas

Gas mostly consists of methane (CH_4) with decreasing amounts of ethane (C_2H_6), propane (C_3H_8), and normal and isobutane (C_4H_{10}). There can be occasional traces of compounds as high as nonane (C_9H_{20}). If the gas is produced from oil then the actual composition will depend will depend upon the physical conditions of the separation (see Chapter 5). In addition to the organic constituents, there are inorganic gases present in natural gas, and these include carbon dioxide (CO_2), nitrogen, hydrogen sulphide (H_2S), helium, and water vapour. The gas which is sold is mostly methane with as little as possible of the other compounds.

Hydrogen sulphide is one of the most poisonous gases known. The reason it causes less fatalities than less poisonous gases, such as carbon monoxide, is because the threshold of detection of H_2S by the human nose is considerably lower than the threshold of toxicity, i.e. one smells it well before it becomes dangerous. The reverse is true for carbon monoxide and by the time a high enough concentration to smell this gas is reached, a fatal dose has probably been taken. Hydrogen sulphide can be used in the manufacture of sulphuric acid and other sulphur products, e.g.

$$2H_2S + 3O_2 \rightarrow 2SO_2 + 2H_2O$$
$$2H_2S + SO_2 \rightarrow 3S + 2H_2O$$

Natural gas is termed 'sweet' when it contains no hydrogen sulphide, and 'sour' when it does.

Brown & Norris (1982) report the use of hydrogen peroxide to remove hydrogen sulphide. Under acidic or neutral conditions and in the presence of a catalyst such as iron the reaction takes 1–2 minutes at room temperature or 2.6 seconds at 140 °F (60 °C)

$$H_2S + H_2O_2 \rightarrow S + 2H_2O$$

The reaction under alkaline conditions is slower.

$$Na_2S + 4H_2O_2 \rightarrow Na_2SO_4 + 4H_2O$$

Carbon dioxide can, if present in sufficient quantities, be removed and used to manufacture 'dry ice', assuming the availability of a suitable market. Hubbard (1982) reports that nearly one third of the gas in the North Sea Brae A field is carbon dioxide which is enough to eat through a normal steel pipe within weeks. The sweetening equipment has caused the platform to weigh almost as much as those designed to work in water twice as deep.

Water must also be removed from natural gas because it causes problems during the transmission of gas through pipelines in cold climates. Water will form a hydrate with a hydrocarbon, normally methane, in which six water molecules form around each hydrocarbon molecule. These are known as clathrate compounds, where the hydrocarbon is physically trapped within a solid ice matrix. These compounds resemble snow and are white snowy crystals with hydrocarbons trapped inside. These hydrates can form at temperatures close to freezing and in sufficient quantities to block gas pipelines at very moderate degrees of frost (Katz, 1971).

Wet gas contains hydrocarbons which are normally liquid at room temperatures and pressures but have a high enough vapour pressure to exist as a vapour. These hydrocarbons, which can include those in the gasolene range, can be removed from the gas by cooling (in the same way that water can be condensed out of the atmosphere on to a cold surface). If these condensable hydrocarbons are present in sufficient number they can be sold profitably. In the early days of the North Sea exploration, when only gas from the southern part was being extracted, many garages in East Anglia were selling 'North Sea petrol'. This was before any oil had been brought ashore and this product was the treated condensate from North Sea gas fields. Dry gas does not contain any such condensable hydrocarbons.

Nitrogen is the one component whose removal is difficult and expensive as, like methane, it needs cooling under pressure in order to liquefy it. If the nitrogen content is not too high the gas can be sold with its associated nitrogen. There is no danger but the gas has a reduced calorific value.

Natural gas is normally transmitted by pipeline. Transportation by tanker involves liquefication of the methane, a process which involves cooling and pressurizing. Thus transport of methane by bulk tanker is from a distant supply point to a distribution centre, whence it can complete its journey as a gas in a pipeline.

Many of the light hydrocarbons, including methane (CH_4), ethylene (C_2H_4), propylene (C_3H_6), butylene (C_4H_8), and benzene (C_6H_6), are used as feedstocks for the chemical industry, where they are converted into synthetic fibres, plastics, synthetic rubber, soaps, detergents, paints, drugs, cosmetics, and agricultural chemicals. Although ethylene, propylene, and butylene, as they as olefins, are not present in natural gas or crude oil, they are easily formed in the refinery by cracking heavier hydrocarbons. Other uses of natural gas include heating, cooking, and lighting.

Liquefied petroleum gases (LPG)

This fraction consists of the normal and branched isomers of propane (C_3H_8) and butane (C_4H_{10}). These gases can be liquefied without any cooling by an increase in pressure. Thus they can be transported as a liquid and used as a gas. As with natural gas, these gases can be used as a source of heat and light, but with the advantage that no pipeline is necessary. Thus they can be used in isolated situations or where mobility is required.

Ligroin

This contains the normal and branched isomers of pentane (C_5H_{12}) and hexane (C_6H_{14}) and has a boiling range from room temperature to 70 °C. It is used for solvents, cleaners, and paint thinners.

Often all the components of crude oil which boil below 100 °C will be grouped together and classified as 'light gasolene'. All the light gasolene hydrocarbon fractions can be obtained as condensate, a mixture which is gaseous under reservoir conditions but which is a liquid when produced at the surface. It will contain all the normal and branched paraffins and aromatics up to, and including, the gasolene range (Hitcham, 1972).

Gasolene

This fraction has a boiling range from 70 to 180 °C and consists of hydrocarbons from C_7 to C_{11}. Below a boiling point of 132 °C for the alkanes and cycloalkanes, and below 180 °C for the aromatics, nearly all the possible members of these series were found to be present in the Ponca City Crude (API project 6; Mair, 1964). A few possible members of the branched and cycloalkanes were either missing or present in only trace amounts. Of the paraffins, the normal compounds were the most abundant, with the singly branched 2-methyl isomers the most abundant of the branched compounds. In the cycloalkanes and the aromatics the component with the single methyl substituent was present in greater abundance than any other isomer including the parent compound.

One of the principal uses for the gasolene-range hydrocarbons is as petrol as a fuel for internal combustion engines. When normal paraffins are used in petrol engines the initial explosion is followed by secondary explosions, known as 'knocking'. This phenomenon is overcome by the use of lead tetraethyl as an anti-knock additive. However, this compound ($Pb(C_2H_5)_4$) is being phased out so as to reduce lead pollution, and organic anti-knock agents are being produced by catalytic reforming of petroleum hydrocarbons in the refinery.

When this phenomenon was first noticed, various chemicals were investigated to discover which caused most and least knocking. 2,2,4-trimethylpentane ('iso-octane') caused least knocking and n-heptane caused most knocking. These were labelled 100 and zero respectively on the 'Octane scale'. Later work has shewn that compounds exist with an octane rating greater than 100, i.e. they cause even less knocking, and that the more compact the molecule the less knocking it causes when burnt in a petrol engine. n-Octane (a long straight-chain molecule) has an octane rating of -19, whereas 2,2,3-trimethylbutane has an octane rating of 112. Aromatic compounds have even higher octane ratings. (The name 'iso-octane' is used in the petroleum industry for 2,2,4-trimethylpentane. This usage is chemically incorrect as isoalkanes are the 2-methylalkanes).

Because the highly branched paraffins, which are especially valuable to aviation and motor fuels due to their anti-knock properties, are not present in straight-run gasolenes, they have to be produced by catalytic reforming at the refinery.

Very few cycloparaffins have been discovered in crude oil, but some which have been shewn to occur in gasolene-range fractions of oils are illustrated in Figure 3.5. The normal paraffin content of gasolenes varies from under 10% to over 60%, while the aromatics can vary from less than 5% to 30% by weight of the gasolene fraction. The branched and cycloparaffins shew similar variations (13–32% and 8–41% respectively). Gasolenes shew decreases in n-paraffin and complementary increases in cycloparaffins and aromatics with increase in molecular weight.

Commercially, at the refinery all those compounds with a specific gravity greater than 0.825 (40° API (see p.94 for definition)) are removed from the gasolene fraction and are added to the gas oil distillate.

Kerosene

Kerosene is produced by the distillation of crude oil; it has a boiling range of 180–270 °C and consists of compounds containing between 11 and 15 carbon atoms per molecule. Kerosene was originally used for lighting purposes but is now an ingredient in jet fuels. Kerosene-range compounds can also be obtained in the refinery by the cracking of heavier molecules.

Kerosene contains many cyclic molecules with the aromatics varying between 10 and 40%, which is higher than the average crude oil. There are also many more condensed naphtheno-aromatics and multiring cycloparaffins in kerosene than in the gasolene fraction. Most of the types of compounds in kerosene are smaller versions of those found in gas oil and will be discussed below.

Gas oil

This fraction is composed of molecules with 15–25 carbon atoms and a boiling range of 275–400 °C. It consists of those fractions which distil between the kerosene and lubricating oil distillates. Gas oil is used as jet and diesel fuel, and those fuels need the opposite specification to those of an internal combustion engine. Diesel engines are compression-ignition engines where the compression of the air by the piston raises the temperature of the air so that the fuel ignites when injected into the cylinder. Long-chain paraffins which would cause knocking in petrol engines are much more suitable for diesel engines. Whereas iso-octane is the standard for the internal combustion engine, n-hexadecane (cetane, $C_{16}H_{34}$) is the standard for the diesel engine. High octane ratings are low cetane ratings and vice versa.

In straight-run gas oil there are very few long-chain hydrocarbon materials, and its major components are cyclic compounds. This is the opposite situation to the preceding fractions. Bicycloparaffins and dinuclear aromatics occur in appreciable amounts in kerosene, while the tricycloparaffins and trinuclear aromatics are important in gas oils. Mixed-type hydrocarbons (naphtheno-aromatics) are present in both kerosene and gas oils and include such compound classes as indanes and tetrahydronaphthalenes. Most of the aromatic compounds in these fractions have fused rings, but biphenyl and substituted biphenyls have been identified in gas oil fractions (Adams & Richardson, 1953). Trinuclear aromatics of the phenanthrene structure have been reported by Martin & Semkin (1953), while ten homologues of the less common anthracene have been isolated from a Kuwait crude (Carruthers, 1956). Some four-ring compounds have also been isolated from the gas oil fractions of crude oil (Moore *et al.*, 1953).

In general, as the boiling points of the fractions of crude oil rise, there is a marked decrease in the quantity of paraffins and a corresponding increase in cycloparaffins and aromatic compounds. These compounds become complicated in structure, having many rings and side chains. Among the aromatic compounds, many heteroatomic molecules, including benzothiophene and naphthobenzothiophene derivatives, have been found.

Lubricating oil

The lubricating oil fraction normally contains compounds from C_{26} to C_{40}, but in extreme cases this range is widened (C_{20} to C_{50}). Detailed compositions of the lubricating oil fractions have proved very difficult to obtain. However, it is known that these fractions contain n-paraffins in the C_{22}–C_{40} range as well as a wide range of branched paraffins, cycloparaffins, and mono-, di-, and trinuclear aromatics of similar type to those in the gas oil fraction, but with a higher molecular weight. Of the cycloparaffins, the most abundant are those with one to five rings, but small amounts of compounds with up to ten rings have been shewn to be present. The aromatics are also predominantly of the fused-ring type with phenanthrene derivatives more common among the three-ring aromatics than the anthracene-based compounds. Much of the aromatic content is to be found in naphtheno-aromatic compounds.

Highly paraffinic crudes will have a high wax content due to the high percentage of normal, iso-, and anteisoalkanes in them. This wax will affect the viscosity of the oil. Pour point and cloud point are methods by which the quantity of paraffin wax with a crude oil can be measured. These tests are empirical and useful for specifying the behaviour of oils at low temperatures, and are determined by allowing a sample of oil to cool in a standard apparatus. The temperature at which turbidity is observed is the cloud point, while the temperature below which the oil will not flow in a definite manner is the pour point. Both these temperatures are determined by the precipitation of paraffin wax. The rate of cooling is important, and hence the need for standard conditions. Oils which are completely wax-free, such as some naphthenic oils, will shew no cloud point, but will have a pour point, albeit very low (ASTM Standard D 97–47).

Heteroatomic compounds containing nitrogen, sulphur, and oxygen tend to be concentrated in the high-boiling fractions. Although some are found in the gas oil fractions, the majority are in the lubricating fraction and residuum. It is the concentration of these heteroatomic compounds which gives these fractions their colour and smell. The composition of a finished lubricating oil will be markedly different from that of the lubricating distillation fraction because some of the wax, non-hydrocarbon compounds, polynuclear aromatics, and multiring cycloparaffins are removed during refining. In general, the presence of medium chain length paraffins gives the lubricating fraction better lubricating properties. The cyclic compounds are detrimental to lubricating oils. However, the use of synthetic chemical additives allows fractions from naphthenic and aromatic crudes to be used for lubricants.

With the exception of the *n*-paraffins, no pure compounds have been isolated from the lubricating oil fractions of crude oils although many compound types have been identified by mass spectrometry.

Lubricating oils are classified by viscosity, refractive index, and density, although of these viscosity is the most important and commonly used.

Residuum

This consists of what is left at the end of a fractional distillation of a crude oil which has produced the fractions described above. The residuum is composed of asphalts, asphaltenes, and resinous compounds. In some very paraffinic crudes the very high molecular weight alkanes will have too high a boiling point to distill over at atmospheric pressure and will remain in the residuum.

The residuum is used for the production of wood preservatives, roofing asphalts, and road asphalts.

Natural solids

There are several groups of compounds which can be removed from crude oil other than by distillation and these groups are the same as those which are often found in isolated deposits. These groups include the paraffin waxes, the asphalts, and resins.

Paraffin waxes

Paraffin wax is composed of the same compounds as the paraffin which occurs in petroleum, i.e. n-paraffins of higher boiling point with some iso-, anteiso-, and cycloparaffins. They can be obtained from petroleum by solvent extraction or by precipitation.

Paraffin wax also occurs as isolated deposits, such as ozerkerite where it is mixed with grit and sand. These deposits occur near enough to the surface to be mined and the wax is purified by melting and filtering so as to remove the inorganic matter. These paraffin waxes, and the similar products that can be obtained from petroleums, consist of compounds higher than $C_{23}H_{48}$ (m.p. 48 °C) and are predominantly straight-chain alkanes with some branched material. Levy *et al.* (1961) describe a paraffin wax containing 39% n-paraffins, 32% isoparaffins, 27% naphthenes, and 1% aromatics.

These compounds are believed to originate from the waterproofing components of the leaves of terrestrial plants. The commercial use of paraffin wax includes candles, paper waxing, anti-rust coatings, polishes, medicines, cosmetics, sealing compounds, electrical insulations, cartridges, coatings for fruit and vegetables, inks, carbon papers, models, and leather dressings.

Resins and asphalts

Resins and asphalts are usually distinguished on the basis of their separation procedure. Precipitation by propane separates resins and asphaltenes from the remainder of the crude, while addition of n-heptane separates the soluble resins from the insoluble asphaltenes. The resins include minor amounts of free acids and esters but the bulk of the resins is composed of molecules similar to, but less aromatic than, the asphaltenes. Hence they have less cohesion and the resins are more soluble than the asphalts.

Asphalts and resins can be classified on the basis of their physical properties which, as will be seen, are related to the size of the molecules which constitute them.

Asphaltic petroleum Asphaltic petroleum is a liquid and is really a petroleum with a very high asphalt content. Its specific gravity is less than 1.0 and all the asphaltene compounds are dissolved, or suspended, in the petroleum hydrocarbons.

Asphalts or neutral resins

These are very rich in heteroatomic molecules, with oxygen being a major heteroatom, and which have a relatively low molecular weight (500–1200). They are semi-solid to solid materials with melting points between 66 to 93 °C. They are between 10 and 70% soluble in petroleum naphtha but between 70 and 98% soluble in carbon disulphide. Their specific gravity is between 1.0 and 1.1 and they are soft, tarry substances which flow when heated. Trinidad Lake asphalt, with the mineral matter removed, is a good example of this class.

Asphaltenes

These are hard solids with melting points well above 150 °C, e.g. Grahamite (m.p. 315 °C), and they are brittle solids which break to give bright shiny surfaces. They are 0–60% soluble in petroleum naphtha but 50–60% soluble in carbon disulphide, and are readily precipitated from oils by *n*-heptane, an occurrence which forms the basis of methods of quantitative assessment of asphaltene content of oils. Their specific gravity is up to 1.2.

Asphaltic bitumens

Examples of bitumen are Wurtzite and Albertite, which have specific gravities greater than 1.5 and are completely insoluble in petroleum naphtha and less and 10% soluble in carbon disulphide. These compounds will swell and decompose before a melting point is reached.

Earlier in this chapter there was a discussion concerning the basic units of the asphalts and although their detailed chemical structure is unknown, they are believed to consist of chains of repeating units. These repeating units contain naphtheno-aromatic sub-units with heteroatoms nitrogen, sulphur, and oxygen present in both the units and the side chains. The greater the length of the chains of repeating units and the greater the degree of linking between these chains, the higher will be melting point and specific gravity and the lower the solubility. In addition, the asphalt nuclei will be more aromatic than the resin nuclei. The asphaltene units group themselves together into particles (molecular weight 3,000–10,000) and into micelles molecular weight > 30,000). The resin units are in the molecular weight range 800–3,000.

As one passes through this somewhat artificial classification, changes in properties (specific gravity, melting point, and solubility) are believed to be due to increasing molecular weight of the components. This increase is due to the greater size of the sub-units and the greater association of the units into particles and micelles with increased bonding between the units.

The main uses of the asphalts is for the production of furnace oils and road asphalts, although they find minor use as binders, fillers, adhesives, and water-insulating materials.

CLASSIFICATION OF CRUDE OILS

Introduction

At first sight it might not be unreasonable to expect that a fairly uniform crude oil could be obtained from an oil-bearing stratum, and any small differences in chemical and fractional composition could be explained as due to local factors such as differential absorption on clays. In many cases and over wide areas the crude oils from the same producing sand can be fairly uniform. However, there are oilfields which are close together but which produce significantly different oils. These differences can be as pronounced as one oil being paraffinic whilst the other is naphthenic or asphaltic.

There is no direct correlation between the age of an oil and its composition and the oldest oils are not always the most paraffinic. The amount of heat to which an oil has been subjected will be the dominant factor in deciding whether that oil is light or heavy, and in general the higher the temperature the oil has reached the more paraffinic it will be. There is, however, a time effect and the oil must stay at a sufficiently elevated temperature for long enough for it to be altered. The depth at which the oil is discovered is likewise no guide to the quality of the oil, and the important factor is the maxiumum temperature to which it has been subjected, although this will be related to other factors such as geothermal gradient, igneous activity, and local heating which can cause variations. In addition, the depth at which an oil is found may not be the maximum at which it has been, either in the reservoir or in the source rock.

Classification systems can be of great value in assessing the relationship between oils, which can in turn assist exploration. No successful single method of classification of petroleums has yet been devised. Various systems involving physical properties, such as specific gravity, or the detailed composition of the oil are used with varying degrees of success.

Oils are often described in terms of their 'base'. This word is not used in the acid–base sense, but more to describe the principal component of the oil. The use of the word base is a historical accident due to the fact that the early analyses of oil were carried out by chemists who were mainly involved with the preparation of ointments, salves, lotions, cosmetics, etc. In such preparations the active ingredient is held in an inactive base. Examination of the first oils showed that these oils contained paraffin wax which could be precipitated by chilling, and thus the early investigators, because of their previous experience, considered that these oils were similar to ointments with a base of wax. Later, when naphthenic and asphaltic oils were analysed, the term 'base' was used in the description of these oils, and however inaccurate the original reason for the use of the word, it is well known and often used.

Classification of crude oils is of great importance to the petroleum geologist so that he can compare crude oils from different depths, horizons, and ages, and to assist him in solving the problems of generation and migration. Classification is also of great assistance to the refiner who wishes to know the amounts of the successive distillates (gasolene, kerosene, gas oil, lubricants, etc.) that can be obtained from the crude oil. The refiner is also interested in the physical properties of the crude oil and its fractions as indicated by such tests as viscosity, cloud point, pour point, specific gravity, refractive index, etc. It is important to make the greatest economic use of crude oil. A knowledge of the type of compounds that any oil is likely to contain, as indicated by its classification, will aid decisions as to the best way of refining that oil, and which combinations of oils will be needed to be put together so as to operate a refinery to its optimum and to produce the correct range of products for a particular market (Smith, 1940; Sachanen, 1945; Dow, 1974).

Physical classification

Early classification systems were based on the likeness of crude oils to organic compounds; for example a paraffinic crude would have straight-chain hydrocarbons dominating in the gasolene fraction, would be of low specific gravity both as a crude and in its fractions, would contain paraffin wax, and would leave little residue when distilled under cracking or non-cracking conditions. Other types of oils classified according to this system were intermediate, asphaltic, naphthenic, and aromatic. All these terms were misleading as an asphaltic-based crude could contain little asphalt but much more naphthenic compounds; whereas a naphthenic-base crude could contain little naphthenic material. The terms were meant to imply that the crudes resembled their namesake, especially in physical properties such as specific gravity, rather than that the crudes were predominantly composed of the compound types mentioned in the name. Paraffins have lower specific gravities than naphthenes, which are lighter than aromatic compounds, and thus paraffinic crudes are the lightest crudes, followed by naphthenic and aromatic.

The standard measure of specific gravity within the oil industry is degrees °API, where

$$°API = \frac{141.5}{\text{specific gravity}(60\ °/60\ °F)} - 131.5$$

The European 'degrees Beaumé' can still occasionally be seen:

$$°Beaumé = \frac{140}{\text{specific gravity}} - 130$$

Both these scales were designed so that hydrometers could be constructed with linear scales. The original Beaumé scale was found to be in error by a constant amount and was replaced in 1921 by the API gravity scale.

The sepcific gravity referred to in the denominator is the specific gravity of the oil. Specific gravity measurements involve the weighing of a known volume of oil and an equal volume of water, the ratio of which gives the specific gravity. Because volume is temperature dependant, both weighings are made at the same temperature, which has been standardized at 60 °F, and these temperatures are recorded, hence the 60 °/60 °F in the denominator.

US Bureau of Mines classification

This classification system is based upon the specific gravity of two distillation fractions of the crude oil, and the specific gravity limits for these two fractions define nine types of crude oil. The crude oil is subjected to a standardized analytical distillation, first at atmospheric pressure and then in vacuum at a pressure of 40 mm of mercury; this is the Hempel distillation. Fractions are collected over intervals of 25 °C and the two key fractions on which the classification is made are: I, 250–275 °C at atmospheric pressure in the kerosene range; and II, 275–300 °C at 40 mm Hg pressure in the lubricating oil range. Table 3.5 lists the specific gravity data and type classifications for the two key fractions, while Table 3.6 shews the use of the key fractions for the

Table 3.5 US Bureau of Mines Hempel distillation, key fractions

Key fraction	Specific gravity	Classification
I	40 °API and higher,	paraffin
	30.1–39.9 °API	intermediate
	lower than 30 °API	naphthene
II	30 °API and higher	paraffin
	20.1–29.9 °API	intermediate
	lower than 20 °API	naphthene

Table 3.6 Key fractions to characterize crude oils

Class		Key fraction I	Key fraction II
1.	paraffin	paraffin	paraffin
2.	paraffin intermediate	paraffin	intermediate
3.	intermediate– paraffin	intermediate	paraffin
4.	intermediate	intermediate	intermediate
5.	intermediate– naphthene	intermediate	naphthene
6.	naphthene– intermediate	naphthene	intermediate
7.	naphthene	naphthene	naphthene
8.	paraffin– naphthene	paraffin	naphthene
9.	naphthene– paraffin	naphthene	paraffin

characterization of crude oils into nine types. Classes 8 and 9 (Table 3.6) have not been recognized in any known crude oil.

In addition to this classification it would be specified whether the oil was wax-bearing or wax-free. In general, those oils whose fractions are paraffinic or intermediate in gravity contain paraffin wax. The presence or absence of wax is determined by the measurement of the cloud point of fraction II. The cloud point is a standard method of precipitating the wax by cooling the oil to make the fraction opaque. If the temperature at which this occurs is below 5 °F, wax is absent, but if the solution becomes opaque above that temperature, wax is present. Wax content is normally quoted when stating the base of the oil. A test which is related to cloud point is the pour point, which is the temperature at which the fraction ceases to flow in a definite manner. Both these tests are related to the precipitation of wax (IP test No. 15/42).

The carbon residue of the undistilled residuum from the crude at 275 °C and 40 mm Hg indicates the asphalt content of the crude. The value of the carbon residue can be obtained by the destructive heating of the residuum to coke, and it is expressed as percentage of the original crude. It is roughly proportional to the amount of hard asphalt that was dissolved in the original oil (Conradson method, IP No.13/42).

Correlation index

The Hempel distillation provides a reasonable classification of crude oils but even so it tends, like previous classification methods, to draw together different crudes under the same heading and to merge small differences when it is often necessary to be able to note these differences and even emphasize them. Whilst a close study of data might reveal these small differences, such a study would be long and tedious.

The correlation index has been devised by the US Bureau of Mines (Smith, 1940) and it is based upon the specific gravity and boiling point of the fractions of the crude oil which are obtained by the Hempel distillation. The correlation index (CI) is obtained from the following formula:

$$CI = \frac{49,640}{T°(abs)} + 473.7G - 456.8$$

where G is specific gravity of the fraction concerned. The correlation index of the n-paraffins is zero and it increases with the increase in cyclic character of the hydrocarbons present in the fraction. Cyclohexane has a correlation index of 51.4, whereas for benzene it is 100 and the polycyclic aromatics even higher.

The simplest way of obtaining correlation indices from experimental data is not by use of the formula but by the construction of a reference framework (Figure 3.12). On this the specific gravity is plotted against the reciprocal of the boiling point in degrees absolute. The normal alkanes fall approximately on a straight line of correlation index value zero. Lines are drawn parallel to the

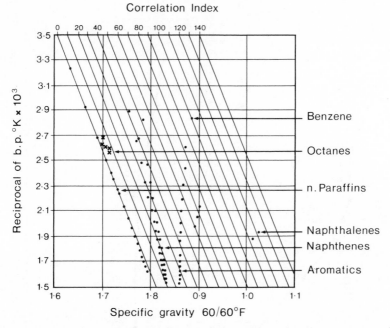

Figure 3.12 Reference frame for correlation indices of crude oil fractions (Smith 1940) (after Gruse & Stevens, 1960). (Reproduced by permission of Bureau of Mines, US Dept. of Interior)

paraffin line and that which passes through benzene has the correlation index value of 100. Thus *n*-paraffins are on one side of the diagram, passing onwards to the branched and cycloalkanes, to the substituted and condensed aromatics at the other extreme. Once the average boiling point and specific gravity of a fraction have been obtained the correlation index is read directly off the diagram.

The correlation index value rises as the cyclic character of the hydrocarbons present in fraction increases, and the following significances can be drawn. Fractions with indices from 0 to 15 are predominantly paraffinic, whereas those whose index is between 15 and 50 are either naphthenic or a mixture of all oil types. Those fractions whose correlation index values are greater than 50 will be mainly composed of aromatic rings.

The simplest use of correlation index as a means of comparing crude oils is shewn in Figure 3.13, where the correlation index of each fraction is plotted against the fraction number. Different crude oils can be identified and the predominant types of compounds in them identified.

One problem with the correlation index method, or any method based on physical properties of petroleum, is that it does not define the relative proportions of hydrocarbons, both ring and open-chain compounds, which are present. As mentioned earlier, most fractions with a correlation index between 15 and 50 can be either naphthenic or a mixture of paraffins, naphthenes, and

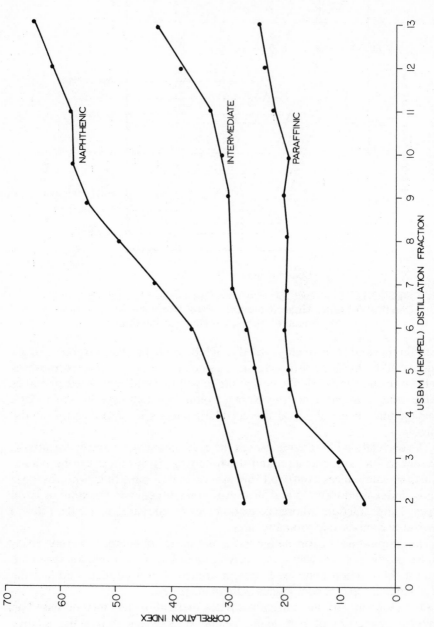

Figure 3.13 Use of correlation index to define oil types. (Reproduced by permission of Bureau of Mines, US Dept. of Interior)

aromatics. To try and overcome this problem, the correlation index may be combined with other information to give an improved classification of a crude oil and to allow better comparisons with other crude oils. Figure 3.14 (Gruse & Stevens, 1960) illustrates how the wax and asphalt content can be incorporated.

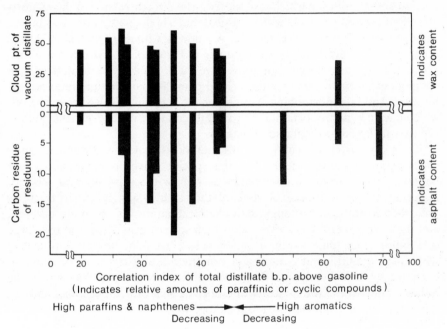

Figure 3.14 Improved oil identification scheme (From *Chemical Technology of Petroleum* by Gruse & Stevens. Copyright © 1960. Used with permission of McGraw-Hill Book Company)

In Figure 3.14 the horizontal axis represents a progression from paraffins with low CI values to aromatics with high CI values. The position of a crude oil on the horizontal scale is fixed by the average correlation index for its fractions which boil between 200 °C at atmospheric pressure and 275 °C at 40 mm Hg. It directly expresses the paraffinic or cyclic nature of the fraction examined. The vertical scale above the horizontal indicates the wax content of the heavy gas oil and light lubricating fractions of the oil. It is related to the average cloud point of the fraction of the oil distilling between 275 °C at atmospheric pressure and 275 °C at 40 mm Hg. The vertical scale below the horizontal is the value of the carbon residue of the residuum (Conradson method) which indicates the amount of asphalt in the crude. Thus this diagram uses three components to classify and identify an oil. These components are the paraffin, the aromatic, and the asphalt contents of the oil.

There are, however, certain inherent weaknesses with such a diagram. The horizontal scale should really be based on the naphthene–aromatic ratio, whereas the correlation index, which is used instead, is greatly influenced by the paraffins that are present, i.e. those which have not distilled over at temperatures lower than the boiling points of the fractions used for the index.

The cloud point test does not include the very high boiling point paraffins which are left behind in the residuum. These paraffins do not distil over because of their high boiling points and they will eventually be cracked to coke and will form part of the carbon residue. Thus they will give a false value for the carbon residue of the residuum which is supposed to be directly related to the quantity of asphalt present. A crude with no asphalt but a high quantity of paraffin wax will have a significant carbon residue. A crude oil with no asphalt should have no component below the horizontal scale in Figure 3.14, but if it has a high proportion of wax it will not only have such a falsely indicated asphalt component on the diagram, but its cloud point and correlation index will be lowered. This false positioning of crudes on the diagram will also occur with oils which have a high proportion of non-volatile components because the method is primarily based on distillation fractions.

The most serious objection to this method is that it is dependent upon specific gravity measurements, and high specific gravity may be due to both naphthenic and aromatic compounds as well as the asphaltic and resinous materials. In the US Bureau of Mines method all high specific gravity fractions are called naphthenic although their chemical nature may be quite different and they can contain any of the other high specific gravity compounds. The correlation index tends to give a slightly better indication because it is dealing with small quantities at a time and can trace the distribution of different components. The two extremes are probably fairly accurate but it is in the middle range where the majority of crudes lie that difficulties arise, and oils quite different in character are grouped together.

Approximate summary

Gruse & Stevens (1960) suggested that because of the great problems involved to obtain accurate analyses and because quite often a quick estimation of the commercial value of the oil and the products obtainable from it is sufficient, the following simple tabulation would be of value;

1. specific gravity
2. gasolene and kerosene distillation
3. sulphur content
4. asphalt content

and to this list the cloud and pour points could reasonably be added.

The specific gravity gives an approximate indication of the relative proportions of the light and heavy, volatile and high-boiling hydrocarbons. Casing head gasolene has a specific gravity of 0.62 while lubricating oils can be 1.0 and some residue as high as 1.06. Crude oils vary between 0.65 and 1.06 depending on the relative amounts of every component present.

The gasolene and kerosene distillates at once define the value of the crude in terms of being a source of these commercially important products. The sulphur content of the crude is an indirect and inverse indication of the refining value of

Table 3.7 Sachanen's classification

Class	Composition
paraffinic	paraffinic side chains > 75%
naphthenic	naphthenic hydrocarbons > 75%
aromatic	aromatic hydrocarbons > 50%
paraffinic–naphthenic	paraffin side chains 60–70%
	naphthenes at least 20%
asphaltic	resin and asphalts > 60%
paraffinic·naphthenic– aromatic	paraffins, naphthenes and aromatics approximately equal
naphthenic–aromatic	naphthenes and aromatics at least 35% each
naphthenic–aromatic– asphaltic	naphthenes, aromatics, and asphalts at least 25% each
asphaltic–aromatic	asphalts and aromatics at least 35% each

the crude. Because sulphur can attack refinery equipment and can give poisonous and obnoxious compounds in the refined products, it has to be removed. This process is expensive and these are reasons why the sulphur content and the value of a crude are inversely related.

The asphalt content is an indication of the ability of making the valuable heavy lubricating oils or the less valuable fuel oils and paving asphalts. The asphalt content can be assessed from the carbon residue of the residuum or by precipitating, by the addition of n-heptane, the asphaltines and weighing them. The cloud and pour points can give an indication of the amount of paraffin wax present in the oil.

Thus a distillation of the oil plus a couple of other tests can give an indication of the value of the oil in terms of its potential for producing valuable refined products.

Chemical classification

Sachanen's chemical classification

Sachanen (1950) suggested a chemical classification which was an attempt to use the chemical nature of the oil as the basis of the classification. The determinations involve a wide variety of physical mesurements in order to determine the relative amounts of paraffinic (including alkyl side chains), naphthenic, aromatic, and resinous-asphaltic compounds in the oil. The total percentage composition of the oil is derived from data from each distillation fraction from gasolene to heavy lubricating oil plus the residuum. Table 3.7 gives the classes of crude oils as defined by this method and their compositional characteristics. This method was an attempt to classify oils by their chemical composition rather than by physical properties. Thus paraffinic oil would, under this scheme, actually contain a high percentage of paraffins and not just have physical properties like those of paraffins. However, this method involves a large number of analyses of a complex nature and since the advent of mass spectrometry and other modern analysis techniques it is easier to use these to analyse oils. Thus this classification system is not widely used.

Tissot and Welte's classification

Tissot & Welte (1978) have proposed a classification system based upon the oil's content of the various structural types of hydrocarbons. These types include alkanes, naphthenes, aromatics including heteroatomic molecules, asphalts and resins. The sulphur content of the crude is also determined. The hydrocarbon groups are determined by modern analytical techniques on the part of the oil with a boiling point above 210 °C at atmospheric pressure. The various hydrocarbon groups include related material so that the paraffin content includes *n*- and isoalkanes but excludes the alkyl side chains on cyclic compounds. Likewise the naphthenic group includes all molecules with one saturated ring but no aromatic groups, whereas the aromatic group includes all molecules which have one or more aromatic rings irrespective of other components. This should be compared with Sachanen's scheme where the different hydrocarbon groups in a molecule would each contribute. Thus methylcyclopentanophenanthrene (Figure 3.7) would only contribute to the aromatic component in Tissot and Welte's system but to paraffinic, naphthenic, and aromatic components in Sachanen's method. Tissot and Welte also group the asphalts and resins in with the aromatics, whereas Sachanen separately used their concentration.

Tissot and Welte's classification is illustrated in Figure 3.15 and it shews how six classes of oils are defined by this method. The six classes of oils can be defined as follows:

Paraffinic oils These will have a saturated open-chain hydrocarbon content of greater than 50%. In general these oils tend to be light oils some of which are very fluid with low viscosity but some, due to their high wax content, are more viscous and have high cloud and pour points. Their specific gravity is normally below 0.85 and their resinous–asphaltic component less than 10% with a very low sulphur content.

Naphthenic oils In these oils saturated cyclic compounds account for 50% or more of the crude oil, although there are few oils in this class. Many of them are degraded oils with less than 20% of the *n*- and isoalkanes which were originally paraffinic or paraffinic–naphthenic oils prior to biodegradation.

Paraffinic–naphthenic oils These contain more than 50% saturated molecules with a sulphur content of normally less than 1%, often in benzothiophenes, while the aromatics account for 25–40% of the hydrocarbons and the asphalts and resins vary between 5 and 15%. Because of the chemical composition of these oils they are slightly more dense and viscous than paraffinic oils.

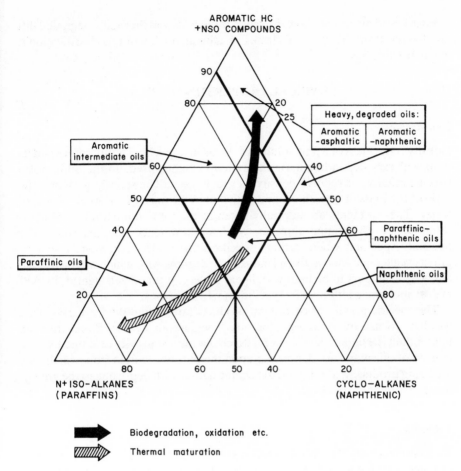

Figure 3.15 Characteristics of the principal classes of crude oils (after Tissot & Welte, 1978. Reproduced by permission of Springer-Verlag, Heidelberg)

Aromatic–intermediate oils These oils contain 40–70% aromatic compounds, many of which are of steroid origin. Sulphur is normally less than 1% and occurs in thiophene derivatives which can be more than 25% of the aromatic compounds present. Resins and asphalts constitute 10–30% of the oil and these oils are heavy, with a specific gravity above 0.85.

Naphthenic–aromatic oils These oils are biodegraded oils which were originally paraffinic and paraffinic–naphthenic. They have low sulphur contents, normally less than 1%, and contain significant amounts of resins (often 25%).

Aromatic–asphaltic oils These oils are mostly biodegraded aromatic–intermediate oils, but there are a few true aromatic oils. They are heavy, viscous oils whose sulphur content is normally greater than 1% and can reach 9%. Their resin and asphaltene content varies between 30 and 60%.

Figure 3.15 also shews how the biodegraded oils and thermally degraded oils are derived from the other oil classes. A tabular version of this classification is given in Table 3.8.

PHYSICAL PROPERTIES OF OILS

Density

Density is the mass of a unit volume of the oil at a specified temperature and its units will vary depending upon the systems of units used. In the oil industry specific gravity, or relative density, is widely used. The specific gravity is the ratio of the mass of a given volume of the oil to the mass of an equal volume of water. These weights are normally taken at the same temperature, which is often 60 °F. The temperature at which both weights were measured are normally specified when quoting specific gravity. The commonest specific gravity scale to be used in the oil industry is 'degrees API' which was devised to allow hydrometers to be built with linear scales. As an oil gets lighter, its API gravity increases, although its specific gravity decreases.

The specific gravity of petroleum products range from 0.6 for casing head gasolene to just over 1.0 asphaltic oils. The gaseous hydrocarbons are even lighter than the liquid ones. Paraffin wax has a specific gravity of about 0.9. The density of oils may vary depending upon the part of the field from which they are being produced, as oils tend to become lighter with depth due to the greater quantities of light paraffins.

Viscosity

Viscosity is a very important oil property because the majority of petroleum products are fluids and of these the greatest number are liquids. Viscosity determines the flow of oil and gas through the reservoir media and thus the amount and rate of oil and gas produced.

Viscosity is the inverse measure of the ability of a substance to flow and the greater the viscosity of the liquid the less readily it flows. Some oil products have very low viscosities and are highly mobile, whereas others are more viscous and some are semi-solids which exhibit plastic properties. A substance has plastic properties if it is able to resist a small force for an indefinite length of time, but at a higher stress causes it to flow. Table 3.9 shows some typical viscosities of common fluids. Poise are units of grams per centimetre per second however the normal unit is the centipoise. The stoke is a unit of kinetic velocity and is obtained by dividing the absolute viscosity, in poise, by the density in grams per cubic centimetre. This gives the stoke the units of $gram^2/cm$.

Viscosity will depend upon temperature and at higher temperatures the viscosity of petroleum liquids is lowered. However, gases, if there is no increase in volume, increase their viscosity with increasing temperature.

Table 3.8 Tissot and Welte's crude oil classification

Concentration in crude oil fraction with bp. 210 °C, 760 mm Hg			Crude oil type	Sulphur content
saturates	> 50%	paraffins > naphthenes Paraffins > 40%	paraffinic	< 1%
		paraffins < 40% naphthenes < 40%	paraffinic–naphthenic	< 1%
aromatics	< 50%	naphthenes > paraffins naphthene > 40%	naphthenic	< 1%
		paraffins > 10%	aromatic–intermediate	> 1%
saturates	≤ 50%	paraffins ⩾ 10% naphthenes ⩽ 25%	aromatic–asphaltic	> 1%
aromatics	⩾ 50%	paraffins ⩾ 10% naphthenes ⩾ 25%	aromatic–naphthenic	< 1%

Table 3.9 Viscosities of common fluids (Gruse & Stevens, 1960, p. 183)

Material	Viscosity	
	(poise)	(Stokes)
air	1.8×10^{-4}	1.5×10^{-1}
water	1×10^{-3}	1×10^{-2}
gasolene	0.6×10^{-2}	0.9×10^{-2}
mercury	1.55×10^{-2}	1.2×10^{-3}
kerosene	2×10^{-2}	2.3×10^{-2}
lubricating oils	$8 \times 10^{-2} - 1200 \times 10^{-2}$	$8 \times 10^{-2} - 1200 \times 10^{-2}$
asphalts and pitches	$10^3 - 10^7$	$10^3 - 10^7$

Dissolved gas will also lower the viscosity of petroleums. Changes in viscosity with changes in the amount of dissolved gas are mirrored by changes in specific gravity. As the dissolved gas content increases, both specific gravity and viscosity fall. In general, viscosity and specific gravity of oils are directly related to each other and vary with the composition of the oil. The larger the average molecular weight of the oil, the higher will be the specific gravity and viscosity of that oil. The heavy asphaltic oils sometimes require heating before they will flow.

Thus viscosity can have an important effect on the production of crude and the lower the viscosity the greater the percentage of oil in place, all other things being equal, that can be produced. Anything which allows gas to come out of solution will lower the viscosity of the oil and impair the recovery of the remaining oil. The inverse of this is one principle on which gas injection schemes work. The injected gas will lower the viscosity and allow the oil to flow more readily. At the same time the gas will precipitate the heavy asphaltenes out from the oil which will also lower the average molecular weight of the crude and thus lower the viscosity. However, the precipitation of these asphaltenes may block some of the reservoir rock pores and cause production problems through reduced porosity.

The simplest and most used ways of measuring viscosities use capillaries and are based upon Poiseuille's law:

$$u = \frac{\pi r^4 p}{8 \eta l}$$

where r is the radius of the tube, l is its length, P is the pressure difference between the ends of the tube, η is the coefficient of viscosity, and u is the quantity of liquid discharged in unit time. Such capillary viscometers are the simplest and most accurate, if correctly used, methods of measuring viscosity. They are normally calibrated with water or reference oils which have a viscosity range, at 20 °C, between 2 and 30,000 centipoise. These methods involve the timing of the flow of a fixed amount of liquid through the capillary and the typical times of flow for petroleum liquids are 100 to 1000 seconds. For any given instrument there will be value of viscosity below which it will not be able to measure accurately or reproducibly.

Another type of viscometer which is in widespread use is that with two rotating cylinders, where the fluid is between the two cylinders. One cylinder is rotated whilst the torque required to keep the other cylinder stationary is measured. These instruments are of particular benefit where the coefficient of viscosity varies with the rate of shear applies. A commercial variant of this type is governed by the equation $F = C\eta W$, where F is the restraining torque, η is the viscosity, W the angular viscosity of the rotating member, and C is an apparatus constant.

Yet another common type of viscometer is the falling ball apparatus whose equation is

$$\eta = Kt(\rho_2 - \rho_1)$$

where ρ_2 is the density of the ball, ρ_1 is the density of liquid, η is the viscosity, K is a machine constant, and t is the time of fall of the ball. Both these last two types are calibrated with liquids of known viscosity.

Certain viscosity measuring instruments were developed empirically and these measure the time required for a given quantity of fluid to flow from a reservoir through a capillary. Because the capillary is short, Pouseuill's law does not apply. The most commonly used instruments were the Redwood in Great Britain (IP No. 70/53), the Engler in Germany, and the Saybolt in the USA. Normally, the viscosity is quoted as the number of seconds taken for the fluid to flow.

Surface tension

The surface tensions of crude oils and their products are confined to a relatively narrow band and therefore this property is of little use in the characterization of oils. It is of importance when considering the migration of oilfield fluids in the ground. Certain non-hydrocarbon compounds, such as fatty acids, will lower the surface tension of oils but there is an optimum concentration of these acids which is equivalent to the amount required to produce a monolayer of fatty acids on the exposed surfaces. The formation of emulsions of liquids and liquids or of liquids and gases is due to the effect of lowered surface tension. If the two fluids are entirely miscible, their interfacial tension will be reduced to zero. If they are not miscible, then the lowering of the surface tension will allow the formation of an emulsion.

Dissolved gases lower the surface tension and this is believed to be due to a dilution effect. In cases of oil–water systems, the pH of the water will control the degree of change of surface tension and the higher the pH, the greater will be the decrease in the surface tension of the associated oil.

The range of surface tensions for oil products is between 20 and 38 dynes per square centimetre. There is an increase in surface tension with the increase in average molecular weight of the compounds in the crude oil. Surface tension will decrease with increasing temperature and increasing pressure. Decreases in the viscosity and specific gravity differences between the oil and water will lower the interfacial tension. Measurement can be made by using IP method 90/44P.

Colour and fluorescence

Crude oils vary in colour, as seen in transmitted light, from light yellow, green, red, brown, and black. The darker browns and blacks are caused by the asphaltic–resinous material in the crude oil. The majority of the colours are due to molecules with aromatic character which have large π electron systems, such as found in the condensed ring compounds or the polyunsaturated alkanes with aromatic groups attached where the π electrons are conjugated. The extent of the conjugation affects the colour and as the size of the conjugated electron system grows the colour of the compound moves from the ultraviolet, eventually crossing the visible spectrum.

Oils get darker in colour as specific gravity increases due to the large aromatic compounds being more dense. The colour of gasolenes produced by distillation, either with or without cracking, is believed to be due to the presence of fluvenes, which are oxygen compounds.

Many crude oils will appear green in reflected light and this is due to a phenomenon known as fluorescence. The colour of an object depends upon the wavelengths of the original white light that are not absorbed by that object. Some compounds absorb light, especially ultraviolet light, and re-emit it as visible wavelengths; this is fluorescence. The property is of value for examining specimens for oil shows, although it must be treated with some caution. Fluorscence is very sensitive to traces of foreign matter that may quench fluorescence, and the fluorescence of a sample will decrease rapidly with age.

Paraffins and naphthenes are weakly fluorescent and while benzene is more fluorescent, it is the larger aromatic compounds which produce the stronger and longer-wavelength fluorescence. The aromatic oils are the most fluorescent of all the oil classes. Some of the admitted wavelength can be reabsorbed by the substance and readmitted as even longer-wavelength radiation.

Much use of this property is made in the logging of wells to locate oil shews in cores, cuttings, and drilling muds. Fluorescence is observed under ultraviolet radiation, which in the petroleum industry is normally at a wavelength of between 2537 and 3650 ångstrom units. The technique allows the detection by the human eye of one part of oil in 100,000 parts of carbon tetrachloride.

Cloud and pour points

These are empirical tests which give an indication of property of oil at low temperatures and indirectly provide an estimation of the presence or absence of paraffin wax. About 35 ml of the oil is placed in a small glass bottle together with a thermometer, and cooled in a freezing bath under standard conditions.

The bottle is periodically removed, tilted on its side, and visually inspected. The temperature at which turbidity is noticed is the cloud point and it is due to the settling out of solid paraffin waxes. Wax-free naphthenic oils shew no cloud point. The temperature at which the oil will not flow in a definite manner is known as the pour point and is normally 2–5 °C below the cloud point.

If there is no paraffin wax present in the oil, the oil will have no cloud point and its pour point will depend upon the viscosity of the oil (IP test No. 15/42).

Aniline point

The aniline point is the minimum temperature at which equal volumes of aniline and the test liquid are completely miscible. It is measured by heating and stirring equal volumes of the two liquids until they are homogeneous. This mixture is cooled with stirring until the two liquids separate, the temperature of such a separation being the aniline point.

For oils of given type, the aniline point increases with increasing average molecular weight, whereas for oils of the same average molecular weight the aniline point increases with increasing paraffinicity. This means that it is possible to use this test to determine the relative amounts of paraffinic and aromatic components within a crude oil. It is useful as a rough estimation of the aromatic content (IP No. 2/53).

If the aniline points of a sample of oil boiling below 275 °C are taken before and after it has been sulphonated, the percentage of aromatics can be estimated (Table 3.10). Figure 3.16 shews how the aniline point increases upon sulphonation. The distillation of the sulphonated sample can give indications of the relative amounts of paraffins and naphthenes. Figure 3.17 and Table 3.11 shew the relationship between mid-point temperature, the temperature at which 50% of the sample boils over, and the composition of the sample.

Table 3.10 Increase of aniline point upon sulphonation and percentage aromatics

Increase in aniline point (°C)	Percent aromatics
5	6
10	11.6
15	16.9
20	21.9
25	26.8
30	31.4
35	35.9
40	40.4

Optical properties

The majority of crude oils rotate the plane of polarized light, and that rotation is normally to the right. There are a few cases where the light is rotated to the left

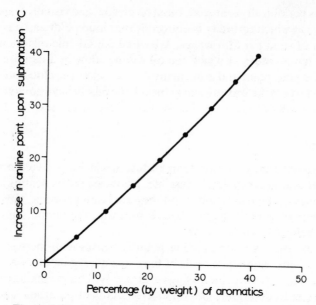

Figure 3.16   Relations between aniline point and aromatic content

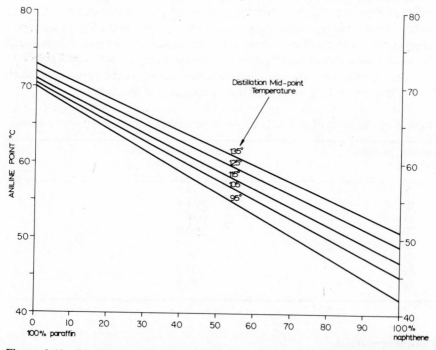

Figure 3.17   Paraffin–naphthene composition as related to distillation mid-point temperature and aniline point of sulphonated samples

Table 3.11 Aniline point and distillation mid-point related to composition

Distillation mid-point (°C)	Aniline point (°C)	
	100% paraffins	100% naphthenes
95	70	42
105	70.5	45
115	71	47
125	72	49
135	73	51

and even fewer where there is no rotation at all. It has been shewn that this power to rotate polarized light is concentrated in a few fractions of the oil, and in particular some of those compounds with molecular weights between 350 and 400.

It is considered that the natural product compounds which survive in petroleum are responsible for the optical activity and thus optical activity is evidence of the low-temperature history of petroleum. Certainly the optical activity is concentrated in the mid-boiling range compounds and high temperature (above 200 °C) seems to destroy optical activity owing to racemization (see Chapter 1).

Optical activity can be exhibited by any compound which satisfies rules relating to the symmetry of the molecule, and certain non-aromatic polycyclic compounds which are optically active have been found in crude oils. Because these compounds are probably less temperature-sensitive than the natural product compounds, it is the presence of the natural product molecules rather than the optical activity of the oil which is evidence for the low-temperature history of oil. In spite of the presence of these compounds, there are many natural product molecules, especially the sterols and their derivatives, present in oils which will contribute to the optical activity of the oils. As far as is known, the optically active components in oils cannot be synthesized inorganically and thus the optical activity of oils is evidence for the biological origin of oil.

Flash and fire points

The flash point is the temperature at which the vapours rising off the surface of a heated oil will ignite with a flash of very short duration when a flame is passed above the surface of the oil. This test is normally carried out in Pensky–Martins closed tester, in which a small flame is periodically applied to the surface of the oil while it is being heated. The flash point is recorded as the temperature at which a flame first passes over the surface. The fire or burning point is the lowest temperature at which that flame becomes self-sustaining and continues to burn once the test flame is removed.

These tests are important as measurements of hazards involved with the storage and handling of petroleum and its products. (IP test No. 34/47).

Refractive index

The absolute refractive index of a substance is the inverse ratio of the speed of light in that substance to the speed of light in a vacuum. The range of refractive indices for petroleum varies from 1.3 to 1.49 and these values are easily determined using the Abbé refractometer.

Since the refractive index is dependent upon the density of the oil, the heavier oils have higher refractive indices. Thus the property offers a quick and approximate method of determining the character of the oil from minute amounts which can be extracted from drill cuttings or cores.

Calorific value

The calorie is defined as the quantity of heat which will raise the temperature of one gram of water from 3.5 to 4.5 °C. The British Thermal Unit (BTU) is the heat required to raise 1 pound of water by 1 °F, and is equivalent to 252 calories. The calorific values of crude oils are inversely related to specific gravity of the oils. This is probably because the paraffins, which are the least dense compounds in oils, are fully saturated and burn readily with a smokeless flame. Thus all the components (carbon and hydrogen) are burnt to give heat, carbon dioxide, and water. Benzene and other aromatics burn with a smokey flame and thus not all the carbon is used to produce heat.

The average calorific value for a crude oil is 18,300–19,500 BTU (4612–4914 k cal) per pound compared with 10,200–14,600 BTU (2570–3679 k cal) per pound of bituminous coal (IP test No. 12/53T). The Scottish proverb 'Guid gear gangs in wee bouk'* seems very appropriate.

REFERENCES

Adams, N. G., & Richardson, D. M. (1953) Isolation and identification of biphenyls from West Edmond crude oil. *Anal. Chem.,* **25,** 1073.
Baker, E. G. (1962) Distribution of hydrocarbons in petroleum. *BAAPG,* **46,** 76.
Bass, D. W. (1960) Composition of crude oils in N. W. Colorado and N. E. Utah. *BAAPG,* **47,** 2038.
Brown, R. A., & Norris, F. D. (1982) Hydrogen peroxide reduces sulphide corrosion. *Oil and Gas J.,* **80,** 36.
Carruthers, W. (1956) The constituents of high-boiling petroleum distillates. Part III. Anthracene homologues in a Kuwait oil. *J. Chem. Sci.,* p. 603.
Dow, W. G. (1974) Applications of oil correlation and source rock data to exploration. *BAAPG,* **53,** 1253.
Gruse, W. C., & Stevens, D. R. (1960) *Chemical Technology of Petroleum,* McGraw-Hill.
Hedberg, H. D. (1968) Significance of high wax crude oils to the genesis of petroleum. *BAAPG,* **52,** 736.
Hitcham, B. (1972) Low molecular weight aromatic hydrocarbons in gas condensates. *Geochim. Cosmochim. Acta,* **36,** 1043.

*'Valuable material occupies small bulk'

113

Hood, A., Clerc, R. J., & O'Neal, M. J. (1959) The molecular structure of heavy petroleum compounds. *J. Inst. Pet.*, **45**, 168.

Hubbard, G. (1982) Sweetening sour gas-Marathon style. *Oilman*, **July 1982**, 9.

Katz, D. L. (1971) Depths to which frozen gas fields can be expected. *J. Pet. Tech.*, **23**, 419.

Levy, E. J., Doyle, R. R., Brown, R. A., & Melpolder, F. W. (1961) Identification of components in paraffin wax by high temperature gas chromatography and mass spectroscopy. *Anal. Chem.*, **33**(6), 698.

Mair, B. J. (1964) Hydrocarbons isolated from petroleum. *Oil & Gas Journal.* Tulsa, Okla., 14 September, p. 130.

Martin, C. C., & Semkin, A. (1953) Determination of aromatic and naphthene rings in aromatics from petroleum fractions. *Anal. Chem.*, **25**, 206.

Moore, R. J., Thorpe, R. E., & Mahoney, C. L. (1953) Isolation of methylchrysene from petroleum. *JACS*, **75**, 2259.

Newton & Leach (1929) *Oil & Gas J.*, **27**(42), 100.

Sachanen, A. N. (1945) *Chemical Constituents of petroleum*, Chap. 10. Rheinhold, New York.

Sachanen, A. N. (1950) Hydrocarbons in petroleum. In *Science of Petroleum*, Brooks, B. T., Dunstan, A. E. (eds), New York, Oxford University Press.

Seifert, W. K. (1969) Effects of phenols on interfacial activity of crude oil (California). Carboxylic acids and the identification of carbazoles and indoles. *Anal. Chem.*, **41**, 562.

Seifert, W. K. (1973) Steroid acids in petroleum. Animal contribution to the origin of petroleum. In *Chemistry in Evolution and Systematics.* Swain, T. (ed.), London, Butterworths, p. 633.

Seifert, W. K., & Howells, W. G. (1969) Interfacially active acids in a Californian crude oil. Isolation of carboxylic acids and phenols. *Anal. Chem.*, **41**, 554.

Smith, H. M. (1940) Correlation index in crude oil analysis. *US Bureau of Mines, Tech. Paper 610*, p. 34.

Smith, H.M. (1966) Crude oil: qualitative and quantitative aspects of crude oil composition. *US Bureau of Mines Inf. Circ. 8286.*

Tissot, B. P., & Welte, D. H. (1978) *Petroleum Formation and Occurrence: A New Approach to Oil and Gas Exploration.* Springer-Verlag, Berlin-Heidelberg-New York.

Chapter 4

Physical Properties of Reservoir Rocks

POROSITY

The porosity of a material is defined as that fraction of the bulk volume of the material which is not occupied by the solid framework of the rock, i.e. it is the ratio of pore or void space to total volume. This ratio is normally expressed as a ratio but occasionally as a decimal or even a fraction. Thus,

$$\text{porosity} = 100 \times \frac{(\text{bulk volume} - \text{grain volume})}{\text{bulk volume}}$$

$$= \frac{\text{pore volume}}{\text{bulk volume}}$$

In oil reservoirs the porosity represents the percentage of the total void space which is available for occupancy by oil or gas and it therefore determines the storage capacity of the rock and consequently the maximum volume of oil or gas that can be present in that rock.

Porosity may be classified as either absolute or effective. Absolute porosity is the ratio of the total void space in the sample to the bulk volume of that sample regardless of whether or not those void spaces are interconnected. Thus while a rock may have considerable absolute porosity, the lack of interconnections between the voids will mean that it will have very low effective porosity and no fluid conductivity. Examples of this are lava, pumice stone, and other rocks with vesicular porosity.

Effective porosity only takes the interconnected pores into consideration and is thus the ratio of the connected pore volumes to the bulk volume. Effective porosity is an indication of the ability of a rock to conduct fluids, but it should not be used as a direct or absolute measure of the fluid conductivity of a rock.

Not all the interconnected pores may be available for the storage or transmission of oil. Many of the pore spaces may be dead-ends with only one entry to the main pore channel system. Porosity measurements take these into account, but they may be misleading as to the rock's capacity for storage or transmission of oil. These dead-end pores may be filled with water which the oil has not been able to displace. Even if such pores are oil-filled the oil may not move out of them under the influence of normal production. Likewise, water drive or gas injection will not move oil from these dead end pores. Only solution gas drive will tend to force such oil out of these pores.

114

Effective porosity may vary laterally and vertically within a reservoir and as many cores as possible should be tested for porosity so as to obtain the most accurate figure for the overall void volume of the reservoir.

Effective porosity is affected by a number of lithological factors, including the clay type, contents and hydration, the heterogeneity of grain sizes, the packing and cementation of the grains, and any weathering that may have affected the rock.

The porosity of the vast majority of reservoirs is below 30% with the most frequently found being rocks with porosities between 10 and 20%. The highly cemented sandstones have the lower porosities, whereas the soft unconsolidated rocks will have the higher values. Silica and clay in the sandstone both reduce the porosity, although the clay by a smaller amount as it is normally porous to a degree. Carbonate reservoirs generally have lower porosities, but higher permeabilities than sandstone reservoirs. Table 4.1 gives an indication of the reservoir quality as judged by porosity.

Table 4.1

Porosity (%)	Reservoir quality
0– 5	negligible
5–10	poor
10–15	fair
15–20	good
20–25	very good

Measurement of porosity

The determination of porosity requires fairly basic, simply operated equipment whose operation can be left to suitably trained but otherwise inexpert personnel. There are various qualitative methods of estimating porosity which can be used when cores are not available. These methods require more highly trained and skilled staff to obtain satisfactory results. The methods include examination of well logs and well cuttings. The electric log is a measurement of the electric potential of rocks and high values of the spontaneous potential (SP) are given by porous beds. Of the radioactivity logs the neutron log is primarily influenced by the presence of hydrogen and thus high readings indicate that the stratum is porous and filled with some hydrogen-containing fluid, e.g. oil, gas, or water. The examination of well cuttings by experienced analysts using binocular microscopes will give reasonable assessments of the amount and type of porosity present. However the results are obtained, their application and the interpretation of the data requires the skilled services of someone who understands the principles of porosity measurement, its limitations, and the geological significance of the value obtained.

The measurement of porosity involves cleaning the samples to remove drilling mud and any small pieces of rock which may have been loosened during the cutting of the core. The organic matter which is within the core must be removed by soxhlet extraction of the core using a light hydrocarbon solvent such as benzene, toluene, or a gasolene-range fraction. The sample is air dried at 110 °C and kept in a desiccator until the measurements are made. The measurements required are the bulk volume and either the pore volume or the solid volume.

The measurement of bulk volume is normally by the volumetric displacement of mercury, and because of the very toxic nature of mercury all those operations should be performed in the correct type of fume cupboard, all apparatus kept in a spill tray, gloves worn at all times, and flowers of sulphur available to deal with any mercury spills. The two commonest methods of measuring bulk volume are by use of the Westman balance and the steel pycnometer.

The Westman balance (Figure 4.1) has a container which is three-quarters

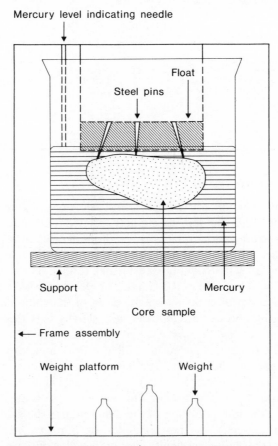

Figure 4.1 Westman balance (From *Oil Reservoir Engineering* by Pirson. Copyright © 1958. Used with permission of McGraw-Hill Book Company)

filled with mercury and the empty frame and cage are placed on the mercury and weights added to the frame until the pointer just touches the surface of the mercury. The cage is then removed and the extracted and dried core is placed in the cage, which is then replaced on the mercury. Additional weights are placed upon the pan until the pointer once again just touches the surface of the mercury. If the additional weight that was required to bring the pointer back to the surface of the mercury was W grams and the density of the mercury is ρ, then the bulk volume of the core is W/ρ.

The Steel pycnometer (Figure 4.2) needs to be cleaned and degreased to prevent mercury contamination and mercury droplets adhering to the surface. The vessel is filled with mercury and the stopper carefully replaced so as to expel slowly any excess mercury. The outside of the pycnometer is cleaned of mercury by brushing with a camel-hair brush.

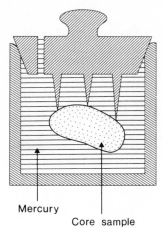

Mercury

Core sample

Figure 4.2 Steel pycnometer (From *Oil Reservoir Engineering* by Pirson. Copyright © 1958. Used with permission of McGraw-Hill Book Company)

The filled and cleaned pycnometer is placed in a clean weighed porcelain bowl and the stopper carefully removed. The core is then placed on the surface of the mercury and the cap carefully replaced. The core displaces its own volume of mercury, which is collected in the dish. Any mercury adhering to the sides of the pycnometer is brushed into the dish which is reweighed. If the weight of the expelled mercury is W grams and the density of the mercury ρ, then the bulk volume is W/ρ.

In order to obtain the porosity, the pore volume or the solid volume have to be ascertained. The effective air porosity can be obtained by the pore gas expansion method (Figure 4.3). This apparatus measures the volume of the solid part of the core so that the pore volume can be obtained by the subtraction of this value from the bulk volume.

If V_1 = the volume of air in the chamber above the stopcock

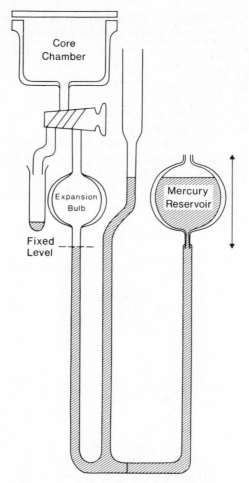

Figure 4.3   Porosimeter—Pore gas expansion

V_2 = volume of the space between the stopcock and the fixed mark below the Bulb

   V_s = the volume of the solid part of the core

   P_a = atmospheric pressure

and P_1 = final pressure in the apparatus

Then $V_s = V_1 - V_2 \times \dfrac{P_1}{P_a - P_1}$

If H is the difference in the level on the manometer corresponding to pressure P_1, then the above equation becomes

$V_s = V_1 - V_2 \times (P_a - H)/H$, which is the equation of a straight line.

By using a selection of ball-bearings of varying sizes, i.e. by varying the value of V_s, and measuring the manometer reading, H, a calibration curve can be drawn.

The cover plate is removed from the core chamber and the core is inserted. The stopcock is turned so that the chamber communicates with the expansion bulb below, and the mercury reservoir is raised until the mercury is just above the stopcock. The stopcock is turned so that the excess mercury runs to the waste trap. The cover plate is now replaced and the seal made airtight. The mercury is lowered about 15 cm and the stopcock turned so as to connect the core chamber and the expansion bulb. The mercury level is adjusted to the fixed mark by adjusting the height of the reservoir. The height of the mercury (H) and the barometric pressure (P_a) will give the value of the solid part of the core from the calibration curve.

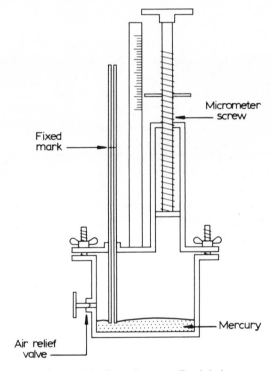

Figure 4.4 Porosimeter—Boyle's law

The Boyle's law porosimeter (Figure 4.4) also measures the volume of the solid part of the core. With the apparatus empty the micrometer is screwed out as far as it will travel and then screwed in two turns, and the piston scale reading noted (d_1). The relief valve is now closed, the piston screwed in until the mercury reaches the fixed mark, and the piston scale reading recorded (d_2).

If V is the volume of air space when the scale reading is d_1, the cross-sectional area of the piston is a, the atmospheric pressure is P, and the pressure within the vessel with the mercury at the fixed mark is $(P + p)$, then:

$$PV = (P + p) [V - a(d_2 - d_1)]$$

A ball-bearing of known volume is then placed in the apparatus and the previous operations repeated with the micrometer screw scale readings being d_1 and d_3. If the volume of the ball-bearing is S, then

$$P(V - S) = (P + p) [V - S - a(d_3 - d_1)]$$

The ball-bearing is replaced by the core of unknown solid volume C and the scale readings become d_1 and d_4. Then

$$P(V - C) = (P + p) [V - C - a(d_4 - d_1)]$$

From these three equations the following relationship is derived:

$$\frac{S}{C} = \frac{d_2 - d_3}{d_2 - d_4}$$

and thus the volume of the solid part of the core is obtained. From this and the bulk volume the porosity may be obtained. This method can be made more accurate if higher pressures are employed and a pressure transducer coupled to a digital voltmeter used to measure the pressure. The absolute value of the pressure need not be known and the appropriate voltmeter readings substituted in the above equation.

All methods which involve the compression or expansion of the gas within the system are subject to the effects of temperature variation. As the volume of the gas changes so will its temperature, which will then slowly come back to ambient. This temperature variation will cause a pressure change in addition to that caused by the volume change. Thermal insulation of these systems is to be recommended.

It is also possible to measure the effective brine porosity of a core using a brine solution with a chloride content approximately the same as that of the connate water from which the core was taken. If the salinity of that water is not known it is usual to make use of a 5% solution of sodium chloride. If the core contains significant amounts of anhydrite then a solution of calcium sulphate is used to saturate the core.

The cleaned and dried core is weighed and transferred to a vacuum container which has a vacuum pump and a reservoir of brine attached. The system is evacuated for about half an hour, after which time the valve to the vacuum pump is closed and that connected to the brine is opened. Brine is allowed in until the core is well covered, but no air is admitted into the system. After half an hour the air is admitted and the sample is left overnight to saturate.

The core is then transferred to a weighed stoppered bottle and weighed. The weight of the brine absorbed by the core is thus obtained and, with a knowledge of the density of the brine, the volume of the brine, and thus of the pores of the core, can be calculated. Bulk volume is measured in the normal way.

Pore volumes and porosities can be obtained by mercury injection methods. The porosity of the sample, at atmospheric pressure, is obtained. The pressure of the mercury is raised, until the mercury is forced to fill all the pore spaces of the core. The volume of mercury used is measured and this equals the volume of the pores. This instrument is the only porosimeter where both bulk volume and pore volume can be measured in the one machine.

The measurement of absolute porosity requires that all the voids, whether or not they are interconnected, are measured. The sample is cleaned and dried and its bulk volume is measured as described earlier. The sample is then carefully crushed so as to reduce to its individual grains and no pores left. The volume of these grains is then measured. The grain volume is obtained from the weight and density of the grains.

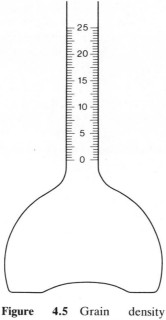

Figure **4.5** Grain density measurement

The apparatus (Figure 4.5) consists of a clean dry flask which has the neck graduated from 0–25 ml. Using a long-stemmed funnel, so as not to wet the neck of the flask, kerosene is poured into the flask until it is between the 0 and 1 ml marks. Any air bubbles must be removed by tapping the flask on a soft surface. Once the kerosene has reached room temperature the flask is weighed (W_1 grams) and the level of the kerosene in the neck is noted (V_1 ml). The grains are weighed and slowly introduced into the flask via a long-stemmed dry funnel, so that they are wet by the kerosene and sink to the bottom of the flask. The volume of sample used should be enough to raise the level of the kerosene to between the 20 and 25 ml marks.

If the new weight is W_2 grams and the new volume is V_2 ml, then the density of the grains (GD) is given by:

$$GD = \frac{W_2 - W_1}{V_2 - V_1}$$

From the weight of the grains and thus density, the volume of the grains obtained. The volume of all the voids is then calculated from the bulk volume and thus the absolute porosity can be determined.

Geological factors affecting porosity

Both primary and secondary porosity are created and altered by geological processes. Primary porosity results from the initial formation of the sediment from individual grains. When these grains are cemented together there will be gaps between them and it is these gaps that give rise to primary porosity. Sediments which are produced by the precipitation of chemicals out of solution are not formed of grains in this way and thus carbonates have much lower primary porosities than do sandstones.

The degree of bonding between the sand grains will also affect primary porosity. The smaller the amount of cement between the grains, the greater will be the porosity, but too little cement will mean friable rocks. In general sediments which have high porosities (25–30%) are very soft and friable with sand grains often produced with the oil at the wellhead, whereas those rocks with low porosities are hard, strong rocks.

The geological situation is not a static one and alterations to the rocks will take place during geological history. These changes, which can affect the porosity of the rock, may be drastic or they may be more moderate. An example of a drastic change is fracturing, whilst the leaching out of some components or a chemical such as dolomite are examples of the less severe alterations. The products of the breakdown of the kerogen, e.g. carbon dioxide, organic acids etc., may cause changes in the pH of the formation waters which may cause solution of deposition of minerals. In general acid conditions are favoured for the solution of carbonates and the deposition of silicates and vice versa. The result of all these changes upon the empty volume of the rock is known as secondary porosity. Secondary porosity is much more important in evaporitic rocks than in sandstones. Not all of the changes may necessarily result in increases in the porosity of the rock.

Sandstones are the most commonly porous rocks and their initial porosity is due to the sand grains abutting each other. If all the sand grains were of a uniform size and the grains were perfect spheres and packed together in the most efficient way, then the size of the grains would be immaterial with regards to porosity. Irrespective of grain size, sandstones made from uniform spheres would have the same porosity. Unfortunately the sand grains within a sandstone are never uniform and the finer particles manage to fall in between the larger grains and fill some of the pore space. Thus the greater the range of sizes of sand particles from which the sandstone is made, the smaller will be the initial porosity. Porosity declines very rapidly as the percentage of very fine grains increases. Initial porosity of sandstones can also be affected by the shape of the grains and the amount of cement. In sandstones the primary porosity is the major contribution to the porosity of the rock, but secondary influences can have some effect.

In limestones, primary porosity is normally much less than in sandstones and results from either voids between the clastic grains of the detrital rock materials, or voids between crystalline limestones along cleavage planes due to the varying sizes of crystals, or from voids due to the removal of organic matter. Although limestones often occur in very great thicknesses they normally have very little primary porosity to make them suitable as oil and gas reservoirs. The exception to this is where chalky or oolitic facies are present.

The factors which produce secondary porosity are fracturing, solution, remineralization, and compaction. The first two generally increase porosity, remineralization can increase or decrease porosity, while compaction tends to reduce porosity.

In sandstones, compaction has very little effect except at very high pressures where the packing of the sand grains becomes less random and eventually the grains are either crushed or plastically deformed. Both of these latter effects will reduce porosity. Cementation has its greatest effect on the initial porosity of sandstones. However as it alters the size, shape, and continuity of the pore channels due to the solution or deposition of quartz and/or calcite and dolomite, any factor which alters the cementation of sandstones will change their porosity. The movement of concentrated brines may bring the correct minerals into contact with the sandstone to cause further deposition and loss of porosity. Alternatively, some minerals may be dissolved, which can increase porosity. Fracturing is not so important in sandstones as a cause of secondary porosity except where sandstones are interbedded with shales, limestones, dolomites, etc.

In limestones, secondary porosity is much more important than primary porosity. The solution of minerals by percolating surface waters which contain dissolved carbon dioxide and organic acids will cause secondary porosity. These solutions dissolve away the more soluble ions as they pass along primary pores, fissures, fractures, joint planes, bedding planes, etc. The rock minerals have different solubilities, with aragonite being the most soluble and followed, in order of decreasing solubility, by calcite and dolomite.

Recrystallization or dolomitization can also cause an increase in porosity. Some carbonate reservoir rocks are almost pure limestone ($CaCO_3$) and if the pore waters within these wells contain significant amounts of magnesium, the calcium in the rock can be exchanged for magnesium in the solution. This reaction was fully described in Chapter 2. Because the ionic volume of magnesium is considerably smaller than that of the calcium which it replaces in the crystal lattice of the rock, the resulting dolomite will have a greater porosity than the calcite from which it was derived. The replacement of calcium by magnesium could result in a maximum 12% increase in pore space. This process can tend to be localized and often the majority of the porosity in a mixed dolomite–limestone rock will be in the dolomite.

As with sandstone reservoirs, the effect of compaction upon carbonate reservoirs is to reduce their porosity, though in this case it is more likely to be due to recrystallization under pressure than by crushing.

Fracturing and jointing often occur in these brittle rocks as the beds are folded and anticlines produced, and are the cause of secondary porosity. Nearly all reservoirs in dolomites and limestones have some fracture porosity. The fracture planes combine with whatever porosity already existed to form an interconnected system. Jointing of these limestone blocks can be caused by mineralogical changes, tectonic stress, or contraction during the consolidation of the sediment. Jointing consists of a series of small fractures which run parallel to one another with others crossing at right angles to give a consistent pattern. In general, the main joint fractures run in a vertical direction and are often found on the crest of anticlines which leads to solution porosity at these points caused by the readily available meteoric waters.

Fracturing will be able to make almost any brittle rock, irrespective of how dense it may have been, into a reasonable reservoir rock. Because carbonates are very brittle, only small tension forces are required to produce significant amounts of jointing. Heavy viscous oils, which cannot flow through rocks of low porosity and permeability, may accumulate and be produced from the fractures in such rocks.

PERMEABILITY

Permeability is defined as the ability of a rock to allow a fluid to flow through its interconnected pores. If the pores are not interconnected then the specimen will have no permeability, irrespective of the absolute porosity. Similar factors will affect permeability as affected porosity, i.e. grain size, grain distribution and packing, angularity, and the amount of cementation and consolidation. Permeability is also affected by the type of clays present, especially where fresh water is also present. Some clays, particularly montmorillonites, swell in fresh water, even to the extent of completely blocking pore spaces.

Darcy's law relates the volumes of fluids passing through a porous medium to their permeability and a generalized form of Darcy's equation is:

$$Q = -\frac{KA}{\mu}\frac{dP}{dL}$$

where Q is the rate of flow of a liquid of viscosity μ through a medium of permeability K, which has a length of L centimetres and a cross-sectional area of A square centimetres. dP/dL represents the pressure gradient of the fluid across the medium and in the direction of the flow of the fluid.

The permeability of the substance through which the fluid is flowing is measured in Darcys (D), although most permeabilities are low enough to be quoted in millidarcys (mD). The permeabilities of reservoir rock normally range between 5 and 1000 mD although commercial production has been obtained where cores have a measured permeability as low as 0.1 mD. In these cases it is believed that the rock has a much larger permeability due to fracturing which is not visible in the small sample examined in the laboratory.

As with porosity, permeability can vary both laterally and vertically within a reservoir rock. Permeability is normally measured parallel to the bedding planes of the reservoir rock because it is along this lateral permeability that the main fluid movement paths reach a borehole. Permeability across the bedding planes, vertical permeability, is also measured and usually found to be less than lateral permeability.

If a sediment were made up of uniform spheres, all of the same diameter, there would be no difference in permeability in any direction. The value of the permeability would depend upon the diameter of these spherical grains, with the sediment made from the larger grains having the greater permeability.

However, sediments are never made up entirely of spheres and they contain many flat, platy minerals. In addition to this, the forces of compaction act vertically downwards and these tend to distort the grains and to reduce vertical porosity. When non-spherical grains are settling out of solution they tend to align themselves with their long axes horizontal. However, in uncompacted sediments this alignment is far from perfect and the grains are at angles to each other. As compaction proceeds and water is expelled the grains tend to take up a preferential horizontal alignment parallel to and overlapping with each other. This leads to a greater permeability in the horizontal direction and solutions which can dissolve rock minerals will preferentially flow in this direction between the plates. Hence permeability will be enhanced even more in that direction.

High vertical permeabilities are caused by fractures and jointing which split vertically across the bedding planes and are commonly encountered in carbonates and other brittle rocks. Table 4.2 gives a rough guide to the quality of a reservoir as determined by permeability measurements.

Table 4.2

Permeability (mD)	Quality
0.1– 1.0	poor
1.0– 10	fair
10– 100	good
100–1000	very good

So far the discussion of permeability has concerned an unspecified fluid, but some fluids are more compressible than others and gases are more compressible than liquids. Klinkenberg (1941) noted that the permeability of a rock to a gas is greater than its permeability to a liquid, and he stated that 'the permeability of a medium to a gas is function of the mean free path of the molecules'. The mean free path of a gas molecule is the average distance travelled by that gas molecule between collisions with other molecules or the container, and it is affected by temperature, pressure, and the type of gas. As the pressure increases the mean free path of the molecules decreases until, at high enough pressures, the measured permeability should approach that of the liquid state of the same material. This is the same as the other physical properties of gases at high pressures approaching the properties of the corresponding liquid.

Mathematically, this can be expressed as

$$K_a = K_\infty(1 + b/P)$$

where K_a is the air permeability at pressure P, while K_∞ is the permeability which is the extrapolation of K_a to an infinite pressure. This should be equal to the equivalent liquid permeability. b is a constant related to the size of the pore openings.

The infinite-pressure value of the permeability (K_∞) can be considered as the equivalent liquid permeability of the reservoir. It is determined by measuring the air permeability at a variety of pressures and plotting these against the reciprocal of the corresponding pressure. If such a curve is extrapolated to infinite pressure ($1/P = 0$), the value of the permeability at infinite pressure (K_∞) will be the permeability of that medium to liquids.

Measurement of permeability

The permeability of reservoir rocks is determined in the laboratory by testing small, normally one-inch diameter, core plugs. The cores are cut, normally parallel to the bedding plane, solvent-extracted and stored in a desiccator. The commonest type of core holder is a rubber bung with a hole in the centre which is slightly smaller than the core to be held so that no gas can leak down the side of the core when it is in place. The bung is placed in the core holder in the instrument so that the required fluid can be passed through the core. The permeameter has a source of fluid to be pumped through the core, a means of measuring the fluid pressure and a flowmeter to measure the quantity of fluid passing. Figure 4.6 shews a permeability-measuring apparatus. For this type of permeameter the permeability (K) is given by

$$K = \frac{Q_a\mu L}{(P_1 - P_0)A}$$

where Q_a is the observed rate of flow in cubic centrimetres per second at an outlet pressure difference of $P_1 - P_0$ atmospheres, L is the length of the sample in centimetres, A is the cross-sectional area of the sample in square centimetres, and μ is the viscosity of the gas in centipoise. The permeability will then be in Darcys.

Figure 4.7 shews the layout of a permeameter actually used for permeability measurements. This also has the rubber core holder, gas supply with regulator and pressure gauge (or mercury manometer), as well as flow meters. The

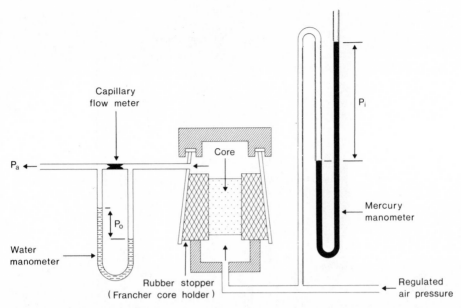

Figure 4.6 Permeability measurement (From *Oil Reservoir Engineering* by Pirson. Copyright © 1958. Used with permission of McGraw-Hill Book Company)

pressure gauge reads the inlet pressure in millimetres of mercury and the flow meters read in cubic centimetres per second. There are three rotameter flow meters of different capacity, and provision for a soap-bubble flow meter.

Once the core is inserted, the value of the highest-range rotameter is opened and pressure applied to the system. If the flow is too low the other rotameters can be opened in turn. The soap-bubble flow meter can be used for cores with very low permeabilities, as the gas flow will be too small to operate any of the rotameters. For this apparatus (Figure 4.7) permeability (K) is given by

$$K = \frac{1520 \, QLP_a\mu \times 10^3}{H(2P_a + H)A}$$

where K is the gas permeability in millidarcies, Q is the rate of flow of air (from a bottle of compressed air) in cubic centimetres per second at 15 °C, L is the length of the core in centimetres, A is the cross-sectional area of the core in square centimetres, H is the head of mercury in millimetres, P_a is the atmospheric pressure in millimetres of mercury, and μ is the viscosity of the air at 15 °C.

Every manufacturer has his own variations but all permeameters use the principles described above. Often the pressure is measured by a gauge or pressure transducer and the flow rate by an electric flowmeter. These modifications allow higher pressures to be used.

All measurements so far described have used dry gas at pressures only a little above atmospheric pressure, and measurements so obtained will have values higher than when a liquid or a gas at high pressure is used. A series of

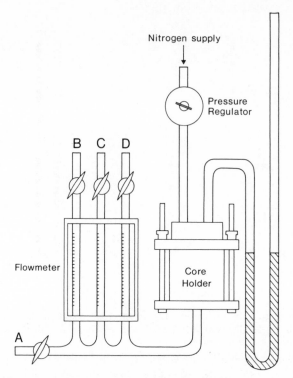

Figure 4.7 Permeability measurement

permeability values, at different pressures, when extrapolated to infinite pressure will give the Klinkenberg value of permeability, which is almost the value of the permeability that is obtained with a liquid which does not cause swelling of the clay constituents of the rock.

As an alternative to obtaining a Klinkenberg value for the liquid porosity, it is possible to measure the permeability of a core to a reservoir fluid. The cleaned and dried cores have to be saturated with the fluid concerned and this is achieved as described for effective brine porosity. The measurement apparatus follows the same principles but has variations to take account of the fact that liquids, rather than gases, are involved. Where cores contain clay minerals which are likely to swell if brines are used other fluids, such as kerosene or oil, should be used.

Unlike porosity there are no well logs which allow accurate estimations of permeability to be made. There are, however, downhole methods of obtaining rough estimates of the permeability of a formation. If when drilling the quantity of mud which returns to the surface is less than that being pumped down, the situation is called lost circulation. Lost circulation is an indication that the drill has entered porous and permeable strata. Sudden decreases in the volume of mud returning mean that the drill has entered a formation with high permeability with a fluid pressure less than that of the drilling mud. If, conversely, the fluid pressure in this highly permeable strata is greater than that of the drilling mud then water will flow into the mud and greater volumes of more drill mud will return to the surface than are pumped down.

A sudden increase in the rate of drilling indicates that a softer and more friable rock has been encountered. Such rocks generally have higher porosities and permeabilities than hard rocks. In a production test a decrease in bottom hole pressure is measured against production rate. In a highly permeable formation the rate of decrease of bottom pressure with increasing production rate will be low, but it will be high in strata of low permeability.

Geological factors affecting permeability

The presence of porosity in a rock does not ensure that the rock will have a significant permeability because permeability depends upon the shape, size, and degree of interconnection of the pores. The packing and arrangements as well as their cementation will affect permeability. Various efforts have been made to take these factors into account when measuring permeability and although several equations have been produced, none of them are in regular use (Pirson, 1958).

The difference in permeability measured parallel and at right angles to the bedding plane is a consequence of the origin of that sediment because grains settle in water with their long axes horizontal. Subsequent compaction of the sediment increases the ordering of the sand grains so that they all lie in the same direction. Within the bedding plane the largest values of permeability will be found parallel to the dominant direction of the old depositional shoreline.

Porosity and permeability can be affected by clay materials within the rock because these swell in contact with water. If the core was oil saturated, the clays may not be swollen so an air permeability measurement would be accurate, but permeability measured with water could be false due to the swelling of the clays. If, however, the core was water-saturated, the cleaning and drying processes may cause disintegration of the clays which can block pore spaces.

Effective and relative permeabilities

So far the permeability of the porous medium to only one fluid has been considered, but in reservoir rocks when they are in the ground there are normally at least two, and sometimes three fluids present. These are oil, gas and water. The ability of a rock to conduct one fluid in the presence of other fluids is the effective permeability of a rock to air, gas, oil, and water are designated K_a, K_g, K_o, and K_w. The relative permeabilities are the ratios of effective permeabilities to the absolute permeabilities, and are denoted k_a, k_g, k_o, and k_w respectively.

Materials can either be water-wet or oil-wet, i.e. in the presence of a mixture of oil and water a water-wet substance will preferentially have the water adhering to its surfaces. Because the majority of reservoir sands are deposited in aqueous conditions it is not surprising that most reservoir sands are water-wet. The wetting fluid, when at low saturation, will have little mobility because of its adhesion to the reservoir surfaces. The non-wetting fluid, which occupies the balance of the pore spaces, will have a higher mobility. Gas is very mobile because it occupies the centre point of the pore spaces and its flow is unimpeded by the surface of the pores.

An example of the permeabilities of oil and water in a water-wet sand are shewn in Figure 4.8. In this example the water does not flow until the water content reaches about 30%. Below this value the water adheres to the rock surface, but the oil, which fills the remainder of the pores, is more mobile and the rock has a high relative permeability to the oil. As the oil content decreases, so does the relative permeability of the rock to the oil.

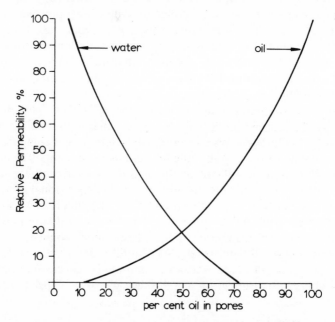

Figure 4.8 Permeability variation in water-wet rocks

The relative permeability of the rock is zero until a minimum value of water content is reached. This minimum value represents the point when all the surfaces of the pores are covered by a layer of water about one molecule thick. After that minimum value, the permeability of the rock to the water increases. Where the two lines cross, the permeability of the rock to both the oil and the water is the same and each fluid can move with equal freedom through the rock under those conditions. As the water saturation increases, the permeability of the rock to the oil decreases until the oil ceases to flow (zero permeability) and only water will move in the sand under these conditions.

Effective and relative permeabilities can be measured in the laboratory and such measurements involve cells designed to cope with two- and three-phase flow of immiscible fluids. The different fluids are kept separate from one another with independent flow systems, pumps, and flowmeters, and they are only allowed to mix in the core.

RELATIONSHIPS BETWEEN POROSITY AND PERMEABILITY

The only real relationship between the porosity and permeability of a rock is that for a rock to have permeability it must be porous. However, the degree of permeability will not be related to the porosity. If sands were made only of sand grains which were perfect spheres of identical radius, perfectly packed, the porosity would be independent of the radius of the sand grains. However, the larger the sand grains the greater will be the permeability of the sand. The total void space, as a ratio of the bulk volume, when spheres of uniform size are perfectly packed is always the same whatever the size of the spheres. However, the larger the spheres, the larger will be the pores between the spheres, whilst leaving the pore/bulk ratio the same. Because fluids flow more easily through larger holes than small ones, sandstones made of larger grains are more permeable.

There has been much work on various reservoir rocks to investigate relationships between porosity and permeability. In general it has been found that if the permeability is plotted against porosity on a log-normal scale (i.e. permeability in millidarcys on the log scale and porosity in percent on the normal scale), straight-line relationships exist between sands within one formation. No relationship seems to exist for unrelated sands. Figure 4.9 is a hypothetical example of these relationships. Because the relationships only exist between sands in the same formation these diagrams can be of no diagnostic value.

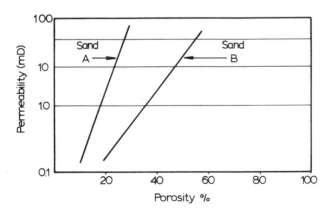

Figure 4.9 Porosity–permeability relationships

Recent work has shewn that the action of drying core plugs before both porosity and permeability are measured can give misleading results. Scanning electron microscope studies have shewn that many rock pores are partially filled with fibrous clays which dehydrate and shrink when conventionally dried.

132

This will result in much larger values of porosity and permeability than exist under reservoir conditions. Cores should therefore be kept wet all times and the pore fluid removed by the passage of a series of miscible liquids.

REFERENCES

Klinkenberg, L. J. (1941) The permeability of porous media to liquids and gases. *API Drill. and Prod. Practice.*
Pirson, S. J. (1958) *Oil Reservoir Engineering*. McGraw-Hill Book Co. Inc., New York.

Chapter 5

Hydrocarbon Thermodynamics

PHYSICAL PROPERTIES OF HYDROCARBON FLUIDS

The ideal ideal gas law can be derived from the kinetic theory of gases and relates the temperature, pressure and volume of a ideal gas by the relationship:

$PV = nRT$

> where P = absolute pressure on gas
> V = volume occupied by gas
> T = absolute temperature of gas
> R = universal gas constant
> n = no of moles (mass/molecular weight)

However, very few gases actually exhibit behaviour anything like that described by the gas equation given above except over a very limited range of conditions. The gas equation was formulated by ignoring the volume occupied by the molecules as well as the attraction they have for each other. Various modifications, such as Van der Waals', to the ideal gas equation have been made, but most are complicated. A most useful method is to insert a correction factor, Z, into the ideal equation so that it becomes $PV = ZnRT$. The compressibility factor Z is a dimensionless quantity and is a function of the pressure, temperature, and the gas involved. It can be determined experimentally by measuring P, V, and T. However, it is often more convenient to obtain reasonably accurate approximations for the compressibility factors by use of either reduced variables (see below) or pseudo-critical variables (see p. 134). If a chart is not available for the gas involved, a value of Z can be estimated with reasonable accuracy by use of the law of corresponding states, as follows.

The ratios of P, V, and T to the critical values P_c, V_c, and T_c (the pressure, volume and temperature corresponding to the critical state, i.e. the point when the liquid and gaseous phases become continuously identical) are called the reduced pressure, reduced volume, and reduced temperature, i.e. $P_r = P/P_c$, $V_r = V/V_c$, and $T_r = T/T_c$. To a fairly good approximation, especially at moderate pressures, all gases obey the same equation of state when described in terms of their reduced variables, P_r, V_r, T_r instead of P, V, T. Thus, if two different gases have the same values for two reduced variables they will have approximately the same value for the third, i.e. they are said to be in 'corresponding states'. This is equivalent to saying that the compressibility factor is the same function of the reduced variables for all gases.

From this it follows that if we have a set of Z/P curves for methane, or any one of the lower paraffins, we can convert the values of pressure and

133

temperature to the critical equivalents and all the compressibility factors for all the lower paraffins for which we have critical data become available. This would be very convenient, as all one would need would be a set of curves and the tabulated data on critical pressure and temperatures for the paraffins.

However, the law of corresponding states is only an approximation and has been shewn by experiment not to be strictly obeyed. If the reduced pressure is plotted against the reduced temperature, one isochore (constant-volume curve) for each value of reduced volume should be obtained, but for each isochore there is a thin spread of lines. In fact there is one for each compound in a regular sequence in order of molecular weight. Likewise, on a pressure–compressibility diagram there is a regular sequence for compounds at the same reduced temperature. Thus the prediction of the properties of any member of a series can be made more accurately if experimental data are taken into account, and the law of corresponding states should only be applied as a first approximation when no satisfactory experimental data are available.

In practice the systems that are being dealt with are complex mixtures rather than simple ones and it is possible to make a rough estimate of the properties of a natural gas by calculating the pseudo-critical constants from an analysis of the gas and using compressibility-reduced pressure charts. Preferably these should be for each component, but if necessary one can manage with the chart for methane. The pseudo-critical pressure $= \Sigma\ y_i P_{ci}$ and the pseudo-critical temperature $= \Sigma\ y_i T_{ci}$ (where y_i is the mole fraction of the ith component whose critical temperature is T_{ci} and whose critical pressure is P_{ci}). We know that there is an error in assuming that the law of corresponding states holds exactly, but this error can be reduced by using a Z/P_r chart constructed from actual natural gas mixtures. Even with such charts it is necessary to know the critical data for the gas under consideration. If the gas has been analysed the pseudo-critical variables can be calculated, but if no analytical data are available it is possible to estimate a value for these variables if the density of the gas is known. There is a linear relationship between pseudo-critical pressure and gas density while the density–pseudo-critical temperature curve is smooth and shallow. Both these curves are widely available in the literature.

THE PHASE BEHAVIOUR OF HYDROCARBON SYSTEMS

The single-component system

The effect of changes in pressure and temperature on the volume of a substance is shewn in Figure 5.1. This illustrates the changes that occur in each phase. Consider the 100 °F isotherm starting on the right of the diagram in the gaseous region. Small changes in pressure result in large changes in volume until the 'dew point' line is reached, at which point liquid starts to form. No change in

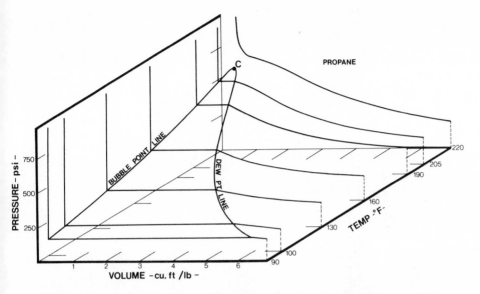

Figure 5.1 Pressure–temperature–volume diagram for a single compound

pressure is needed for the volume decrease to the 'bubble point' line, after which very large pressures are required to make small decreases in the volume of the liquid. All the isotherms below the one that passes through the point C shew the behaviour of the hydrocarbon to be the same with the exception that the volume range in which the mixture of gas and liquid occurs gets smaller, and at higher temperatures there is no liquid region as such at all. It is however more convenient to use a series of two-dimensional diagrams, each of which is a projection of the solid figure. The diagrams are pressure and temperature; pressure and volume; and volume and temperature.

Pressure–volume

Figure 5.2 illustrates a pressure–volume or *P–V* diagram. Each isotherm shews the relationship between the pressure and the volume of a pure substance for a fixed temperature. Let us consider the isotherm for 100 °F starting at point X. For the section X–M the gas will approximately obey the ideal gas law, i.e. $PV = $ constant, and the curve is a rectangular hyperbola. As the gas pressure is increased the volume drops in accordance with the relationship $P \propto 1/V$ until the point M is reached. At that point a liquid phase begins to separate from the gas and the system is then entirely gaseous but in equilibrium with an infinitesimal amount of liquid phase. This point is called the dew point. In the gaseous phase of a pure substance the state corresponding to the dew point is often called the saturated gas or vapour,

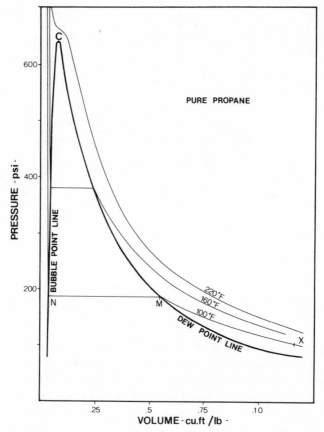

Figure 5.2 Pressure–volume diagram for a single compound

and because only one component is present the composition of the two phases is the same.

Once the dew point is reached the liquid phase will continue to separate and hence the volume will continually decrease, but the pressure remains constant. This pressure is the vapour pressure and it does not depend upon the amount of liquid present. Thus the isotherm continues at constant pressure to the point N where the substance is entirely liquid, but in equilibrium with an infinitesimal amount of gas. This is called the bubble point and the state is that of the saturated liquid. If the pressure is increased past the bubble point there will be very little change in volume because of the low compressibility of liquids. The very large pressures that are required to cause a very small decrease in the volume will result in a nearly vertical isotherm.

It is normal to include several isotherms on one diagram and the locus of the dew points at different temperatures is called the 'saturated gas or dew point line'; likewise the locus of the bubble points is called the 'saturated liquid or bubble point line'; and together they make the 'saturation dome' within which is the two-phase region. The point at which the saturated liquid and gas lines meet is called the 'critical point' (C) and it is the highest temperature and pressure at which the two phases of a pure substance can coexist.

Temperature–volume

Figure 5.3 is a plot at various pressures (isobars) of the volume and temperature of a pure substance. For an ideal gas at constant pressure the relationship between volume and temperature tends to be linear ($V \propto T$). If a gas is cooled at constant pressure there is a point at which it starts to condense as can be seen in a water-jacketed condenser. The condensation will be completed at constant temperature while the volume decreases. Hence the isobar becomes parallel to the volume axis and this represents the two-phase region. Finally, the saturated liquid line is reached, and from this point on one is considering the contraction of a liquid. Since this is very small compared with a gas, the isobar becomes very steep. The saturated gas and liquid lines form a saturation dome as they do in the *P–V* diagram. Both these diagrams are projections of the *P–V–T* diagram onto different planes.

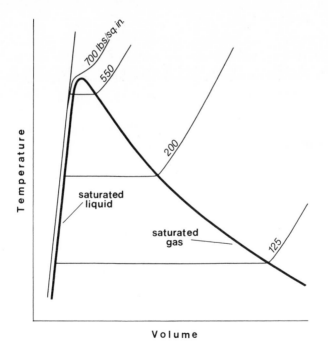

Figure 5.3 Temperature–volume diagram for a pure substance

Pressure–temperature

Figure 5.4 is an example of a pressure–temperature diagram drawn to cover the whole range of possible states. The line OA represents the vapour phase as a

138

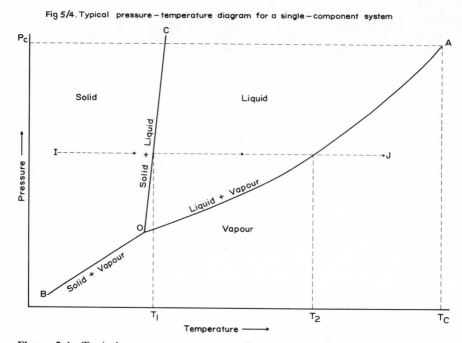

Fig 5/4. Typical pressure – temperature diagram for a single – component system

Figure 5.4 Typical pressure–temperature diagram for a single component system (after Burcik, 1957. Reproduced by permission of John Wiley & Son, Inc.)

function of the temperature and points above OA are liquid, while those below it are vapour, and all the points on the line are the two phases liquid and vapour. The upper limit of the line OA is point A, the critical point, and the lower limit is point O, the triple point where all three phases coexist together.

Petroleum studies do not normally involve use of the solid–liquid and solid–vapour isochores, and so we will move on to Figure 5.5 which is a more detailed look at the upper end of the liquid–vapour isochore OA.

In the gaseous state, an ideal gas at constant volume has a linear relationship between the temperature and the pressure, i.e. $P \propto T$. On the P–V and V–T diagrams in the two-phase region, the isobars and isotherms are parallel to the volume axis and thus on the P-T diagram they merge and become one line. When the temperature of a fixed volume of liquid is lowered the pressure will decrease until the saturation pressure is reached when there is a very rapid increase in volume as the liquid expands into a gas. Hence in the two-phase region the whole system will move to the right and thus the isochores do not meet on the saturation line.

Critical point (for a pure substance)

This is the state at which the liquid and the gas phases become continuously identical. Consider the substance at point A in Figure 5.6 and increase the temperature at constant pressure. The substance will reach point D and a

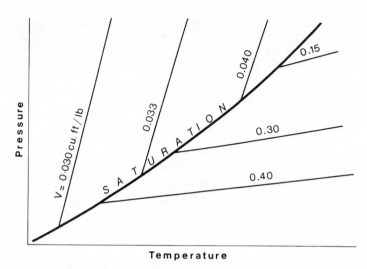

Figure 5.5 Pressure–temperature diagram (section OA of Figure 5.4)

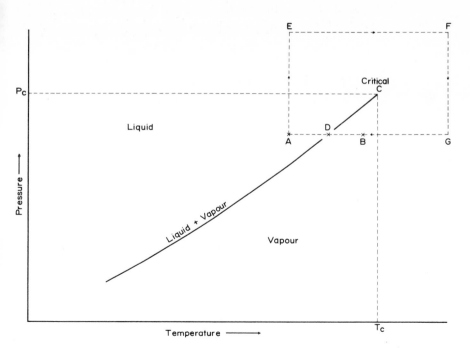

Figure 5.6 Typical pressure–temperature diagram for a single component system in the region of the critical point (after Burcik, 1957. Reproduced by permission of John Wiley & Sons, Inc.)

140

second phase, which is less dense, will emerge, shewing that the substance at A must have been a liquid. Similarly, by cooling at constant pressure the substance from point B to point D, the appearance of a more dense phase will shew that at point B the substance is a gas.

Now consider starting at point A, which is known to be a liquid, and isothermally increase the pressure to above the critical pressure P_c (point E). An isobaric temperature rise to point F, above the critical temperature T_c, followed by an isothermal pressure decrease to point E and an isobaric temperature decrease to point B will mean that we have reached a point where the substance is known to be a gas without an abrupt change of phase. This shews that the vapour and liquid states are very similar forms of the same condition of matter as we can pass between them without any abrupt change of state. The terms 'liquid' and 'vapour' only have meaning near the two-phase region, and in areas far removed from the critical point there is only 'fluid'.

Estimation of critical volume, pressure, and temperature.

This involves the use of the law of rectilinear diameters as the critical state is approached the various physical properties of the liquid and gas phases become more and more alike. For example, the viscosity of the liquid phase decreases while that of the gas increases until at the critical point they are equal. It has been found that the average density of the liquid and vapour when plotted as a function of temperature is a straight line. This is known as the law of rectilinear

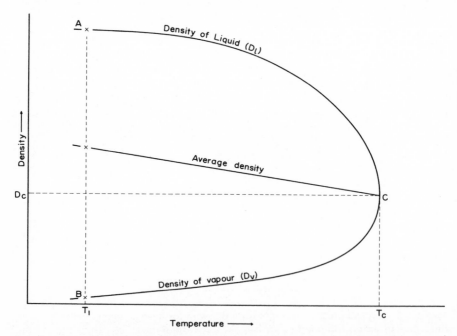

Figure 5.7 Typical density–temperature relationships (after Burcik, 1957. Reproduced by permission of John Wiley & Sons, Inc.)

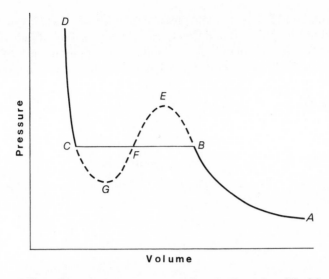

Figure 5.8 Graphical expression of Van der Waals'
equation

diameters and is illustrated in Figure 5.7, which can be expressed as

$$\frac{D_1 + D_v}{2} = aT + b,$$

where D_1 and D_v are the densities of liquid and vapour, and a and b are constants. The relationship in fact is not quite linear, but approaching the critical point it becomes more nearly so, and thus if one can plot the densities of the two co-existing phases as a function of temperature as near the critical point as possible and extrapolate to the critical point, one can obtain the density at the critical point. From that the molecular weight, the molar critical volume can be calculated.

The constants a and b are the same ones that appear in the Van der Waals' equation and may be calculated from that equation:

$$(P + a/V^2)\,(V - b) = RT$$

which is a cubic equation in V, i.e.

$$V^3 - b + \frac{RT}{P}\,V^2 + \frac{a}{P}\,V - \frac{ab}{P} = 0$$

which is shewn graphically in Figure 5.8. This equation fits a suggestion by J. Thomson that the gas–liquid curve has the form ABEFGCD, thereby emphasizing the continuity of state. By very careful experiment the line AB can be pushed in the direction of E and the line DC towards G. In the first case there follows a sudden liquefaction, and in the second, an evaporation that is nearly explosive.

The three roots of the equation correspond to B, the dew point; C, the bubble point; and F, a point of no physical significance. As one approaches the critical point, the points B and C get nearer together until at the critical point there is a horizontal point of inflection at the critical point in the Van der Walls' curve. For this to be so, the first and second derivative of the equation must be equal to zero,

$$\text{i.e.} \quad \left(\frac{\partial P}{\partial V}\right)_{,T} = 0 \quad \text{and} \quad \left(\frac{\partial^2 P}{\partial V^2}\right)_{;T} = 0$$

The Van der Waals' equation may be rewritten as

$$P = \frac{RT}{(V - b)} - a/V^2$$

$$\therefore \left(\frac{\partial P}{\partial V}\right)_T = -\frac{RT}{(V - b)^2} + 2a/V^3$$

$$\left(\frac{\partial^2 P}{\partial V^2}\right)_T = \frac{2RT}{(V - b)^3} - \frac{6a}{V^4}$$

Thus at the critical point

$$-\frac{RT_c}{(V_c - b)^2} + \frac{2a}{V_c^3} = 0$$

and

$$\frac{2RT_c}{(V_c - b)^3} - \frac{6a}{V^4} = 0$$

From these we find that:

$$V_c = 3b$$

$$T_c = \frac{8a}{27Rb}$$

$$P_c = \frac{a}{27b^2}$$

Two-component systems

Single phase

In general, the behaviour of a two-component mixture in the single-phase region is similar to that of a pure substance. A mixture of a given component can be treated as a definite material and considered as a pure substance. The constants needed for a two-component system in the one-phase region need to be experimentally determined. Problems arise when one takes into account the variable composition, and in practice it is better to relate each property separately with composition.

Volume–composition Figure 5.9 shews the variation of volume with composition for mixtures of methane and propane at 220 °F. At low pressures the lines are approximately straight, indicating that the law of additive volumes more or less applies:

$$V_{tot} = n_1 V_1 + (1 - n) V_2$$

where V_{tot} is the total volume of the mixture and n_1, n_2 the volume fractions of components 1 and 2.

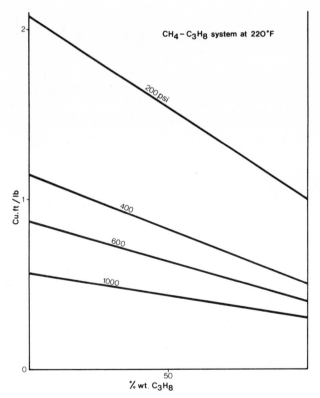

Figure 5.9 Volume–composition diagram for the methane–propane system at 220°F

This law may be used at low pressures for calculating the approximate volume of gaseous mixtures. At higher pressures the lines are not straight and the best data are obtained experimentally.

144

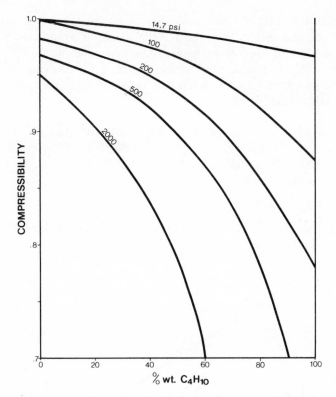

Figure 5.10   Compressibility of methane–butane mixtures

Compressibility–composition. This is illustrated in Figure 5.10, which shews the compressibility of methane–butane mixtures at different pressures. If the law of additive volumes were strictly adhered, to there would be a linear change in Z (the compressibility) with composition. The greater the proportion of the butane, the larger the change in Z. The 2,000 psi isobar stops because at that pressure the liquid phase will start to separate out at higher concentrations of the butane, i.e. gaseous mixtures of methane and butane higher than that critical concentration cannot exist above that pressure.

Pressure–volume and temperature–volume. With constant composition for single phase mixtures these principal properties are very similar to those of pure substances and the curves are of the same shape as those for pure substances.

Two phase

Pressure–volume. Figure 5.11 shews the variation of pressure and volume of a substance through the gaseous, two-phase, and liquid regions. Let us consider

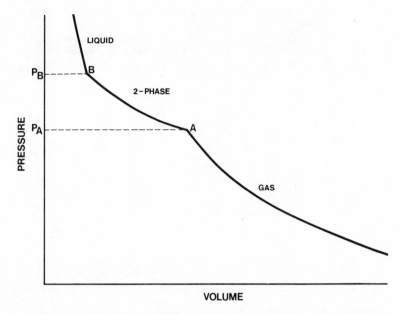

Figure 5.11 Pressure–volume diagram of a binary mixture below the critical temperature (after Burcik, 1957. Reproduced by permission of John Wiley & Sons, inc.)

a component mixture in the gas state just below the critical temperatures of the components and decrease the volume. The pressure will rise until the substance reaches point A, the dew point. So far the curve is as that for a single-component system, but it is at the dew point and beyond that the major differences occur. At the dew point the system is entirely gaseous, in equilibrium with an infinitesimal amount of the liquid phase. The composition of the whole system is that of the gas phase; although the liquid phase composition may be quite different and the liquid is normally richer in the less volatile component. If the volume is decreased further, the rate increase of pressure will change abruptly and more and more liquid will separate until the bubble point (B) is reached. The system at the bubble point is entirely liquid, but in equilibrium with an infinitesimal amount of the vapour which is richer in the more volatile component. The composition of the liquid is that of the whole system.

Thus it is in the two-phase region that the difference between single- and two-component systems manifests itself. Whereas for a pure substance the line AB is horizontal, in the two-phase system it rises from the dew point to the bubble point. The actual shape of the curve depends on the composition.

Finally, a further decrease in volume causes a great increase in pressure, as occurs in a single-component system. There will be a whole set of isotherms for each mixture of a given composition and these are shewn in Figure 5.12. Again, there is a saturation dome with a critical point, but unlike the pure

146

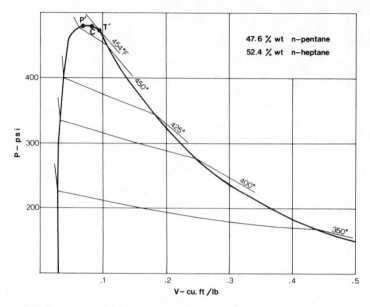

Figure 5.12 Complete pressure–volume diagrams of a binary mixture. (Reproduced by permission of Gulf Publishing Company)

substance, the critical point is not at the highest temperature at which the mixture can exist in two phases.

Retrograde Phenomena Let us consider the *P–V* diagram for a mixture of propane and methane which is 35.3% methane by weight (Figure 5.13). The isotherm T_c passes through the critical point, but the 156.6 °F isotherm, which is tangential to the dew point line at T', is the highest temperature at which the two phases can coexist as a mixture of this composition. The last liquid to disappear on an increase in temperature at this point is not of the same composition, nor with the same properties, as that of the gas phase at T' with which it coexists. Hence T' is not the critical point but the 'cricondentherm'. Likewise point P' is the highest pressure at which the two phases can coexist as a mixture of this composition, and here again the composition and properties of the two phases are different; thus it is not the critical point, but the 'cricondenbar'. The 157 °F isotherm is gaseous throughout its length, but the 130 °F isotherm starts as a gas on the left of the diagram, becomes liquid and vapour in varying proportions between the dew and bubble point lines, and finally continues as a liquid.

Let us consider an isotherm between isotherms T' and T_c, the isotherms representing the temperatures of the critical point and the cricondentherm. As the pressure is increased, the volume will decrease until the dew point is reached, at which point liquid starts to form. As the pressure is further increased, more liquid will separate until the amount reaches a maximum, after which the amount of liquid decreases until the dew point line is reached when the system is completely gaseous (plus an infinitesimal amount of liquid.) The sequence has

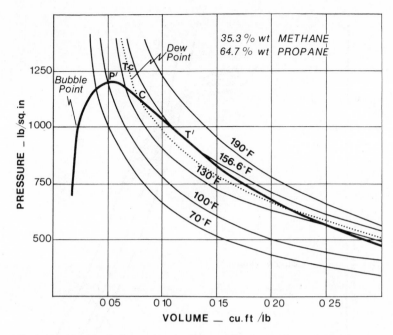

Figure 5.13 Typical pressure–volume diagram of methane-propane mixture

been gas–liquid–gas, which is known as retrograde condensation. (Figure 5.14)

If one now considers a system represented by F (Figure 5.14) and isobarically increases the volume so that it ends in the condition G, we would have had the sequence: liquid–gas–liquid, which is isobaric retrograde vaporization.

Pressure–temperature. In a pure substance the *P-T* plane projection for the two-phase region is a line, because isothermal condensation takes place completely at a definite temperature. However, in two-component systems the situation is different, as shewn in Figure 5.15.

The bubble point line and the dew point line again form a saturation dome, and all within it (AC'B) is a mixture of two phases. All above the dome is liquid and all below is gas, while anything far removed is fluid. Let us consider a substance represented by M, which is in the gas region, and isothermally increase the pressure. The pressure will rise until at point N the system is all vapour in equilibrium with an infinitesimal amount of liquid. Further increase in pressure will cause more and more liquid to condense, so that when point R is reached the system is all liquid in equilibrium with an infinitesimal amount of vapour. By the time the pressure has reached point S, the substance is all liquid.

Sometimes the liquid–vapour volume composition is shewn on such diagrams and in Figure 5.15 the line XC' represents 75%; YC', 50%; and ZC', 25% by volume of the liquid. In the isothermal compression described above, the point O would be 25%; point P, 50%; and point Q, 75% by volume of the liquid.

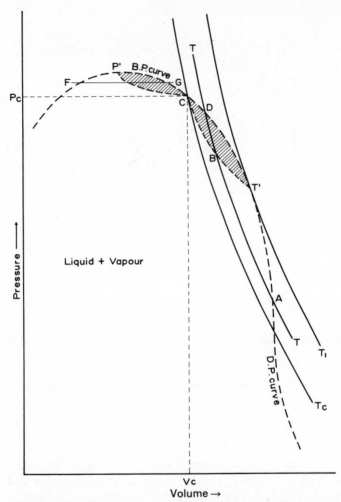

Figure 5.14 Pressure–volume diagram for a two-component system in the region of the critical point (after Burcik, 1957. Reproduced by permission of John Wiley & Sons, Inc.)

This diagram shews that the two phases can coexist above T_c and P_c. The cricondentherm is the highest temperature at which the two phases can coexist and the cricondenbar is the highest pressure at which the two phases can coexist. The diagram can also shew retrograde behaviour. Consider a substance as represented by C and let it be isothermally compressed. When the dew point is reached liquid will form and the amount of liquid will increase to a maximum at E. After this point the liquid will decrease until the dew point line is reached where the system is all vapour in equilibrium with an infinitesimal amount of liquid (point F). A further increase in pressure to point G will result in an all vapour system. This pressure increase occurs at a temperature higher than the critical temperature (T_c) but lower than the critical pressure (P_c). The isobaric temperature rise KJIH occurs at a pressure above the critical pressure

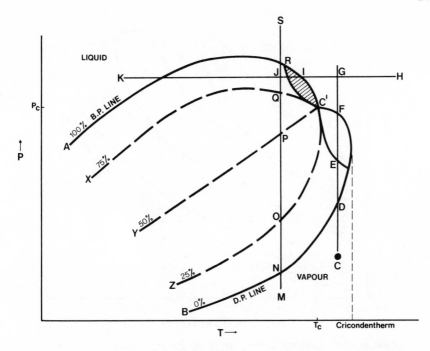

Figure 5.15 Pressure–temperature diagram for a typical two-component system (after Burcik, 1957. Reproduced by permission of John Wiley & Sons, Inc.)

but below the critical temperature. For retrograde phenomena to occur either the pressure or the temperature of the system must be below and the other above the critical value. If both are below the critical value a normal condensation or vaporization occurs whereas if they are both above the critical point no changes of state can occur. The section E to F is an isothermal retrograde condensation and the section J to I is an isobaric retrograde vaporization.

Composite pressure–temperature diagrams.

The diagrams so far have only shewn one composition of the mixture, and there will be an infinite number of these *P*–T diagrams for varying compositions. Figure 5.16 shews saturation domes for some mixtures of a binary hydrocarbon mixture and it is interesting to note that some of the critical pressures of the mixtures are higher than the critical points of the individual paraffins from which they were made. Figure 5.17 shews the loci of the critical points for binary mixtures of the lower paraffins and it is apparent from this that the greater the divergence of the critical temperatures of the components, the higher is the critical pressure of the resulting mixture. It will be noted that the actual critical pressures of mixtures bear no relationship to the pseudo-critical pressure used in calculations.

150

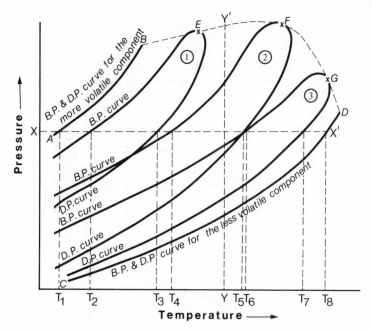

Figure 5.16 Composite pressure–temperature diagram for a two-component system (after Burcik, 1957. Reproduced by permission of (John Wiley & Sons, Inc.)

It should also be noted that this type of behaviour is relatively simple and many more complex types are found. These include azeotropic or constant-boiling mixtures, or systems where the components are only partly miscible. However, the petroleum hydrocarbons generally behave in the simple way described above.

Temperature–composition diagrams.

These can be obtained by a study of the *P–T* diagram for several compositions. At constant pressure the pure components will plot as single points on a *P–T* diagram. All the other mixtures will have two points, one for the dew point and one for the bubble point. This is represented by the line X–X' on Figure 5.16. The result is shewn in Figure 5.18.

Pressure–composition diagrams.

These can be obtained from the composite pressure–temperature diagram in a similar way as for the temperature–composition diagram, by plotting the pressure corresponding to each composition at a constant temperature. Each mixture will have two pressures, the dew and bubble points, while the pure substances will have one point each. Figure 5.19 illustrates such a diagram where A and B are the vapour pressures of the bubble and dew points

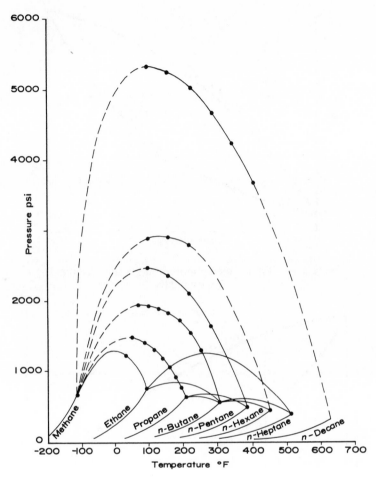

Figure 5.17 Critical loci of binary *n*-paraffin mixtures (after Burcik, 1957. Reproduced by permission of the Gas Processors' Association)

for the more and less volatile components respectively, whereas C and D represent the dew point pressure and the bubble point pressure for a mixture with 25% of the less volatile component. Above the bubble point line is liquid, below the dew point line is vapour, whilst in between them is the two-phase region. The points X and Y are thus at the same temperature and the same pressure and they represent the compositions of the coexisting liquid and vapour phases. W_l is the weight percentage in the liquid phase of the less volatile component and W_v is the weight percentage in the vapour phase of the less volatile component.

Let us consider the system shewn in Figure 5.20 and the substance described by point A. An increase in pressure will cause no phase change until point B is reached, where the vapour is in equilibrium with an infinitesimal amount of the liquid. This is the dew point, pressure P_1. The composition of the vapour is that of the original system, z, but the composition of the liquid is given by x_1 (the

152

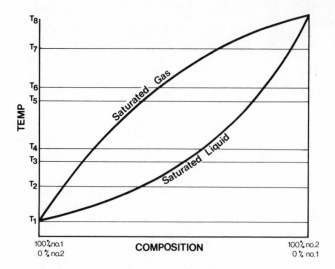

Figure 5.18 Typical temperature–composition diagram for a binary mixture

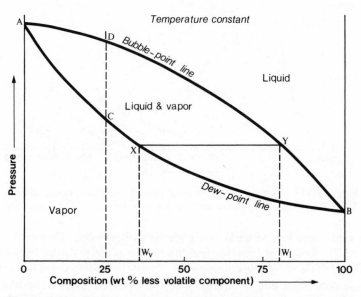

Figure 5.19 Typical pressure–composition diagram for a two-component system (after Burcik, 1957. Reproduced by permission of John Wiley & Sons, Inc.)

more volatile component). As the pressure is increased, more and more liquid condenses and the compositions of the coexisting vapour and liquid at any pressure are given through the dew and bubble point lines, e.g. at P_2 the composition of the liquid is x_2 of the more volatile component and $1 - x_2$ of the less volatile, while the composition of the vapour is y_2 of the more volatile and $1 - y_2$ of the less volatile component. As the pressure is increased to P_3,

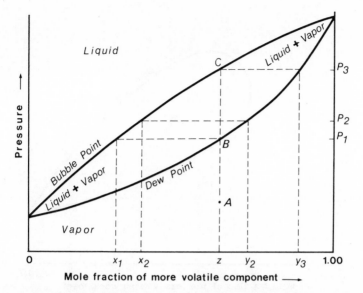

Figure 5.20 Pressure–composition diagram illustrating an isothermal compression through the two-phase region

the bubble point C is reached. The composition of the liquid is the same as the original composition of the vapour (i.e. that of the whole system) and it is in equilibrium with an infinitesimal amount of the vapour, of composition y_3. The extremities of each horizontal line give the composition of each phase, but not the amount of each. This can be calculated as follows.

In the system described by Figure 5.20, n = the number of moles in the system; n_1 = the number of moles of the liquid; n_v = the number of moles of the vapour; z = the mole fraction of the more volatile component in the system; x = the mole fraction of the more volatile component in the liquid; and y = the mole fraction of the more volatile component in the vapour.
Hence:

    nz = number of moles of more volatile component in the system;
    n_lx = number of moles of more volatile component in the liquid;
    n_vy = the number of moles of more volatile component in the vapour

Obviously

$$nz = n_lx + n_vy \text{ and } n_l = n - n_v$$

and by eliminating n_l we get:

$$nz = (n - n_v)x + n_vy$$

or

$$\frac{n_v}{n} = \frac{z - x}{y - x}$$

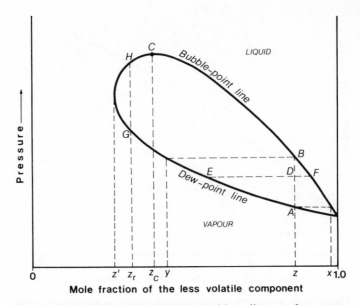

Figure 5.21 Typical pressure–composition diagram for a two-component system at a temperature between the critical temperatures of the two components (after Burcik, 1957. Reproduced by permission of John Wiley & Sons, Inc.)

Thus if the system were composed of 2 moles of butane and 1 mole of hexane and the mole fractions of the butane in the liquid and vapour respectively are 0.42 and 0.92, we obtain:

$$z = 0.667 (\text{i.e. } 2/3); \qquad x = 0.42; \qquad y = 0.92$$

$$n_v = n \frac{z - x}{y - x} = 3 \frac{(0.667 - 0.42)}{(0.92 - 0.42)} = 3 \times \frac{0.247}{0.500} = 1.48$$

thus the number of moles of vapour is 1.48 and since there are only 3 moles in total the number of moles of liquid must be 1.52.

So far, all the pressure–composition diagrams we have examined have been for components at temperatures less than their critical temperature, but Figure 5.21 shews a system at a temperature between the critical temperatures of the two components. No two-phase region exists for systems with a mole fraction of the less volatile component of less than z'. The system of composition of z' is the system whose cricondentherm is equal to the temperature at which the diagram was drawn. The critical point is C, which corresponds to a composition of z_c. The interpretation of this diagram is similar to that of the other diagrams and it also demonstrates retrograde phenomena. Consider a gas of composition z_r, which is between z' and z_c, and compress this gas isothermally. When it reaches point G it crosses the dew point line, and liquid starts to condense until a maximum amount of liquid is obtained. After that the liquid decreases until at point H the dew point line is met and the system is all gaseous.

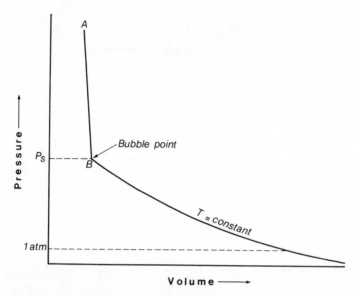

Figure 5.22 Pressure–volume isotherm for a crude oil (after Burcik, 1957. Reproduced by permission of John Wiley & Sons, Inc.)

Multicomponent systems

These are in general similar to the two-component systems, but the pressure–composition and temperature–composition diagrams cannot be drawn in two dimensions. Figure 5.22 shews the pressure–volume diagram for a crude oil of fixed composition. It is very similar to the two-component system except that the dew point is not very pronounced and is very difficult to detect. As the pressure is increased at constant temperature, the volume decreases until the bubble point is reached, which in crude oils is called the 'saturation pressure' since in crude oils it is customary to regard the vapour phase that forms at the bubble point as a gas that is dissolved in the liquid. As the pressure is lowered, more and more gas is liberated and normally at atmospheric pressure the system consists of gas and liquid. To vaporize completely may need very low pressures so that the dew point is almost unattainable.

Figure 5.23 illustrates the pressure–temperature variation of a multicomponent system, which is not unlike that for a two component system and which at low dew point pressures could be a crude oil.

If the surface conditions are at A and the reservoir conditions are B, this would represent a reservoir with a gas cap containing both liquid and vapour and producing both. If the reservoir conditions were F, and the surface conditions were at A the oil would be an undersaturated crude, i.e. it exists in the reservoir as a liquid but produces both gas and liquid. If the reservoir

156

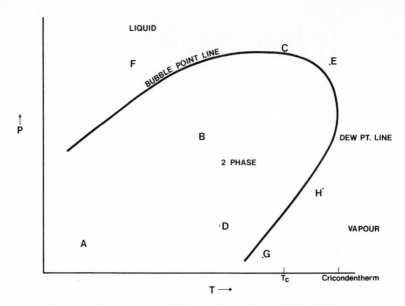

Figure 5.23 Pressure–temperature diagram of a multicomponent system (after Burcik, 1957. Reproduced by permission of John Wiley & Sons, Inc.)

conditions are H and the surface G, we have a dry gas reservoir producing dry gas, but if the reservoir is H and the surface is D the crude exists as a gas, but it is produced as gas and liquid, i.e. it is a condensate reservoir. A reservoir with conditions at E and surface conditions at G is a retrograde condensate reservoir. The crude exists as gas and is produced as gas but during production, i.e. pressure drop, it passes through the two-phase region where both gas and liquid exist.

QUANTITATIVE PHASE BEHAVIOUR
(two-phase ideal solution)

Ideal solutions are those in which the attractive forces are all equal and there is no heating effect when the components are mixed, and the total volume is the sum of the volumes of the components.

Raoult's law states that in an ideal solution the partial pressure of a component in the vapour (P_a) is equal to the product of the mole fraction of that component in the liquid (X_a) and the vapour pressure of the pure component (P_a^0):

$$P_a = x_a P_a^0 \text{ or } P_i = x_i P_i^0$$

hence $P_t = \Sigma x_i P_i^0 =$ total vapour pressure of the solution. This total pressure is also the bubble point pressure, since an infinitesimally greater pressure results in an all-liquid system. If the infinitesimal amount of vapour that exists at the bubble point is a perfect gas, then Dalton's law of partial pressures will apply, i.e.

$$P_i = y_i P_t \quad \text{or} \quad y_i = P_i/P_t$$

and hence by using $P_i = x_i P_i^0$ we can get the bubble point pressure and from $y_i = P_i/P_t$ we get the vapour composition. For example, calculate the bubble point pressures and the vapour composition at the bubble point for mixtures containing 0.5 mole each of n-butane and n-pentane, and 0.25 mole of n-butane and 0.74 mole of n-pentane at 60 °F. The vapour pressures of n-butane and n-pentane at 60 °F are 26.3 and 7.0 psi respectively.

Component	P_i^0	x_i	$P_i = x_i P_i^0$	$y_i = P_i/P_t$
C_4H_8	26.3	0.5	13.5	0.790
C_5H_{10}	7.0	0.5	3.5	0.21
			$P_t = 16.65$	
C_4H_8	26.3	0.25	6.58	0.56
C_5H_{10}	7.0	0.75	5.25	0.44
			$P_t = 11.83$	

These results are displayed in Figure 5.24. Points A, B, C, and D have been calculated; FACE is the bubble point line and FBDE is the dew point line.

Note that for an ideal solution the bubble point line is a linear function of the composition, which for a two component system is

$$\text{BPP} = x_1 P_1 + x_2 P_2 \text{ (this is Raoult's law), and since } x_2 = 1 - x_1, \text{ BPP}$$
$$= x_1(P_1^0 - P_2^0) + P_2^0, \text{ which is linear in } x_1.$$

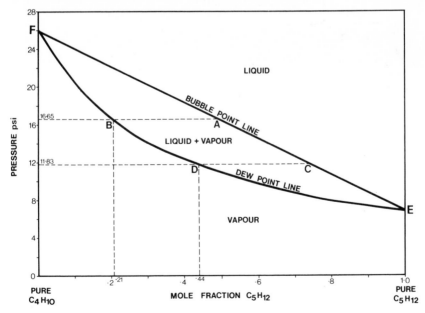

Figure 5.24 Pressure–composition diagram of the n-butane/n-pentane mixture at 60 °F

Calculation of the liquid and vapour composition

For a two-component mixture in the two-phase region, as the pressure is reduced below the bubble point more and more vapour forms and the vapour becomes richer in the more volatile component. Hence for a binary system in the two-phase region the composition of the liquid and the vapour are different and neither is equal to the overall composition.

Let x_1 and x_2 be the mole fractions of the two components in the liquid phase. By Raoult's law:

$$x_1 P_1^0 + x_2 P_2^0 = P_t$$

and for the liquid $x_1 + x_2 = 1$. Thus:

$$\left. \begin{aligned} x_1 &= \frac{P_t - P_1^0}{P_1^0 - P_2^0} \\ x_2 &= \frac{P_t - P_1^0}{P_2^0 - P_1^0} \end{aligned} \right\} \text{ liquid composition}$$

By the application of Dalton's law we obtain:

$y_1 = P_1/P_t$ and since by Raoult's law $P_1 = x_1 P_1^0$,

$$\left. \begin{aligned} y_1 &= \frac{x_1 P_1^0}{P_t} \\ y_2 &= \frac{x_2 P_2^0}{P_2} \end{aligned} \right\} \text{ vapour composition}$$

Assuming ideal behaviour, calculate the composition of the liquid and vapour at 140 °F and 53 psi for a system of one mole n-pentane and one mole of n-butane, given that at 140 °F, $P_{c_4}^0 = 92.6$ psi and $P_{c_5}^0 = 31$ psi.
Let C_4H_8 be component 1; then since

$$x_1 = \frac{P_t - P_2^0}{P_1^0 - P_2^0}$$

$$x_{c^4} = \frac{P_t - P_{c_5}^0}{P_{c_4}^0 - P_{c_5}^0} = \frac{53 - 31}{92.6 - 31} = 0.357$$

Thus $x_{c_5} = 0.643$

Because $y_1 = \dfrac{x_1 P_1^0}{P_t}$,

$$y_{c_4} = \frac{x_{c_4} P_{c_4}^0}{P_t}$$

$$= \frac{0.357 \times 92.6}{53} = 0.623$$

and thus $y_{c_5} = 0.377$

Hence by these equations we get the composition of the liquid and the vapour in the two-phase region, but not the amounts. These can be obtained from the equation:

$$\frac{n_v}{n} = \frac{z - x}{y - x}$$

Thus in the above example we get

$$\frac{n_v}{2} = \frac{0.5 - 0.357}{0.623 - 0.357} = 0.54$$

thus $n_v = 1.08$

Calculation of bubble point pressure

It has already been shewn that in an ideal two-component system the total pressure is given by:

$$P_t = \text{BPP} = x_1 p_1^0 + x_2 P_2^0 = x_1(P_1^0 - P_2^0) + P_2^0$$

At the bubble point the composition of the liquid is equal to the overall composition while the composition of the vapour is given by

$$y_1 = x_1 p_1^0 \, P_t$$

Thus if we require to calculate the bubble point pressure and the vapour composition at that pressure for an equimolar mixture of n-butane and n-pentane, we use the above formulae.

$$\text{BPP} = x_{c_4} P_{c_4}^0 + x_{c_5} P_{c_5}^0$$
$$= 0.5 \times 92.6 + 0.5 \times 31 = 46.3 + 15.5 = 61.8 \, \text{psi}$$

$$y_{c_4} = \frac{x_{c_4} P_{c_4}^0}{P_t} \frac{0.5 \times 92.6}{61.8} = 0.749$$

Therefore $y_{c_5} = 0.251$

Calculation of the dew point pressure for a two-component system

At the dew point the system is essentially all vapour in equilibrium with an infinitesimal amount of liquid. If the vapour behaves in an ideal way, then

$$y_1 = \frac{x_1 P_1^0}{P_t}$$

Hence if we knew x_1 we could calculate y_1. It has just been shewn that

$$x_1 = \frac{P_t - P_2^0}{P_1^0 - P_2^0}, \text{ hence:}$$

hence:

$$y_1 = \frac{([P_t - P_2^0]/[P_1^0 - P_2^0])P_1^0}{P_t}$$

This can be solved for P_t and the composition of the liquid that is present in an infinitesimal amount can be obtained from the expression:

$$x_1 = \frac{P_t - P_2^0}{P_1^0 - P_2^0}$$

Thus we calculate the dew point of an equimolar mixture of n-pentane and n-butane together with the liquid composition at the dew point by use of these latest equations. The composition of the vapour at the dew point is equal to the overall composition, i.e. $y_{c_4} = y_{c_5} = 0.5$.

$$y_{c_4} = \frac{P2t - P_{c_5}^0}{P_{c_4}^0 - P_{c_5}^0} \times \frac{P_{c_4}^0}{P_t}$$

i.e. $0.5 = \dfrac{P_t - 31}{92.6 - 31} \times \dfrac{92.6}{P_t}$

or $P_t = \text{DPP} = 44.99$ psi

$$x_{c_4} = \frac{P_t - P_{c_5}^0}{P_{c_4} - P_{c_5}} = \frac{44.99 - 31}{92.6 - 31} = 0.227$$

Thus $x_{c_5} = 0.773$

Summary of results involving an equimolar mixture of n-butane and n-pentane at 140 °F.
  (i) Bubble point pressure = 61.8 psi.
 (ii) Composition of vapour at bubble point; $y_{c_4} = 0.749$, $y_{c_5} = 0.251$.
(iii) Composition of liquid and vapour at 53 psi:
    $x_{c_4} = 0.357$; $x_{c_5} = 0.643$; $y_{c_4} = 0.623$; $y_{c_5} = 0.377$.
(iv) Dew point pressure = 44.99 psi.
 (v) Composition of liquid at dew point; $x_{c_4} = 0.227$; $x_{c_5} = 0.773$
Figure 5.25 shews these quantities plotted on a pressure–composition diagram.

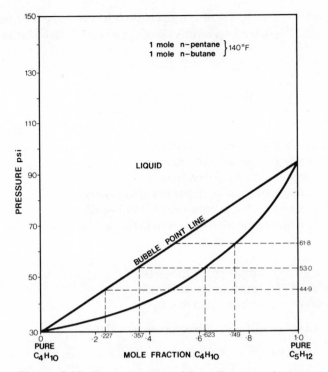

Figure 5.25 Pressure–composition diagram of the *n*-butane/*n*-pentane system at 140 °F

The amounts of liquid and vapour present at 53 psi are given by: $n_v/n = (z - x)/(y - x)$ which earlier shewed that $n_v = 1.08$.

The apparent molecular weight (AMW) is given by AMW = $\Sigma\, y_i MW_i$, and the weight of the liquid or vapour is the product of AMW and the number of moles. Hence at 53 psi:

the weight of liquid = $AMW_1 \times n_1$
= $(357 \times 58 + 0.643 \times 72) \times 0.92 = 61.65\ lb$

the weight of vapour = $AMW_v \times n_v$

= $(0.623 \times 58 + 0.377 \times 72) \times 1.08 = 68.35\ lb$

Thus the total weight of the system is 61.65 + 68.35 lb = 130 lb. (Since we have one mole of *n*-butane (58 lb) and one mole of *n*-pentane (72 lb) this confirms that the calculation of the compositions is correct.

Multicomponent ideal systems

The calculation of the bubble point pressure and the composition of the vapour at the bubble point requires no new principles when applied to multicomponent systems. The application of Raoult's law will give the bubble point pressure, BPP $= \Sigma\, x_i P_i^0$, where x_i is the mole fraction of the ith component in the liquid and P_i^0 is that component's vapour pressure. Thus by Dalton's law we get

$$y_i = \frac{P_i}{P_t} = \frac{x_i P_i^\circ}{P_t} = \frac{x_i P_i^\circ}{\text{BPP}}$$

Let

$n\ =$ total number of moles in the system
$n_1 =$ total number of moles in the liquid
$n_v =$ total number of moles in the vapour
$z_i\ =$ mole fraction of the ith component in the system
$x_i\ =$ mole fraction of the ith component in the liquid
$y_i\ =$ mole fraction of the ith component in the vapour

Then

$z_i n\ =$ moles of ith component in the system
$x_i n_1 =$ moles of ith component in the liquid
$y_i n_v =$ moles of ith component in the vapour

and a material balance for the ith component will be:

$$z_i n = x_i n_1 + y_i n_v$$

Raoult's and Dalton's laws are now applied to the ith component:

$$y_i = \frac{P_i}{P_t} = \frac{x_i P_i}{P_t}$$

and by eliminating y_i we get

$$z_i n = x_i n_l + \frac{P_i^\circ}{P_t} x_i n_v$$

and hence

$$x_i = \frac{z_i n}{n_1 + (P_i^\circ/P_t)n_v}$$

since at the bubble point the system is all liquid and the component of the liquid and the whole system are the same.

$\Sigma\ x_i = 1$. Thus

$$\Sigma \frac{z_i n}{n_1 + (P_i^\circ/P_t)n_v} = 1$$

By the elimination of x_i instead of y_i, an equivalent equation for the vapour composition can be obtained. Once the composition of the liquid has been obtained, Dalton's law will give the vapour composition:

$$y_i = x_i P_i^0 / P_t$$

In practice the equation is solved on a trial-and-error basis with an intelligent guess being made for the relative values of n_l and n_v. When all the compositions have been calculated they are summed to see if their total is unity. The process is repeated until the sum of the mole fractions is one. For approximate calculations, all that is needed are the composition of the feedstock and the boiling points of each component. The vapour pressure at the required temperature is estimated via Trouton's rule and the Clapeyron–Clausius equation. Trouton's rule states that the ratio of molar heat of vaporization (latent heat L) and the boiling point in degrees absolute is a constant. For hydrocarbons this constant is 20 and thus the latent heat can be approximated and put into the Clapeyron–Clausius equation:

$$\log_e P_2/P_1 = \frac{L}{R}\left[\frac{1}{T_1} - \frac{1}{T_2}\right]$$

The use of Trouton's rule approximation for the latent heat of any hydrocarbon together with the boiling point and the vapour at the boiling point (atmospheric) permits the estimation of the vapour pressure of that hydrocarbon at any other temperature.

NON-IDEAL SOLUTIONS

Raoult's law is not applicable to non-ideal solutions as it is really a special case of Henry's law which states that the vapour pressure of a component is proportional to the mole fraction of that component in the liquid phase, i.e. $P_i \propto x_i$ or $P_i = k_i x_i$, and for Raoult's law $k_i = P_i^0$. For an ideal solution we know that $P_i = P_i^0 x_i$ and $P_i = y_i P_t$; thus $y_i = x_i P_i^0 / P_t$. For the non-ideal solution $P_i = k_i x_i$, and if $P_i = y_i P_t$, then $y_i = k_i x_i / P_t = K_i x_i$. The constant K_i is a function of both temperature and pressure. At any given pressure it rises with an increase in temperature, but at fixed temperature it falls with increase in pressure. It is now possible to alter the equations that have been used in the ideal case by substituting the ratio of vapour to total pressure by the constant K. Values for the constants K for each hydrocarbon at various temperatures and pressures are determined experimentally by virtue of the fact that $K = y_i / x_i$. Both the liquid and vapour concentrations are carefully measured. At low pressures they are independent of the composition, but at high pressures, K values must be obtained from mixtures similar to that being investigated.

The previous equations for a binary system become

$$x_1 = (1 - K_2)/(K_1 - K_2)$$
$$x_2 = 1 - x_1 = (1 - K_1)/(K_2 - K_1)$$

$$y_1 = K_1 x_1$$
$$y_2 = 1 - y_1 = K_2 x_2$$

whilst the equations for multicomponent systems become:

$$\sum x_i = \sum \frac{z_i n}{n_1 + K_i n_v} = 1$$

$$\sum y_i = \sum \frac{z_i n}{n_v = n_1/K_i} = 1$$

Once again, in the case of multicomponent systems one lets $n = 1$ and estimates a reasonable value for n_1 and then sees if $x_i = 1$. If the value of $\sum x_i > 1$ then increase the estimated value of n_1, or vice versa.

Calculation of bubble and dew point pressures

The new equations for non-ideal systems given above no longer contain pressures, and thus cannot be used directly to calculate the pressures required. However, because at the bubble point the system is all liquid, except for a minute amount of vapour, $n_v = 0$ and $n_1 = n$; thus

$$\sum \frac{z_1 n}{n_v + n_1/K_i} = 1 = \sum \frac{z_1 n}{0 + n/K_i} = \sum K_i z_i$$

Likewise, at the dew point the system is all vapour, except the small amount of liquid; therefore $n_1 = 0$ and $n = n_v$, and thus

$$\sum \frac{z_i n}{n_1 + K_i n_v} = 1 = \sum y_i/K_i$$

Thus the bubble and dew point pressures are found by a process of trial and error, first making an assumption about the likely pressure and seeing if with this pressure the summations are equal to unity.

HYDROCARBON CHARACTERISTICS

Gas formation volume factor

This is the volume, in barrels, occupied by one standard cubic foot (std cu. ft) of gas at reservoir pressure and temperature, and it may be obtained from the gas laws. Let P_0 and T_0 to be the pressure and temperature of the reservoir and P_1 and T_1 to be the standard pressure and temperature (14.7 psi and 60 °F or 520 ° Absolute on the Fahrenheit scale), then:

$$\frac{P_0 V_0}{Z_0 T_0} = \frac{P_1 V_1}{Z_1 T_1}$$

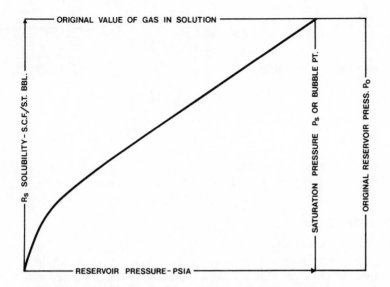

Figure 5.26 Gas solubility curve where the original reservoir pressure was above the bubble point (after Burcik, 1957). Reproduced by permission of John Wiley & Sons, Inc.)

Now for all practical purposes $Z_1 = 1$ and hence

$$V_0 = \frac{Z_0 P_1 V_1 T_0}{Z_1 P_0 T_1} = \frac{Z_0 \times 14.7 \times 1 \times T_0}{1 \times P_0 \times 520} = \frac{0.0283 Z_0 T_0}{P_0} \text{ cu. ft}$$

To convert this figure from cubic feet to barrels one must divide by 5.62, thus

$$v = V_0/5.62 = 0.00504 Z_0 T_0/P_0 \text{ barrels per std cu. ft.}$$

To calculate v one must know Z_0, the compressibility factor, and if this is not available by experimental determination then a value must be calculated. Values for the pseudo-critical pressure and temperature can be obtained either from composition ($P_c = y_i P_{ci}$; $T_c = y_i T_{ci}$) or from plots of P_c and T_c against gas specific gravity. Values of Z_0 can be obtained from published plots of the critical variables against compressibility.

Gas solubility (R_s)

This is the solubility of the gas in standard cubic feet per barrel of oil at 60 °F and 14.7 psi (Figure 5.26). P_0 is the original reservoir and pressure and R_s is the original value of the gas solubility, and as the pressure falls the solubility of the gas falls. For unsaturated crude oils the gas solubility remains at the original value, R_s, between the reservoir pressure and the saturation pressure, i.e. between P_0 and P_s, after it falls.

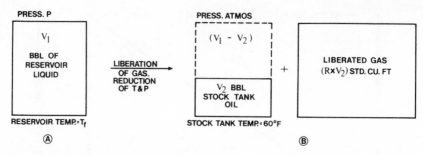

Figure 5.27 Schematic relationships of reservoir, stock tank oil, dissolved and liberated gas (after Burcik, 1957. Reproduced by permission of John Wiley & Sons, Inc.)

Oil formation factor (B_0)

This is the volume, at reservoir conditions, occupied by one stock tank barrel of oil plus the gas in solution. If the pressure on a sample of oil is reduced below its saturation pressure gas will be evolved and the residual volume will be less. This shrinkage in volume of the oil may be expressed either as a fractional change in the volume or a relative volume ratio based on either the original or final oil volume. Consider the system shewn in Figure 5.27; then we have:

$$\text{shrinkage based on final oil volume} = Sh_2 = (V_1 - V_2)/V_2$$
$$\text{shrinkage based on original oil volume} = Sh_1 = (V_1 - V_2)/V_1$$
$$\text{formation volume factor} = B_0 = V_1/V_2$$

Obviously $B_0 = 1 + Sh_2 = 1/(1 - Sh_1)$ and B_0 is the most commonly used expression, but it varies with pressure as shewn in Figure 5.28. If the original pressure is P_0, the original formation volume factor B_0 then as the pressure is reduced the value of B decreases. For the undersaturated crude a fall in pressure from the original P_0 to the saturation pressure, P_s, is accompanied by an increase in volume due to expansion and B increases to a maximum B_s after which further pressure reductions cause the value of B to fall as in the saturated case.

Two-phase formation volume factor (B_t)

The volume occupied in the reservoir by one stock tank barrel of oil plus the free gas which was dissolved in it is the two-phase formation volume factor, $B_t = B_0 + (R_0 - R)v$. B_0 is the liquid volume at reservoir conditions of one stock tank barrel, $(R_0 - R)v$ is the volume of free gas under reservoir conditions that was originally in solution. Figure 5.29 shews the volume changes that occur when the oil goes from reservoir conditions to stock tank conditions. Figure 5.30 shews the changes in B_t as the pressure is reduced.

The portion of the curve P_0–P_s is coincident with that for a plot of P against B_0 (both for undersaturated crudes) since the system is entirely liquid over this range. At pressures below P_s the value of B_t increases due to the evolution of gas.

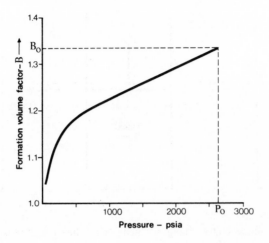

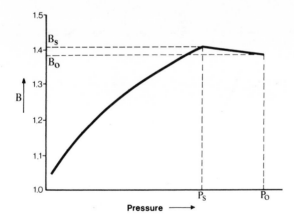

Figure 5.28(a) Plot showing the dependence of the formation volume factor on pressure (for a saturated crude). (b) Plot of formation volume factor versus pressure (for an undersaturated crude) (after Burcik, 1957. Reproduced by permission of John Wiley & Sons, Inc.)

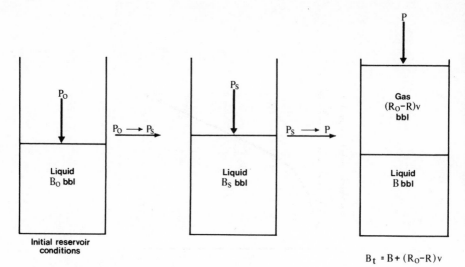

$$B_t = B + (R_O - R)v$$

Figure 5.29 Volume factor changes which occur when pressure is decreased on the B_o barrels of reservoir fluid (after Burcik, 1957. Reproduced by permission of John Wiley & Sons, Inc.)

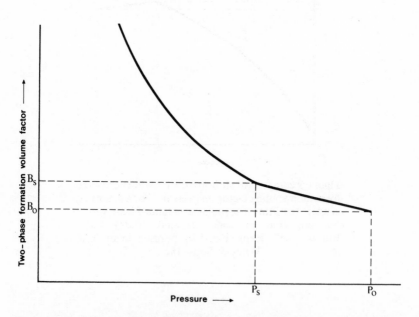

Figure 5.30 Plot of two-phase formation volume factor as a function of pressure for an undersaturated crude (after Burcik, 1957. Reproduced by permission of John Wiley & Sons, Inc.)

REFERENCE

Burcik, E. J. (1957) *Properties of Petroleum Reservoir Fluids*. John Wiley and Sons Ltd., New York.

Chapter 6

Kerogen Formation and Diagenesis

INTRODUCTION

The organic matter in rocks is of three main types (Hedberg, 1964; Stevens, 1956). Firstly, there is the potentially mobile organic matter which is readily soluble in a wide range of organic solvents and which often accounts for between 0.5 and 5.0% of the rock (by weight). This soluble component of sedimentary organic matter includes petroleum, asphalts, bitumens, and waxes. Secondly, there is the insoluble organic matter which has the potential to generate hydrocarbons, either by being buried for long periods of time at a sufficient depth and temperature or by artificial pyrolysis in the laboratory. Lastly, there is the insoluble organic material which is incapable of any further hydrocarbon generation.

Petroleum is a complex mixture of organic molecules, which are either fluids or solids dissolved or suspended in other petroleum fluids and which can migrate and are liable to be altered by heat, pressure, filtration, microbial action, and adsorption on clays. Petroleum originates from the organic matter that is deposited in sedimentary basins. Most of this organic material is of marine origin but significant amounts of terrestrial material can be carried by rivers and aeolian transport, and can thus contribute in varying degrees to the petroleum source material.

Cox (1946) suggests that organic matter can be converted into petroleum and that the following are the required essentials: (a) a marine environment; (b) a maximum temperature of 100 °C for most oil basins, but an absolute maximum of 200 °C as shown by the presence of chlorophyll; (c) a pressure between 2,000 and 5,000 psi; (d) sufficient time, because no petroleum has been formed since the Pliocene; (e) bacteria to degrade the original organic matter; (f) catalysts to allow the conversion of the organic matter to hydrocarbons at the low temperatures involved.

Fresh organic matter is attacked in the unconsolidated sediment by bacteria and transformed into an organic substrate which is rich in nitrogen, sulphur, and oxygen. This is known as the humus complexes. These complexes are converted into kerogen, which is the major source of oil and gas, by a series of processes involving physical, chemical, and biological alteration without any significant effect of elevated temperature. This stage is known as diagenesis. The structure of kerogen is believed to involve a number of sheets composed of aromatic and alicyclic rings with varying proportions of heteroatoms and aliphatic chains. The nucleii are linked by chain-like bridges and both can contain the heteroatoms. In addition, lipid molecules can be entrapped within

170

the kerogen matrix. In an immature kerogen these sheets are well separated, low in aromatic character and contain many functional groups such as, $-OH$, $-NH_2$, $-COOH$, $=C=O$, $-O-$, and $-S-$ (hydroxyl, amino, carboxyl, ketonic, etherial, and sulphidic). In addition, alkyl groups are found attached to the kerogen nucleus. As the kerogen evolves and matures, the sheets become closer together and more parallel. Concomitantly the functional groups may be lost.

At this stage it is worth while noting that kerogen is not the only source of the compounds that are found in petroleum. The individual compounds may be incorporated with little or no further chemical alteration or they may proceed via a different pathway, for example, the generation of alkanes from fatty acids. In fact there are many such compounds which do not proceed via the kerogen route and are trapped in the sediment with no or only minor changes but retaining their original carbon skeleton. These are often called geochemical fossils although the same or similar molecules can be derived from the kerogen.

Most geochemists accept that the reservoired hydrocarbons have a biogenic origin, and the evidence that petroleum is derived mainly from plant sources, but with a small animal contribution, is as follows. There is a molecular similarity between the types of compound found in petroleum and the types of organic compound synthesized by living organisms. This applies, not only to the hydrocarbons, but to the non-hydrocarbons such as the porphyrins and other plant pigments. Within petroleum there have been identified certain compounds which are known to have come from plants and animals and which could not have been synthesized abiogenically.

Many fractions of petroleum have the ability to rotate the plane of polarized light, a property which is usually found in compounds that have been generated biogenically. In all petroleums compounds called porphyrins have been found and these are known to have been derived from chlorophyll, and thus petroleum does contain the remains of living organisms. (Figure 6.1) Chlorophyll is essential to the process of photosynthesis and it gives plants their green colours. Thus the presence of porphyrinic and isoprenoid fragments of chlorophyll molecules in petroleum is reliable evidence for the plant (and animal) origins of petroleum. Stable isotope studies of crude oils shew the carbon $^{13}C/^{12}C$ ratios of crude oils are much more closely related to those of organic matter than they are to carbonates or to atmospheric carbon dioxide. In fact, in petroleum the carbon $^{13}C/^{12}C$ ratios are even more negative than they are for normal organic material and the only organic materials that have similar ratios to those of crude oils are the lipids. This again supports the theory that these lipid compounds, which are found in plants and animals, must provide much of the basic organic material that is eventually to become reservoired crude oil (Table 6.1, Figure 6.2; Schultz & Calder, 1976; Craig, 1953).

The ratio of ^{13}C to ^{12}C is determined using an isotope ratio mass spectrometer. The ratio (∂), in parts per thousand ($^0/oo$) relative to a standard

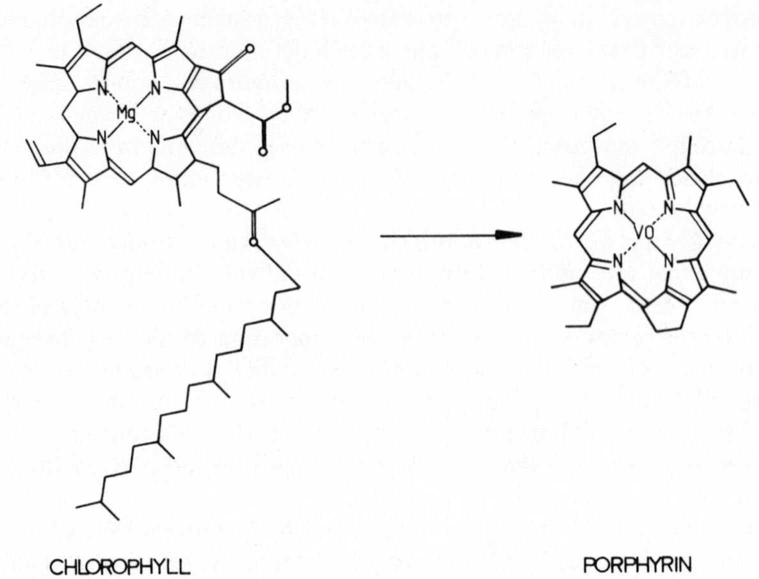

CHLOROPHYLL   PORPHYRIN

Figure 6.1   Chlorophyll and porphyrins

Table 6.1   Typical carbon isotope composition of natural materials

	$\partial\ ^{13}C\ (^0/_{00})$
Peedee belemite (PDB)	0
typical limestone	+5
marine plants	−10 to −18
freshwater plants	−22
land plants	−22 to −26
crude oils	−20 to −32
biogenic gas	−55 to −85
thermogenic gas	−25 to −60
plankton lipids	−30

is given by

$$\partial^{13}C = \left[\frac{(^{13}C/^{12}C)_{\text{sample}}}{(^{13}C/^{12}C)_{\text{standard}}} - 1 \right] \times 1000$$

The most widely used standard is a belemite from the Peedee Formation in South Carolina, USA (PDB) (Figure 6.2).

The presence of nitrogen compounds in crude oils also suggests the organic origin of oil because many of the nitrogen-containing compounds found in crude oils are known in biologically produced situations.

Petroleum-like substances have been found in recent soils and sediments, but these mixtures differ from petroleum in several ways. They differ not only in their total concentration but also in the relative concentrations of the

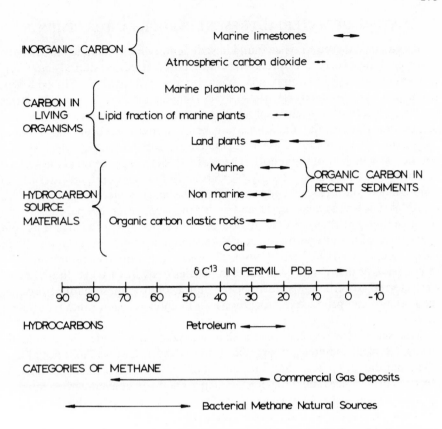

Figure 6.2 Values of some carbonaceous materials on the PDB carbon isotope scale (after Feux, 1977. Reproduced by permission of Elsevier Scientific Publishing Co.)

individual compounds present. Not only are some compounds present in greater relative abundance than in petroleum, such as the odd carbon number alkanes but some compounds such as the low molecular weight hydrocarbons are present in much smaller relative quantities.

It is surely no coincidence that fossil fuels are found exclusively in sedimentary areas that can be related to geological eras of great biosynthetic activity, whilst there is an absence of petroleum from Precambrian rocks, although there is evidence for life in rocks as old as 3.9×10^9 years. If oil had been formed abiogenically, then it follows that it should have been found in rocks whether or not there is evidence of life. The fact that oil is formed only when there is evidence of biosynthetic activity tends to support a biogenic source for petroleum. The various fractions in crude oil are not in thermodynamic equilibrium, which is what one would expect if the crude oil were of an abiogenic origin.

NATURE OF MATERIAL PRESENT IN ROCKS AND SEDIMENTS

The organic compound types found in sedimentary rocks include aminoacids, carbohydrates, lipids, isoprenoid compounds (including steroids and terpenes), heterocyclic compounds, phenols, and hydrocarbons. In addition, there are complex mixtures whose exact composition is unknown and varies between locations, e.g. the asphalts, bitumens, resins, and petroleum, as well as the macromolecules that constitute kerogen and its precursors. It is not surprising to learn that these compounds are subject to biological attack, particularly in view of the reported level of 10^6 bacteria per gram at the surface reducing to 10^3 per gram at a depth of 50 cm. Below this depth polymerization and condensation reactions occur to form the macromolecule kerogen. This stage is known as diagenesis and it is followed, after deeper burial, by compaction and thermal degradation or the early stages of catagenesis, the next stage of the alteration of sedimentary organic matter (Bray *et al.*, 1956; Meinschein, 1961b). Mair (1964) discussed the possible routes for the conversion of many natural product molecules to hydrocarbons. In particular he referred to terpenoids, fatty acids, and alcohols. These natural product compounds can produce hydrocarbons via the non-kerogen route mentioned earlier.

Meinschein (1961a) has shown that sediments and crude oils should be derivable from sedimentary organic matter. Bray & Evans (1965) noted that the older the rock, the greater the amount of hydrocarbon present and the CPI (carbon preference index; the predominance of odd- over even-carbon number of alkane chains; see below) decreases. They also observed that this decrease in CPI was depth-related and that the greater the percentage of hydrocarbon in a rock sample, the closer was its CPI to that of crude oils. Meinschein (1961b) considered that normal and branched alkanes come from different sources, with the normal alkanes being derived, in part, from fatty acids and alcohols, whereas the branched and alicyclic compounds are related to the isoprenoid compounds. The aromatics were considered to be derivatives of the terpenes, sterols, etc., from plant and animal products. Meinschein also noted that over 80% of the benzene–methanol extract of rocks was non-hydrocarbon material.

Aminoacids

Aminoacids are derived from both terrestrial and aquatic organisms, often by the hydrolysis of the peptide links between individual aminoacids in proteins (Figure 6.3). Aminoacids produced by the bacterial breakdown of proteins are chemically more stable than those aminoacids derived from proteins that resist bacterial action. Water-soluble aminoacids are the most easily detroyed, especially under oxidizing conditions. After the initial bacterial attack the destruction of aminoacids continues thermally as the temperature increases with depth of burial. Recent experiments have shown that those aminoacids that are less sensitive to heat are those found in ancient sediments. The types of aminoacid found in oilfield brines and modern sea waters are very alike.

a. Hydrolysis of Peptide Linkage

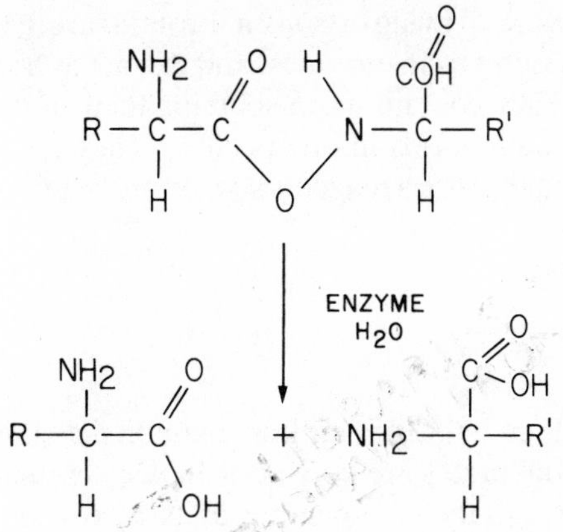

b. Types of Amino Acid

I) Acidic e.g. Asparatic Acid

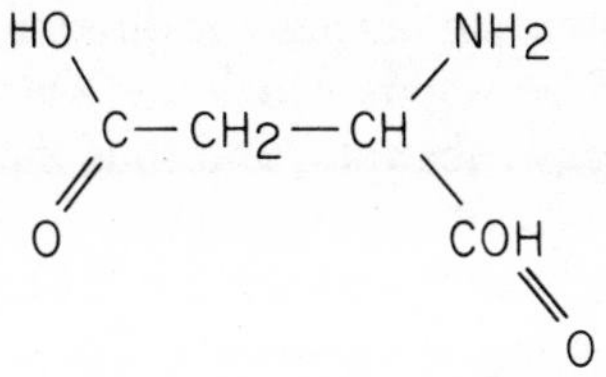

II) Basic e.g. Lysine

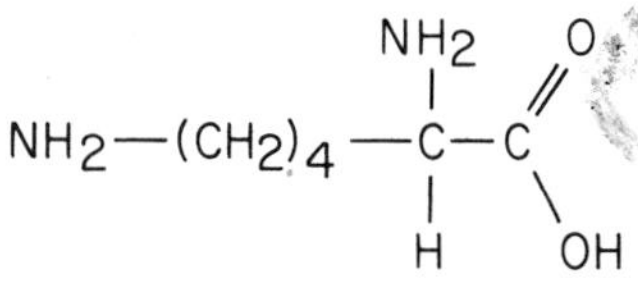

III) Neutral e.g. Alanine

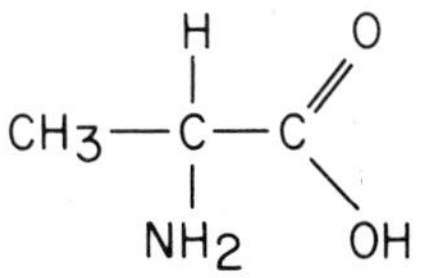

Figure 6.3 Aminoacids, types and peptide linkage

Proteins are long regular polymers of aminoacids and form part of the structural framework of living organism. Proteins are in the contractile element of muscles, are essentials in enzymes, and can act as hormones and regulators of metabolism. They contain more subunits than carbohydrates, often up to twenty different aminoacids in one protein. They are very sensitive to heat, extremes of pH, and various reagents so they are readily decomposed into their constituents.

Carbohydrates

Carbohydrates (Figure 6.4) are far less common than the aminoacids in marine sediments. Carbohydrates are polymeric molecules made up of long chains of either one sugar variety or repeating units of two or three different sugar compounds. Perhaps the most abundant carbohydrate is cellulose, built from long chains of D-glucose. The individual sugar units are held together with oxygen bridges and these can be broken to give the individual sugars or sometimes the groups of two or three sugars that form the repeating units. Carbohydrates generally form less than 5% of the total organic carbon of a sediment. Studies of present-day materials suggest that much of the carbohydrate material found in ancient sediments came from the lower plants, especially algae. Free sugars which are water soluble are readily utilized by bacteria as an energy substrate, and thus these sugars are short-lived in their free form and are accordingly restricted to waters and recent sediments. Any sugars that survive bacterial attack are short-lived in the increasing temperature regime of compacted sediments. Thus individual sugars are unlikely to survive in any sediment for very long; however, the polymeric carbohydrates may in certain circumstances do so. Because they are polymeric, these large carbohydrates tend to be much less soluble, and combined carbohydrates have to be extracted by acid hydrolysis from their stabilized clay–carbohydrate complexes. Cellulose (a glucose polymer) survives to the Tertiary but in general, ancient polymers are of shorter chain length than their modern equivalents. In modern sediments (offshore California), carbohydrates decrease from 300–500 μg per gram of sediment at the surface, to 100 μg/g at a depth of 3 m, and a meagre 10–20 μg/g at a depth of 10 m. This drastic decline in carbohydrate/sugar concentration may be related to one of three likely causes. These sugars may possibly degrade rapidly like other organic matter under the oxidizing conditions prevailing at the time; or perhaps they may be removed by the formation of carbohydrate–clay complexes which may effect a chromatographic segregation which will limit considerably their detection. Yet another possibility is a polymerization sequence which might be expected to produce high molecular weight components of lower solubility and greater stability, isolating once more the sugars originally present from detection. Carbohydrates are important polymers in the tissues of plants and are available for incorporation into sedimentary organic matter.

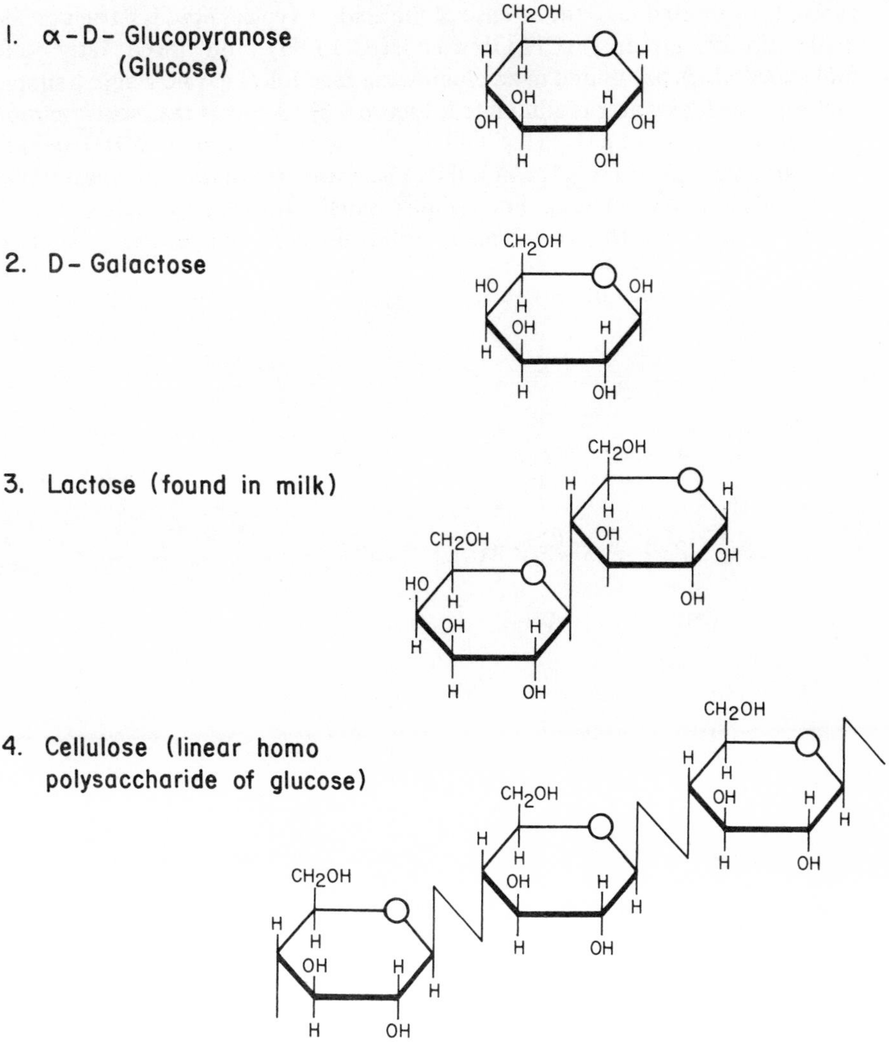

1. α-D- Glucopyranose
   (Glucose)

CH₂OH

2. D- Galactose

CH₂OH

3. Lactose (found in milk)

CH₂OH
CH₂OH

4. Cellulose (linear homo
   polysaccharide of glucose)

CH₂OH
CH₂OH
CH₂OH

Figure 6.4 Carbohydrates, typical monomers and polymers

Lipids

Lipid is a collective term for compounds which, whilst they are soluble in organic solvents, e.g. ether, hydrocarbons etc., are only very sparingly in water. 'Lipid' is derived from the Greek word meaning fat and many are similar in composition to many petroleum compounds.

The most common lipids are the animal fats and vegetable oils, and upon hydrolysis they yield fatty acids and alcohols. Alcohols and fats are odourless, whereas fatty acids have pungent aromas so that when fats and oils become rancid they smell due to the release of the acid. A typical lipid is a triglyceride made up of glycerol ($CH_2OH \cdot CHOH \cdot CH_2OH$) and three fatty acid molecules which are united in an elimination reaction (i.e. one where a simple molecule such as water is eliminated; Figure 6.5). Some of the most common fatty acids are palmitic ($C_{16}H_{33}COOH$), stearic acid ($C_{18}H_{37}COOH$) and the unsaturated oleic acid ($C_{18}H_{30}COOH$). There are quite a few polyunsaturated fatty acids to be found in lipid material. Animal fats contain mostly saturated paraffin chains which are normally solid at room temperature, whereas

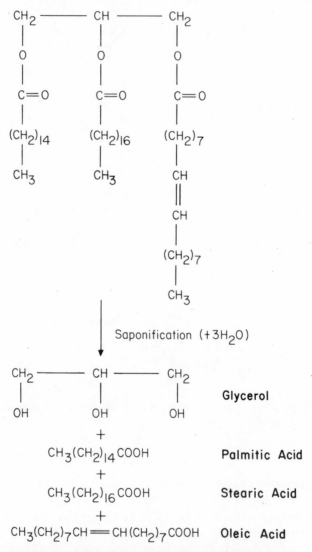

Figure 6.5 Triglycerides. The typical saponification of a triglyceride into its constituents

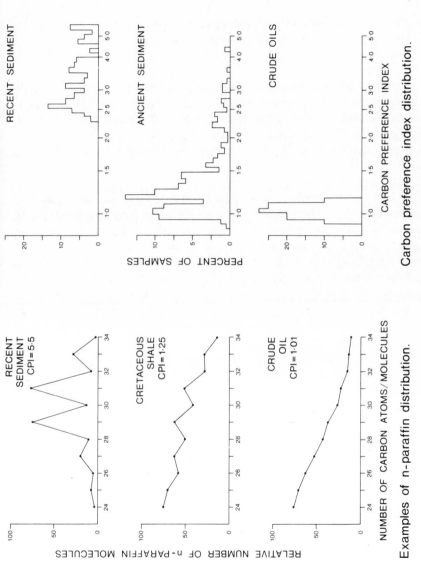

Figure 6.6 Examples of *n*-paraffin distribution (after Bray & Evans, 1965. Reproduced by permission of The American Association of Petroleum Geologists)

vegetable fats are normally liquid and contain unsaturated chains. (Because of the way that animal fats deposit solid fat on arteries there has been a switch to the comsumption of vegetable fats.) These unsaturated fatty acids are very important because the relative amounts of saturated and unsaturated fatty acids in an organism will depend not only upon that organism's species, age, taxonomy, and diet, but also upon the climatic conditions in which that animal lived. The reason is that in cold water it is much easier to keep unsaturated acids fluid within the body, whilst in warm water it is easier to keep the saturated acids in a fluid, mobile, and useful state.

In addition to this, the unsaturated fatty acids are much more reactive and the double bonds can be easily opened with oxygen to form peroxides. Such organic peroxides are very reactive compounds and when the peroxide bonds break open they react with one another to form organic polymers; it is these reactions which are involved, in part at least, in the formation of the giant molecules which make up kerogen.

The fatty acids which occur in the lipids and other similar acids have been isolated from peats, lignites, waxes, petroleum natural waters, as well as meteorites. It has been shewn that such fatty acids can be decarboxylated to yield alkanes with one less carbon atom than the parent fatty acid. Because biologically produced lipids normally contain fatty acids with 14, 16, 18, and 20 carbon atoms, decarboxylation of such acids will give alkanes with 13, 15, 17, and 19 carbon atoms respectively. Brooks & Smith (1967) have studied the diagenesis of plant lipids, especially the plant waxes, during the formation of coal and petroleum. These occur as the waterproofing layer on leaves and are mostly either n-alkanes of high molecular weight or the n-alkyl esters of n-alkanoic acids which are resistant to degradation during diagenesis and they therefore accumulate in sediments. As maturation proceeded Brooks and Smith reported a fall in CPI and average molecular weight, suggesting that fragmentation of wax hydrocarbons occurred as well as the hydrolysis of the esters which give n-alkanes and n-alkanoic acids. The latter can be decarboxylated to give alkanes with one carbon atom less. Perry et $al.$ (1979) have shewn that many marine sedimentary fatty acids are produced by bacteria and thus there is the possibility of large quantities of fatty acids being available in sediments.

Philippi (1974) suggests that the triglycerides which are derived from phyto-and zooplankton can be a source of fatty acids, especially the C_{18} and C_{16} compounds. However, vegetable and animal waxes contain esters of fatty acids and higher alcohols which are in the C_{30}, C_{32}, and C_{34} range.

Alkanes

The distribution of alkanes in rocks and sediments is very variable and will depend not only upon the type of incorporated material but also on the thermal history of that rock or sediment. Long-chain alkanes with odd carbon numbers will predominate in most young sediments and especially in those which are deposited under terrestrial influence (Figure 6.6). However, with increasing

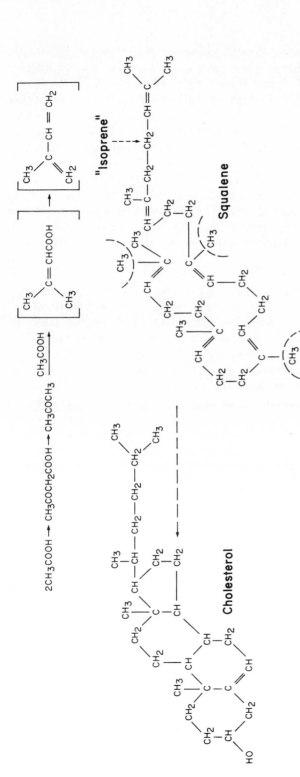

Figure 6.7 Synthesis of cholesterol and squalene (after Cantarow & Schepartz, 1954. Reproduced by permission of W. B. Saunders Company)

depth of burial and increasing age this odd preference disappears by dilution with newly generated alkanes which have no odd or even carbon number preference (Philippi, 1965; Eisma & Jung, 1964; Albrecht *et al.*, 1976). Koons *et al.* (1965) have shewn that although the *n*-alkanes in plant waxes and recent sediments have a predominance of odd carbon numbered members, work on sponges and corals has shewn *n*-alkanes with a CPI much closer to unity (i.e. 1.1–1.4 instead of 2.4–10). They also noted that in recent marine sediments where the organic content is of marine origin, the *n*-alkanes have little or no odd-number preference. Dembicki *et al.* (1976) report the presence of *n*-alkanes with a predominance of even-numbered carbon chains from an organic-rich rock where bacteria may have subsisted on algal remains in a highly saline carbonate environment. Thus the *n*-alkanes present in a rock will depend upon the original organic matter as well as upon changes caused by maturation.

The different relative amounts of terrestrial and marine material which contribute to the range of sedimentary alkanes can be indicated not only by the odd/even preference, but also by the molecular-weight range of the alkanes and of the fatty acids. In predominantly terrestrial material, especially that which has accumulated in a restricted environment, the molecular weight range will be between C_{24} to C_{32}. This range is the same as that found in the alkanes derived from the plant waxes. However, where the terrestrial input to the organic material is a minor constituent, the alkanes will range from C_{14} to C_{22} and will most likely be derived from the lipids of marine organisms. Thus in recent sediments and even in some ancient sediments two ranges of alkanes may appear depending upon the relative magnitudes of the terrestrial and marine input. The lower alkane range will have come from the marine source whereas the higher alkanes will be derived from terrestrial organic material.

The geochemical routes by which fatty acids are converted to alkanes so as to produce the alkane distribution now found in reservoired oil and ancient sediments have exercised the minds of many workers. Martin *et al.* (1963), Bray & Cooper (1968), and Knenvolden (1966) have all speculated about the role that fatty acids play in the formation of petroleum. In many recent sediments there are alkanes with an odd carbon number preference, and fatty acids with an even carbon preference, which could be the source of the *n*-alkanes. However, in ancient sediments the absolute amount of alkane per gram of organic carbon is greater than for recent sediments and these alkanes in ancient sediments do not have a pronounced odd carbon preference and are of much lower molecular weight. In this way alkanes in ancient sediments resemble those in crude oils. The simple decarboxylation of the fatty acids found in recent sediments will not give alkanes similar to those found in ancient sediments because even carbon numbered fatty acids give odd-numbered alkanes. Likewise, the low molecular weight range and large quantity of alkanes in ancient sediments could not be obtained from the quantity of high molecular weight acids found in recent sediments.

Although thermal cracking can explain the lower molecular weight range, the change in carbon preference and the increase in quantity need other theories. Philippi (1974) has proposed a free-radical method for producing a whole molecular weight range of n-alkanes from the biologically available fatty acids and this together with the production of alkanes from the kerogen probably explains the composition and quantity of alkanes found in crude oils and ancient sediments.

The free fatty acids, from which alkanes can be derived, tend to cover the whole range of molecular weights but the lower molecular fatty acids (C_{16}–C_{18}) are the major component of those acids bound to kerogen. Thus when in later stages of maturation these acids decompose to give n-alkanes, the bulk of the additional acids will be of lower molecular weights than those found in recent sediments.

There have been several cases reported of even carbon number preference alkane mixtures, especially from carbonate and evaporite sediments and even occasionally from crude oils (Welte & Waples, 1973). Dembicki *et al.* (1976) have also reported the presence of even carbon number alkanes in the range C_{22}–C_{28} in greater concentrations than the odd alkanes (C_{21}–C_{29}) from the same sample. The rock from which these alkanes with even preferences were obtained also contained algal remains, and Dembicki suggested that the cause of the higher concentration of even carbon alkanes was a highly saline solution where bacteria, which subsisted on the blue-green algal remains, produced these alkanes with an even preference.

A similar distribution in a native bitumen from the Unita basin was reported by Douglas & Grantham (1974). Welte and Waples (1973) have shewn that this type of distribution is normally found with a high phytane-to-pristane ratio, whereas alkane mixtures with an odd carbon preference are normally associated with pristane predominance or equality in the pristane and phytane concentrations. The suggestion has been made that in very reducing environments the reduction of fatty acids and alcohols, including phytanic acid and phytol, predominates over the decarboxylation reactions and the result is an even carbon number preference in the alkanes so produced.

The predominance of odd-numbered carbon atoms in the n-alkanes can be quantified by the use of the carbon preference index (CPI) (Bray & Evans, 1961) or the odd–even preference (OEP) (Scalan & Smith, (1970).

$$\text{CPI} = \frac{1}{2}\left[\frac{C_{25} + C_{27} + C_{29} + C_{31} + C_{33}}{C_{26} + C_{28} + C_{30} + C_{32} + C_{34}} + \frac{C_{25} + C_{27} + C_{29} + C_{31} + C_{33}}{C_{24} + C_{26} + C_{28} + C_{30} + C_{32}}\right]$$

An alternative formula is

$$\text{CPI} = \frac{2C_{29}}{C_{28} + C_{30}}$$

The ratio suggested by Scalan and Smith, the OEP, is not fixed to particular molecular weight ranges but is applicable whatever the molecular weight range of the alkanes and it can thus be used for alkane mixtures which do not include those members needed for CPI calculations.

$$\text{OEP} = \left[\frac{C_i + 6C_{i+2} + C_{i+4}}{4C_{i+1} + 4C_{i+3}} \right]^{(-1)^{i+1}}$$

C_i is the relative weight percentage of the n-alkane which contains i carbon atoms per molecule. The ratio takes account of the average weights of five consequentive alkanes centred about the alkane with $i + 2$ carbon atoms. Because this ratio is not tied to a given set of alkanes it is applicable to more limited alkane mixtures. It is also more sensitive to local odd/even variations rather than overall trends.

Despite the greater flexibility of the OEP, it is the CPI which is probably the most commonly used method of illustrating the odd/even predominance of alkane mixtures. In general the CPI will decrease with the increase in the maturity of the organic matter from which the alkanes have come until the sediment is in a state to produce hydrocarbons, when the CPI will be nearly unity and very similar to that of crude oils. However, Vandenbrouche (1976) has indicated that great caution should be taken when using the CPI as an indicator of maturity of algal matter because alkanes from immature algal matter can have anomalously low CPI values.

As the odd/even preference disappears with increasing maturation, the distribution of the n-alkanes moves to lower molecular weights (Figure 6.6) due to the generation of shorter molecules either by thermal cracking or such mechanisms as the free radical decarboxylation of fatty acids (Powell, 1975; Tissott & Welte, 1978).

Besides the normal alkanes there are, in both sediments and crude oils, a large number of branched alkanes. These have alkyl groups attached to the main carbon chain and the simplest branched alkanes are those that have a single methyl group attached to the second carbon atom of the main chain. These are called the isoalkanes or 2-methylalkanes, whilst those with the methyl group attached to the third carbon atom in the chain are the 3-methylalkanes or anteisoalkanes. These compounds have been shown to occur in bacterial waxes (Tissot & Welte, 1978) and as a component of plant waxes. In the Green River shales of the Unita basin, the iso- and anteisoalkanes form a homologous series, without any odd or even preference, mainly in the range C_{20}–C_{30} but occasionally with compounds up to C_{45}.

Both the normal and branched alkanes are believed to have originated mainly from the waterproof coating on the leaves of land plants as well as from bacterial sources. Lower molecular weight iso- and anteiso- fatty acids have been detected in the lipids of marine organisms and bacteria, while the equivalent alkanes have been discovered in oils and ancient sediments.

The wax found in oils is composed of higher molecular weight normal, iso-, and anteisoalkanes and is the same as occurs in isolated paraffin wax deposits such as ozokerite. The high molecular weight of the components suggests a terrestrial origin as these compounds and their fatty acid precursors are not generated by marine organisms.

Crude oils and ancient sediments also contain complex mixtures of other branched alkanes. Many of these have substituent groups at varying places on the chain and also may have more than one side chain, which may have more than one carbon atom. Most are believed to be the remnants from the breakdown of other natural product species, although some unusual branched alkanes such as the 6-methyl-, 7-methyl-, and 8-methylheptadecanes have been found in blue-green algae (Han & Calvin, 1970).

Other natural product compounds

Much of the natural product material which finds its way into sediments and eventually into crude oils is based upon the 2-methyl-1,3 butadiene or isoprene molecule (Figure 6.7). These five carbon units are the natural building blocks in biological organisms to form terpenes, steroids, carotenoids, and rubber. Figure 6.7 illustrates the biological synthesis of the steroid cholesterol from acetic acid via isoprene units, and the unsaturated hydrocarbon squalene which occurs in most organisms and is the unsaturated variant of the hydrocarbon squalane which is found in many crude oils. Figure 6.8 illustrates many of the compounds which are believed to be derived from isoprene units.

Erdman & Schwendinger (1964) discovered the steroid cholestene in recent sediments, and its reduction product, the hydrocarbon steroid cholestane, has been found in Green River shales. Kaplin *et al.* (1970) discovered hydrocarbons and fatty acids in Precambrian rocks while Gohring *et al.* (1967) reported a series of isoprenoid compounds in crude oils and Cretaceous bituminous shales. Powell and McKirdy (1973) have noted that oils which originate from immature terrestrial matter have low pristane to phytane ratios, while Raschid (1979) has shown that the pristane/phytane ratio increases with an increasing depth of burial of the sediment and with increasing organic maturation. The ratio of pristane to n–C_{17} alkane is also reported to decrease with increasing thermal maturation (Ishiwatari *et al.*, 1977). This decrease will occur because not only are the branched alkanes less thermally stable than the equivalent normal alkanes, and thus more liable to be destroyed with increasing burial and temperature, but very little new branched material compared with normal compounds will be produced during thermal maturation. Most of the natural product compounds are simple derivatives of those compounds which entered the sediment from plant and animal organisms and thermal maturation will not produce any more such compounds. The amount present depends directly on the amount incorporated into the sediment, and will not increase. In contrast, thermal maturation produces many normal alkanes from the decarboxylation of fatty acids, breakdown of the kerogen, etc., than were originally incorporated.

Many authors have reported the presence of natural product species in sediments and crude oils which can be related to plant and animal compounds, and this provides more evidence for the biogenic origin of oil. Barton *et al.* (1956)

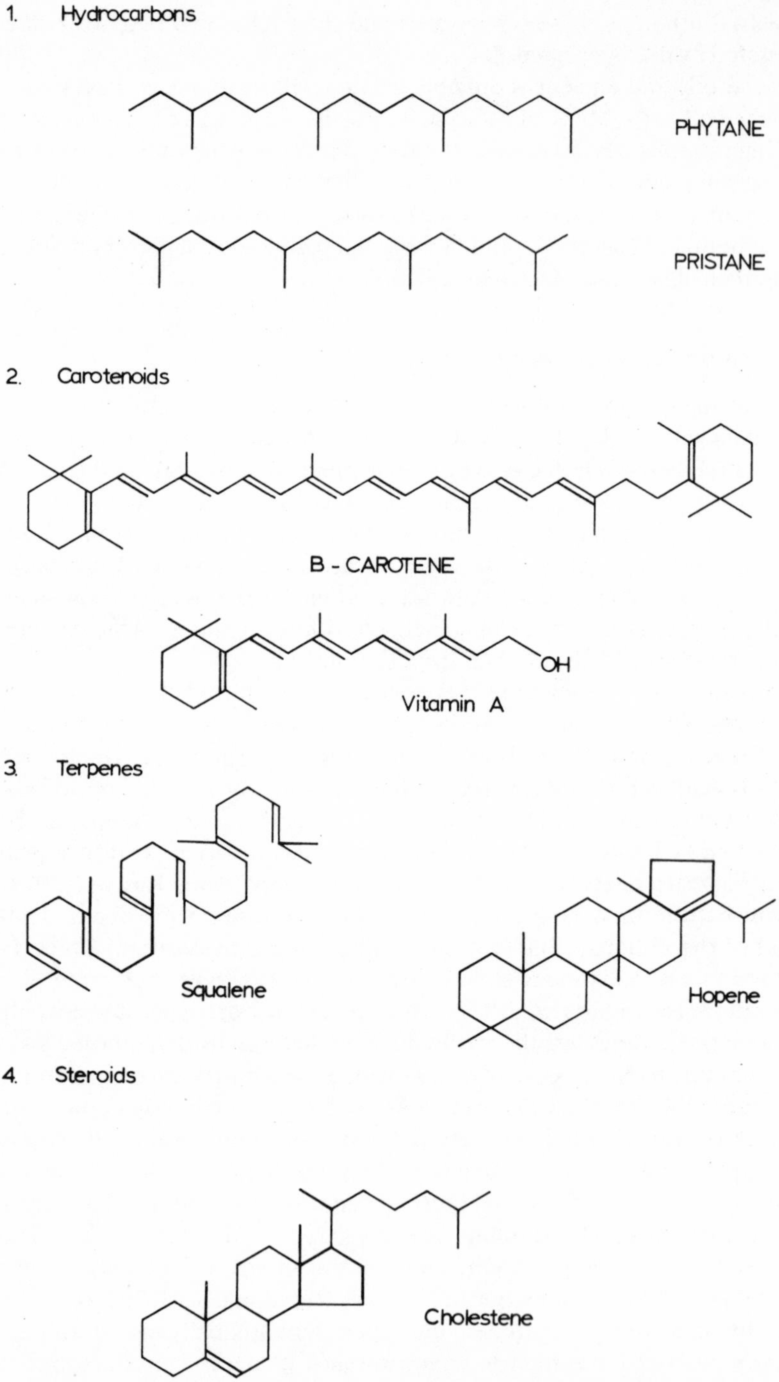

186

1. Hydrocarbons

PHYTANE

PRISTANE

2. Carotenoids

B - CAROTENE

Vitamin A

3. Terpenes

Squalene

Hopene

4. Steroids

Cholestene

Figure 6.8 Natural product compounds based upon the isoprene units

reported oxyallobetul-2-ene, a triterpenoid lactone in petroleum which can be related to the plant tritepenoid betulin. Barghoorn *et al.* (1964) have found alkanes including pristane and phytane in Precambrian sediments, while Waples *et al.* (1974) have reported the presence of 2,6,10,14,18-pentamethyl eicosane, a regular C_{25} isoprenoid hydrocarbon, in Tertiary sediments, and noted that isoprenoid hydrocarbons from C_{16} to C_{25} and C_{30} have all been reported in oils and sediments. The presence of isoprenoid fatty acids in Green River shales has been reported by several workers, including Douglas *et al.* (1966).

The phytol side chain of chlorophyll (Figure 6.1) is believed to be the source of many isoprenoid compounds. Blumer & Thomas (1965) have shown the presence of saturated and unsaturated isoprenoid hydrocarbons in zooplankton. These animals feed on plants and convert the chlorophyll they obtain from the plants into various porphyrins and isoprenoid hydrocarbons. An alternative route to pristane and phytane is shewn in Figure 6.9. If the phytol is in moderately oxygenated top sediment layers it will be oxidized to phytenic acid which after

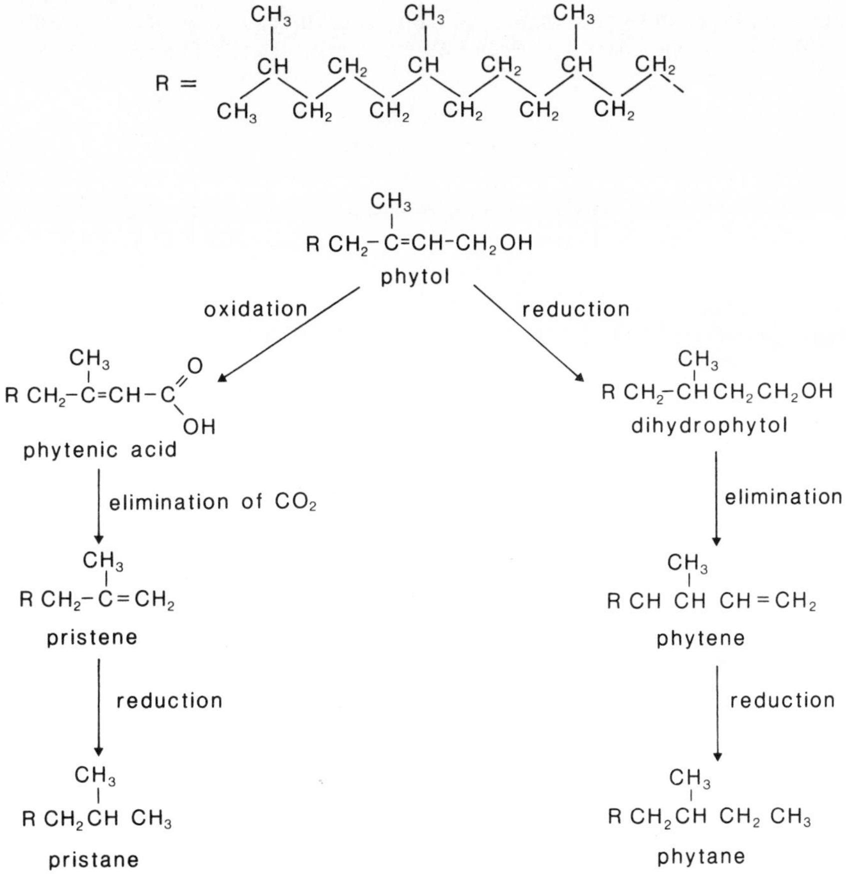

Figure 6.9 Production of pristane and phytane

decarboxylation can be reduced in anaerobic sediments to pristane. However, if the phytol is incorporated directly into an anoxic sediment it will be reduced to dihydrophytol and can be converted to phytane. In either case the phytol is available to contribute to the production of kerogen.

Laboratory experiments (Bayliss, 1968) have shewn that pristane and phytane can be produced from the phytol sidechain of chlorophyll, albeit at high temperatures and pressures. Albrecht & Ourrison (1969) have shewn that the direct heating of pure phytol has given a range of isoprenoid compounds from C_{14} to C_{20}, both saturated and unsaturated. Many sesquiterpenoid compounds have been discovered in insect wax, fungi, and higher plant leaves as well as in higher animals. These compounds which have greatly diversified physiological properties are based on the cyclopentano-perhydro-phenanthrene nucleus, and include vitamins, hormones, bile acids, cholesterol, and many natural drugs and poisons.

Other constituents of plant and animal tissues have been suggested as sources of other compounds found in sediments and crude oils. These plant constituents include the pigments, cell-supporting tissues (carbohydrates, chitin), protective tissues (cutin and waxes), barks, spores, leaves, and seeds.

Chlorophyll is probably the most important plant pigment. It has already been demonstrated that the phytol side chain of the chlorophyll molecule can be broken to give a whole range of open chain isoprenoid compounds. The isoprenoid compound pristane is found in zooplankton, especially copepods, which feed on phytoplankton, although pristane is not found in algae. Pristane is a major component in the body fats of copepods and it is believed to be a buoyancy aid so that the organism can maintain its position without using energy by swimming. Thus it is likely that the copepods manufacture pristane from their foodstuffs, and it is concentrated by its passage through the food chain. Hence zooplankton become a major source of pristane in marine environments.

Squalene has been shewn to exist in both the animal and plant kingdoms and it is important because it can be the source of many polycyclic triterpenoids, e.g. cholesterol from plants and animals, ergosterol from yeast, fucosterol from algae, and fungisterol from fungi.

Carotenes are another type of plant and animal pigment and are the compounds which give the yellow colour to leaves in the autumn. Related to carotenes is vitamin A (Figure 6.8) which, with its derivatives, is important in the biochemistry of vision. The carotenoid pigments are lipid soluble and are found associated with lipids in nature. In fact the yellow colour ascribed to many fats is not due to the fats themselves, but to dissolved carotene pigments. As shewn in Figure 6.8, carotenes are long hydrocarbon chains with alternating double and single bonds with a cyclic structure at each end. The alteration of double and single carbon–carbon bonds allows the π electrons of the double bonds to be delocalized over the whole chain and the length of the chain with its delocalized electrons will determine the wavelength of the light absorbed by the molecule; and hence the colour of the pigment. Carotenes have been shewn

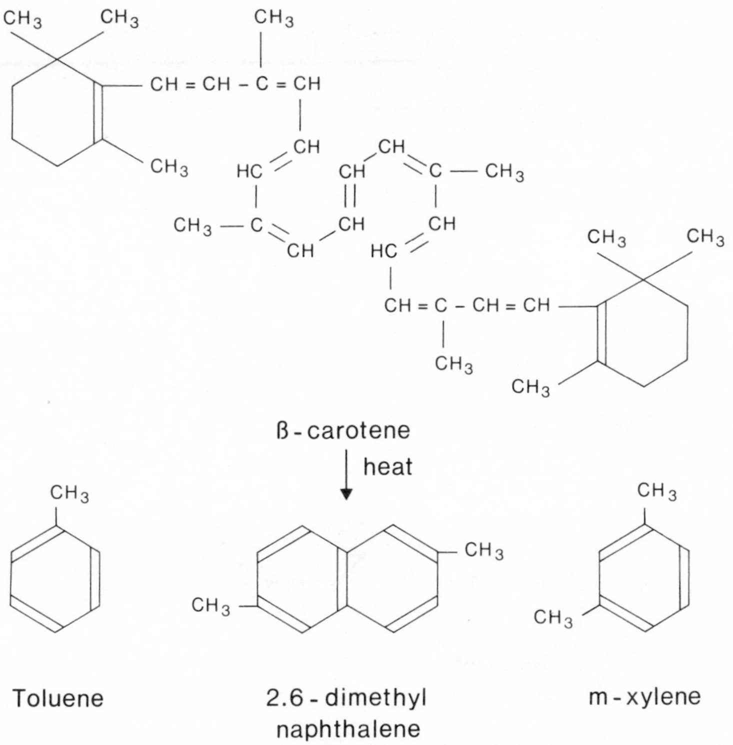

Figure 6.10 Thermal degradation of β-carotene (after Breger, 1960. Reproduced by permission of Pergamon Press Ltd)

to be the source of some of the simple aromatic compounds which are found in sediments and crude oils (Figure 6.10).

Other parts of organisms are available to be incorporated in sedimentary organic matter. Membranes are layers of phospholipids sandwiched between two layers of protein. The phospholipids are glycerol molecules to which are esterified two fatty acid molecules (frequently at least one is unsaturated) and phosphoric acid (lecithins). Cephalins are similar to lecithins but with a base attached to the phosphoric acid. Cutin is deposited on the outside of plant cells and is a biopolymer of hydroxy and epoxy fatty acids, whereas suberin is found on the inside of plant tissues and is a biopolymer consisting of aromatics and polyesters (Kolatturkurdy 1980). Chitin is found in the skeletal material of arthropods and is a polymer consisting of D-glucosamine units, acetylated on the amino group, probably linked by 1,4,β-glucoside bonds. Pectin, which is present in fruits, is composed of D-galaturonic acid units, connected via a 1,4 linkage, probably as α-glycosides, with many of the carboxyl groups present as methyl esters (Figure 6.11).

Aromatic compounds

Aromatic compounds, especially the mobile low molecular weight compounds,

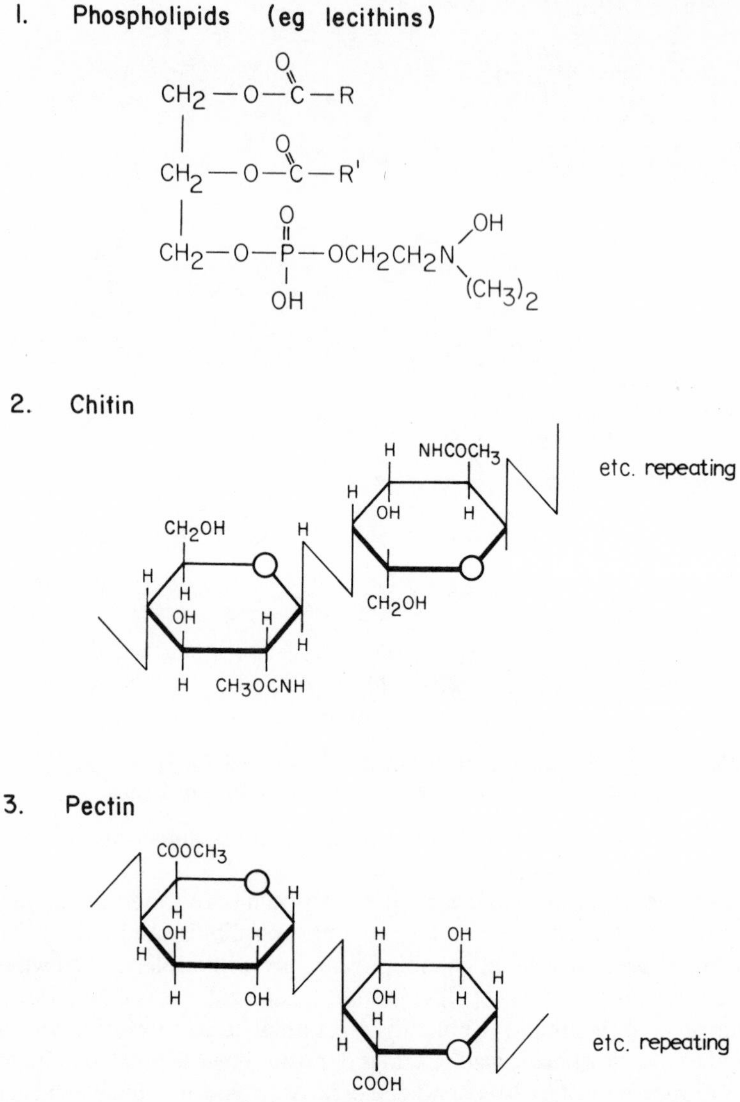

I. Phospholipids (eg lecithins)

$$CH_2-O-\overset{\overset{O}{\parallel}}{C}-R$$

$$CH_2-O-\overset{\overset{O}{\parallel}}{C}-R'$$

$$CH_2-O-\overset{\overset{O}{\parallel}}{\underset{\underset{OH}{|}}{P}}-OCH_2CH_2N\overset{OH}{\underset{(CH_3)_2}{}}$$

2. Chitin

etc. repeating

3. Pectin

etc. repeating

Figure 6.11 Plant components

are often toxic to living organisms and animals do not synthesize them. However, plant tissue contains large amounts of aromatic compounds in large polymers, e.g. lignin and tannin. Thus, of the common aromatic constituents of crude oil, i.e. benzene, toluene, xylene, and naphthalene, none are known to occur freely in living organisms though they have been found in sediments in relatively low quantities compared with crude oils and ancient sediments.

The major source of sedimentary aromatic matter is from lignins which are a large group of complex high molecular weight polyphenol substances abundant in the woody parts of plants. The units from which lignins are built are believed to be alcohols based on phenyl-propane derivatives such as coniferyl alcohol, sinapyl alcohol, *p*-coumarsyl alcohols, as well as derivatives of guaineol, piperonyl, and syringic (Figure 6.12). Lignin can be broken down in aerobic conditions and thus in hot climates the essential components of trees, i.e. cellulose and lignins, are hydrolysed in water to leave a concentration of

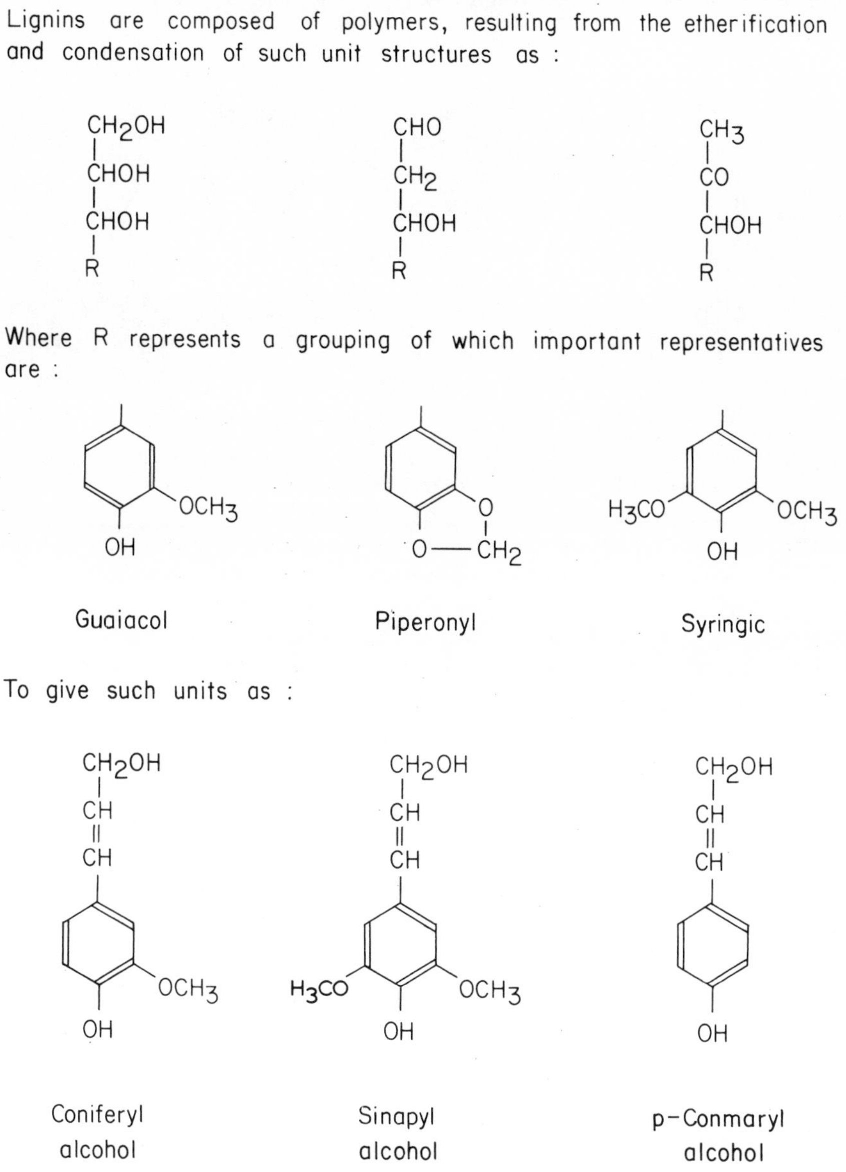

Lignins are composed of polymers, resulting from the etherification and condensation of such unit structures as :

Where R represents a grouping of which important representatives are :

Guaiacol   Piperonyl   Syringic

To give such units as :

Coniferyl
alcohol

Sinapyl
alcohol

p−Conmaryl
alcohol

Figure 6.12 Lignin units

water-insoluble lipids which can be the precursors of petroleum. The lignin molecule is constructed of aromatic rings, with oxygen bridges, and as the lignin is incorporated into the sediment in anaerobic conditions and then matured, the oxygen is lost and the rings become the basis of humic coals.

Tannins are also aromatic polymers which are generally based on gallic and ellagic acids and are found in fungi and algae as well as in the bark of oak trees, in unripe fruit and seed coats (Figure 6.13).

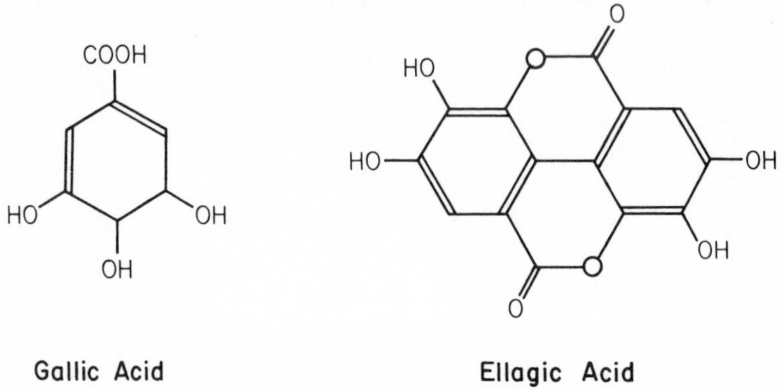

Gallic Acid   Ellagic Acid

Figure 6.13   Aromatic compounds in plants (tannins)

Glycosides are the product of the reaction of carbohydrate (sugar) molecules with aglycones. Many flavouring substances, such as mustard and horseradish, derive their taste from glycosides while several important drugs are also glycosides (e.g. digitalis and phlorizin). The common sugars involved include *D*-galactose, *D*-mannose, *D*-fructose, *D*-arabinose, *D*-ribose, *L*-xylose, and *D*-rhamnose, while many of the aglycones are aromatic species (Figure 6.14).

Another source of simple aromatic compounds is believed to be via the alteration of carotenes (Figure 6.10).

Other heteroatomic compounds

These compounds contain atoms other than carbon and hydrogen, the most common being oxygen, nitrogen, and sulphur. Some of the simpler compounds containing these elements have already been mentioned, e.g. the carbohydrates, fatty acids, proteins, and aminoacids, but there are much larger compounds to be found in sediments and crude oils, such as chlorophyll, steroid acids, pyridines, quinolenes, and plant alkaloids. These compounds are important for two reasons. Firstly, the presence of nitrogen in compounds can affect the storage of petroleum as well as poisoning the catalysts which are used in the cracking reactions in the refinery. However the second, and more important point about their presence, is that they shew that organic species derived from plants have played their part in the production of petroleum. Many of these are pigments with which we have already dealt, e.g. chlorophyll,

NAME	SOURCE	SUGAR	AGLYCONE
Arbutin	Leaves of arbutus & other plants	Glucose	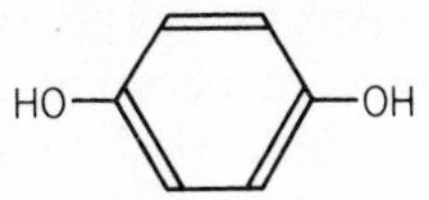 HO—⟨ ⟩—OH Hydroquinone
Coniferin	Bark of fir trees	Glucose	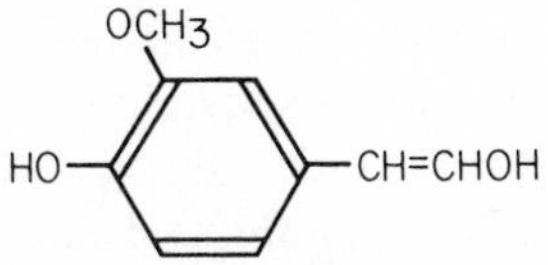 OCH₃ on ring; HO—⟨ ⟩—CH=CHOH Coniferol
Phlorizin	Bark of apple, cherry, pear and plum trees	Glucose	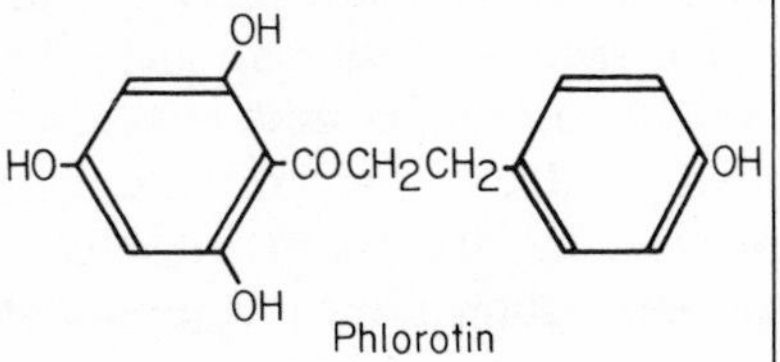 OH HO—⟨ ⟩—COCH₂CH₂—⟨ ⟩—OH OH Phlorotin
Salicin	Bark of aspen and willow trees	Glucose	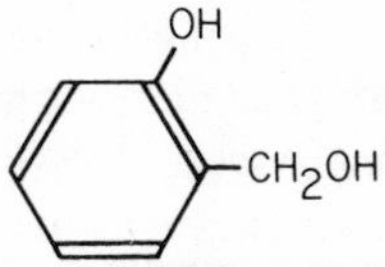 OH ⟨ ⟩—CH₂OH Saligenin

Figure 6.14 Glycosides and their components. (From *A Textbook of Biochemistry* by Mitchel. Copyright © 1950. Used with permission of McGraw-Hill Book Company)

the green pigment, but there are also the related compounds such as haemin, the red-blood pigment.

Figure 6.1 shews some of the stages in the degradation of chlorophyll to give porphyrins and isoprenoid chains. The degradation of haemin will be very similar as it is also based upon the same porphyrin nucleus. It is believed that many of the porphyrin skeletons are able to be incorporated into the kerogen and at a later stage to be regenerated in a modified form. The iron or magnesium in the haemin or chlorophyll is lost at an early stage in the degradation of these molecules and they are replaced by vanadyl groups (VO) or nickel which stabilize the molecules and aid their preservation. These compounds are thermally unstable and their presence in sediments and petroleum can give indications of the maximum palaeotemperature to which the immediate environment of these sediments has been subjected.

HUMIC ACIDS, KEROGEN, AND OIL SHALE

When the organisms, which contain the types of organic compounds described above, die their organic matter can undergo a variety of reactions. In aerobic conditions bacteria are able to consume the organic matter, utilizing part of it as energy, and part for cell generation. The carbohydrates and proteins are the most susceptible to bacterial attack while the lipids and lignins are least degraded. In aerobic conditions most of the organic matter can be destroyed but under anaerobic conditions there seems to be a greater chance of the organic material surviving.

Fermentation is an anaerobic process whereby bacteria use oxygenated organic matter, especially carbohydrates, to produce energy. One of the by-products of such fermentation can be the production of methane (see Chapter 7). Sulphate-reducing bacteria utilize the oxygen from sulphate ions (SO_4^{2-}) and reduce the sulphur to sulphide (S^{2-}). The sulphide can be converted to hydrogen sulphide or metal sulphides in the sediment.

The factors which stop such bacterial attack on sedimentary organic matter are difficult to quantify, but they could include lack of nutrients, presence of toxic products, and hostile environments, i.e. increase in temperature and pressure as the basin subsides. Because of this, and because anaerobic decomposition is not as complete as aerobic oxydation, organic matter which is quickly incorporated into anaerobic sediments will have a greater likelihood of becoming kerogen and eventually oil and gas.

The biodegraded monomers from the biopolymers are condensed in the next stage of diagenesis. Firstly, fulvic acids, which are soluble in acids, will be formed, then the acid-insoluble humic acids, and eventually humin will be produced. Humin is insoluble in alkalis. Thus a process of polycondensation occurs which results in polymers which become less and less soluble as their size increases.

Humic acids are large molecules with molecular weights within the range 10,000–300,000. Humic acids produced in subaerial soils consist of nuclei with reactive groups linked together. The nuclei are often cyclic structures, aromatic, naphthenic, or heterocyclic and the linkages are oxygen, sulphur, peptides, or methylene groups. However, in subaquatic sediments the cyclic nuclei are less phenolic and are composed of proteins, carbohydrates, and lipid derivatives (from plankton) and lignin and cellulose (from land plants). These humic acids are richer in hydrogen than humic acids from soils and contain more aliphatic chains and alicyclic rings. The humic acids are derived from the fulvic acids (molecular weight range 700–10,000) by condensation reactions. Hydrocarbons are not directly involved in these reactions because of their lack of functional groups, but they can be physically trapped in the humin matrices, and their precursors (fatty acids etc.) can be attached to these humic and fulvic acids. The humin can be regarded as the precursor of kerogen.

When the organic material is land-derived and especially when it has been subject to anaerobic, high pH conditions, then the humin will become peat which with deeper and deeper burial will become the various grades of coal, from brown through to anthracite. This process should be compared with the formation of kerogen which comes from the sapropelic, or high-lipid, organic matter found in spores and plankton which have been deposited in anaerobic aquatic muds, normally marine, deltaic, or lacustrine. The land-derived and deposited material will give coals which are found in their place of deposition; however, the water-deposited organic matter will form kerogen whose breakdown will give oil and gas. The type of kerogen and the products generated by its breakdown will depend upon the source of the materials from which the kerogen was generated.

Philp *et al.* (1978a) have carried out geochemical studies on recently deposited algal mats and oozes which they proposed contained possible kerogen precursors. They reported that these algal mats and oozes contain insoluble residues, very similar to kerogen in that they are cross-linked aromatic nuclei with additional components, such as esters, attached. When these insoluble residues were pyrolysed the major products were phytenes, pristenes, sterenes, and triterpenes. Philp *et al.* (1978b) have also heated the cells of certain plants and obtained phytane and five isomeric phytenes. The presence of montmorillonite increased the rate of production of these compounds from the cells of *Nostoc muscorum* but not from *Rhodopseudomonas* spheroids.

Oil shale is a compact, organic-rich argillaceous rock of sedimentary origin, which has an ash content of more than 33%. Upon destructive distillation the organic matter yields an oil, whilst normal solvent extraction of the same shale would only yield a maximum of 2–3% bitumen. An oil shale can thus be considered to be an immature source rock of the conventional sort and destructive distillation, or pyrolysis, produces the same effect as deep burial at high temperatures and pressures for millions of years has produced in conventional petroleum source rocks.

However, there is one significant difference between oil shales and conventional petroleum source rocks, and that is the minimum amount of organic matter each contains before it is of value. While it is generally accepted that for a petroleum source rock the minimum amount of organic matter required to produce significant amounts of reservoired hydrocarbons is of the order of 0.5%, the amount of organic matter in an oil shale has to be much higher for the extraction of the shale oil to be economic. The reason for this difference is that the material has to be pyrolysed *in situ* or to be mined and then fed into the retorts where the immature kerogen will be cracked into mobile products that can be distilled. Often a minimum figure of 5% organic content, that is ten times the figure required for conventional source rocks, is quoted as the minimum organic content which makes oil shale worth exploiting.

Oil shales are pyrolysed (or artificially matured) at a temperature of around 500 °C to produce the shale oil. The oil thus produced will vary depending upon the type of organic matter to be incorporated in the rock. The oil shale with a kerogen derived from algal remains will have the highest convertibility into shale oil, often greater than 60%, whereas oil shales containing a sapropelic-type kerogen have a much lower convertibility which can be as low as 25%. This, of course, is akin to the percentage convertibility of equivalent types of normal kerogens. Similarly, the oil type derived from the different oil shale kerogens varies, with oils from algal kerogens having a high content of aliphatic chains whereas those from the sapropelic-type shale tend to have a high proportion of cyclic compounds. The most notable thing about the composition of all these shale oils is the large amount of heterocyclic compounds, particularly those containing nitrogen (often 40–60%) as well as the large quantities of olefins, which are virtually absent in natural crude oil. This may be due to the fact that the retorting occurs in anhydrous conditions in which there will be a shortage of available hydrogen, whereas kerogens which mature naturally in the earth do so in the presence of interstitial water and thus in a hydrogen-rich environment. The hydrogen-lean situation produces unsaturated and cyclic compounds whereas the hydrogen-rich environment produces saturated compounds.

A typical oil shale is the Mahogany bed of the Green River Shale from the Eocene of Wyoming, which has a bulk volume composition of $C_{215}H_{330}O_{12}N_5S$ and was deposited in a shallow lake under anaerobic conditions. Upon distillation this shale produces 66% oil, 9% gas, 5% water, and 20% carbon dioxide. The oil is dominated by heterocyclic compounds (45–60%), with smaller amounts of cycloalkanes (20–25%), straight-chain alkanes (10–15%), and aromatics (10–15%). Most of the sulphur in this case is found in the gaseous components.

The commercial exploitation of shale oil decreased (until 1973) due to the then relative cheapness of natural crude oil. In Great Britain the production of shale oil ceased when the government subsidy was removed. This was because the processing of oil shale was labour intensive, involving both mining and retorting operations, as well as the disposal of the ash which remains after the distillation. However, the recent and rapid increase in the price of natural crude oil, together with greater mechanization and automation of the oil shale handling, may see a resurgence of the shale oil industry in the not too distant future.

Williams (1982) notes that rising costs and a slump in the price of crude oil have caused some slowing of the re-emergence of the oil shale and other synthetic fuel projects. However some schemes are progressing and Union Oil of California should start producing shale oil in 1983. The shale is retorted at 700–900 °F and the shale oil is treated to remove arsenic, nitrogen, sulphur and other impurities. The average yield of 34 API° gravity oil is 34 US. gal. per ton of shale.

Two other problems were met with the production of shale oil which will still need to be taken into account in any new shale oil processing operation. Because of the amount of sulphur and nitrogen contained in the organic matter of the shale, the products of retorting contain a significant percentage of these elements. Compounds with these elements are often carcinogenic, poison refinery catalysts, and attack the retorting and refining equipment. Besides this, the destructive distillation of oil shales leaves an inorganic residue once the kerogen has been cracked and the oil removed. This ash is very light and has expanded so it will occupy a volume 50–100% greater than the original oil shale. Thus the disposal of this ash can create problems, as is evident in the Lothians of Scotland where there are many hillocks, now grass covered, which are the spoil tips from the early oil shale industry, which ceased in that country in the 1950s.

HYDROCARBON OCCURRENCES

Trask & Wu (1930) studied recent sediments and tried to extract soluble hydrocarbons, but finding none of these assumed that such hydrocarbons had not been formed and were not present in recent sediments. However, Smith (1952) did find petroleum-like hydrocarbons in recent sediments taken from the Gulf of Mexico. He pointed out that these mixtures of hydrocarbons differed considerably from petroleum in the relative and absolute amounts of the compounds present. There were fewer compound types and the proportions of different compounds were not the same as in petroleum. Kidwell & Hunt (1958) also found petroleum-like substances in a sand lens within a clay and the organic matter decreased with depth from 1.45% at the surface to 0.6% at 200 ft, but the percentage of hydrocarbons increased with depth.

Hoan & Meinschein (1977) have reported a variety of fatty acids which have been isolated from fossil fruits, indicating that these acids were present in organic matter at the correct geological time to be available as petroleum precursors. It has been noted that the hydrocarbons obtained from recent sediments differ from those in crude oil and ancient sediments. Baker (1960) and Bray et al. (1956) state that the hydrocarbons from recent sediments only include relatively simple mixtures of aromatic compounds and the n-alkanes showed a significant odd carbon number preference which does not occur in crude oils or in organic matter in ancient sediments (Figure 6.6).

Hunt (1961) reports that the amounts of alkanes and naphthenes were less in recent sediments, while recent sediments lack compounds in the C_3–C_{14} range which can account for up to 50% by volume of many crude oils. Thus, as the sediment gets older the hydrocarbons form a higher percentage of the organic material with a diminished molecular weight and a carbon preference index approaching unity.

Dembicki et al. (1976) have found an even-number carbon preference in the range C_{22}–C_{28} alkanes in an organic-rich rock, 310×10^6 years old. This rock also contained algal remains and it has been postulated that the even-number preference could, in this case, be due to the highly saline carbonate environment, and the bacteria that existed in those conditions subsisted on the blue-green algal remains.

Abell & Margolis (1982) have extracted hydrocarbons from the Plio-Pleistocene sediments of palaeo-lake Turkana in Kenya. Three groups of alkanes were recognized. One group (C_{15}–C_{19}) was derived from algae, a second group (C_{20}–C_{30}), with no odd/even preference, was derived from the action of microbes on soil organics and the third group (C_{21}–C_{35}), which had a high odd/even preference, was derived from plant waxes.

Bray *et al.* (1956), Meinschein (1961a), and Hunt (1972) have all identified hydrocarbons from soil samples which are closely comparable with those found deposited in recent aquatic sediments. Again, while these hydrocarbon mixtures are similar to petroleum, they are definitely not petroleum. Some people have advanced theories in which groundwater could carry the hydrocarbons found in soils and recent sediments down to reservoirs. Buckley *et al.* (1958) have shewn that pore water can contain dissolved hydrocarbon gases such as methane, ethane, propane, and butane.

So far all the discussions have involved sedimentary rocks as these are essential for the origin of the oil. However, both igneous and metamorphic rocks can provide secondary reservoir porosity, and migrating petroleum will be able to fill these pore spaces and hence oil can be reservoired in, and produced from, such non-sedimentary rocks without being formed in such surroundings. However, in such a case the rocks involved will have to be near sedimentary rocks and in communication with the sedimentary rocks whence the oil would have come.

Oil has been found in commercial amounts on all five inhabited continents and their continental shelves. Antarctica is the only continent from which oil and gas are not produced in commercial quantities. Oil has not yet been found under the open seas away from the continents, but that may only be due to the lack of exploration in these areas due to technological difficulties. Exploration in these deeper waters is now being contemplated. In general, and with the exception of very small fields, oilfields are extremely localized and the majority (two-thirds) of all known oil is in the Middle East.

Weeks (1958) stated that the upper half of the Tertiary has accounted for 35% of the world's discovered reserves and 87% of all oil and gas reserves are in Mesozoic–Tertiary sediments. Petroleum has been found from the surface down to 25,000 feet, with pressures up to 20,000 psi and temperatures of up to 150 °C.

SEDIMENTARY SETTING FOR PETROLEUM GENESIS

Sedimentary rocks which contain organic matter suitable for conversion to petroleum were originally deposited in aqueous conditions and the organic matter incorporated therein was also water-borne. The organic matter present in this water is either in solution, present as a colloidal solution, or as particles in suspension. In modern oceans the majority of the organic matter is in the upper layers where photosynthesis occurs, and the organic content decreases over the first few hundred metres until it reaches an approximately constant value irrespective of depth. The particulate matter will fall slowly under the influence of gravity to the sea bed. A similar fate will also befall the organic matter that is in colloidal suspension, but flocculation will need to occur before this can happen. The organic matter that is in solution may well only reach the bottom of the ocean if it is absorbed onto clay particles as these fall through the water. During the passage of all this organic matter through the water column it will be attacked by bacteria, etc., which will rework suitable substrates. Likewise, when the organic matter settles on the sea bed, bottom dwelling organisms will subsist on some of it. Thus it is only the relatively resistant organic matter that survives to be incorporated into the sediment. It may be rewarding to consider the most favourable conditions for the preservation of organic matter which may eventually be converted to fossil fuel.

The first essential is that the correct starting material is provided. The right type and amount of organic matter need to appear in the water column above the sediment. Plant and animal matter contribute and this material may be either of marine origin from aquatic plants and organisms or it can be terrestrial plant and animal debris swept into the sea by rivers at deltas, or even the airborne spores and pollens from terrestrial plants which are blown by prevailing winds out to sea where they settle.

The production of alkanes from natural products like fatty acids and proteins has been demonstrated and even more complicated molecules with ring structures such as terpenes and steroids can be related to other natural product compounds. Chlorophyll is also a likely fossil fuel precursor since porphyrins derived from chlorophyll have been found in petroleum (Groenings, 1953). The carotenes can by thermal degradation provide some of the simpler aromatic compounds (Figure 6.10). Stevenson (1960) states that there are more unmodified plant remains in aquatic sediments than in soils because if plant debris had remained in soils under aerobic conditions it could have been metabolized by micro-organisms. However, in the aquatic sediments the debris is soon transported into an anaerobic environment where it is more likely to survive.

Day & Erdman (1963) have investigated the thermal degradation of β-carotene and have shewn that aromatic compounds can be formed, while Draffen et al. (1971) have shewn how cholesterol can be the source of hydrocarbons. The isoprenoid alkanes have been shewn to be important components of both oil and sedimentary organic matter. Pristane occurs in animals and marine organisms such as in copepods and in benthic planktonic algae and zooplankton but not in contemporary plants (Blumer et al., 1963, 1964, 1965, 1971; Clark & Blumer; 1967; Mold et al., 1963.) Phytane has been observed in zooplankton; however, besides animals such as zooplankton which feed on plants and produce several saturated and unsaturated isoprenoid hydrocarbons from the phytyl side chain of chlorophyll, there may be a way of producing pristane, phytane, and other isoprenoid hydrocarbons depending on the oxic conditions of the sediment into which the chlorophyll is incorporated (Figure 6.9) (Brooks et al., 1969). Han & Calvin (1969) consider that the isoprenoid compounds found in sediments can originate from the chlorophyll side chain and that squalene can give a range of C_9–C_{25} isoprenoid hydrocarbons by a variety of bond cleavage reactions.

Breger (1960) discussed the suitability of many natural product compounds as hydrocarbon source materials. He considered that carbohydrates and proteins do not give petroleum upon degradation, but that lignin was one possible source

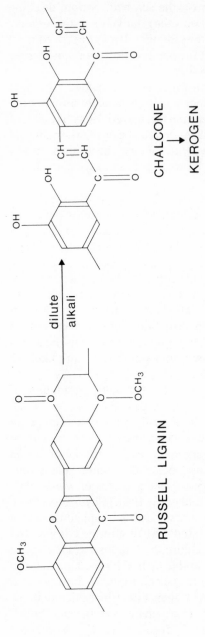

Figure 6.15 Formation of kerogen from lignin (after Breger, 1960. Reproduced by permission of Pergamon Press Ltd)

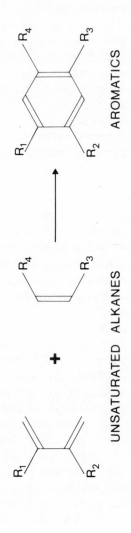

Figure 6.16 Diels Alder reactions

of kerogen (Figure 6.15), while pigments were varied as to their source potential. The flavinoids do not produce hydrocarbons while the porphyrins survive almost unchanged and the carotenes give aromatic compounds. Other routes by which aromatic compounds can be formed include Diels Alder type reactions (Figure 6.16) and disproportionation reactions where two molecules are produced, one saturated and the other an aromatic compound. However, much of the sedimentary organic matter is incorporated into kerogen which will in turn produce more petroleum hydrocarbons. Some compounds will be physically trapped in the kerogen and will be later released unchanged; some compounds will be weakly bound to the kerogen and will be only slightly altered when the kerogen breaks down; while other compounds will be totally integrated into the kerogen and will not be released from the kerogen in the form in which they entered it.

Laplante (1974) considered that there were two major types of kerogen; one which upon degradation will produce oil and another which, because of its lower initial hydrogen content, will have gas as its major thermogenesis product (Figure 6.17). Hunt (1979) has a different picture of these two fundamentally different kerogen types while agreeing on the major products each yields upon maturation (Figure 6.18). Tissot et al. (1974) considers that each type of kerogen will produce oil or gas depending upon its subsequent history. It has been suggested that the gas in the North East Netherlands (composition 81.3% methane, 33% other hydrocarbons, and 4.4% nitrogen) is due to the heating of the coal measures, while Eckleman et al. (1962) have found the $^{13}C/^{12}C$ ratios of some oils to be closer to the carbon isotope ratios of terrestrial organic matter than to the ratios of marine matter. Breger & Brown (1962) have suggested that the Chattanooga shale, which contains up to 25% organic matter, does not have a marine origin.

Thus at one time, and for both geological and geochemical reasons, aquatic plant and animal life were considered to offer the most promising source of petroleum, whereas the sources of gas were considered to be predominantly terrestrial in origin. However, there is now evidence that not only can oil be generated from organic matter of both marine and terrestrial sources, but that the final breakdown product of sedimentary organic matter, irrespective of its origin, will be gas. Thus there will be on the maturation path of all types of sedimentary organic matter an 'oil window' and a 'gas window'. The relative positions of these 'windows' and the amounts of products generated in those windows will depend upon the type of kerogen involved.

Although all organisms and plants contain some hydrocarbons, these are not enough or of the correct type to account for the composition of reservoired hydrocarbons. Even those petroleum-type compounds for which an obvious precursor can be found, such as the alkanes and fatty acids, will not supply the entire range of compounds found in oil. Therefore other sources of hydrocarbons are required and these extra sources are the kerogen, from the precursors via a different mechanism and by the alteration of primary produced compounds.

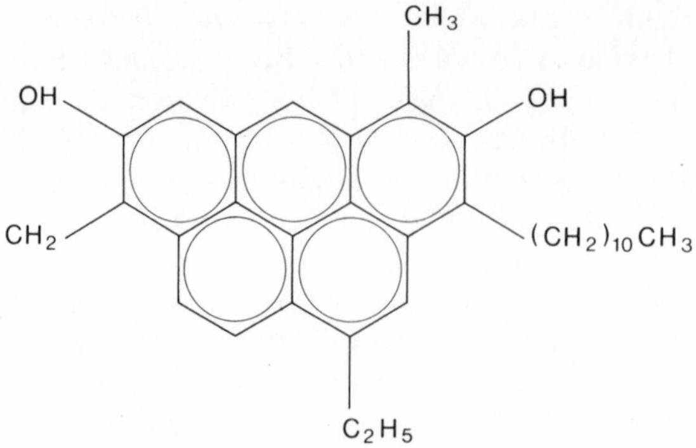

OIL KEROGEN

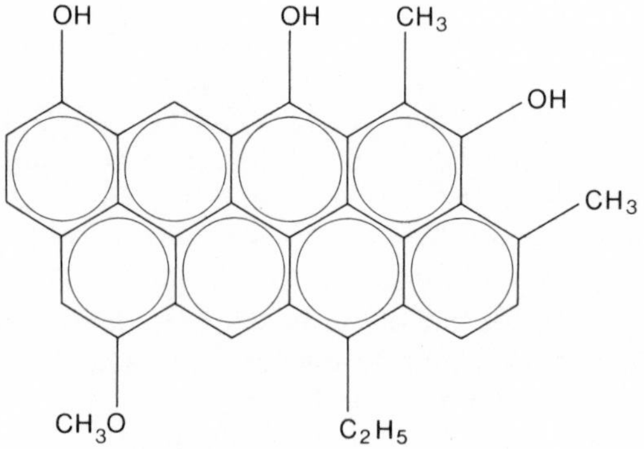

GAS KEROGEN

Figure 6.17 Kerogen types (Laplante, 1974. Reproduced by permission of The American Association of Petroleum Geologists)

So far, only the type of organic matter has been discussed; however, the amount of organic matter within a sediment needs to be considered. Welte (1965) considered that the lowest amount of organic matter that must be present in a potential source rock must be at least 0.5% for clastic sediments and 0.3% for carborates and evaporites.

Even if the correct amount of organic matter is available and it is of the correct or suitable type, there is still no guarantee that the end result will be reservoired oil and/or gas. Other conditions need to be satisfied.

Conditions for the preservation of source material

Stratified water column and anoxic sediments

The absence of organisms which can destroy the deposited organic matter is

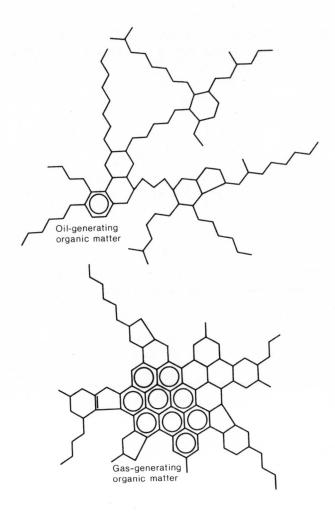

Oil-generating
organic matter

Gas-generating
organic matter

Figure 6.18 Kerogen types (From *Petroleum Geochemistry and Geology* by John M. Hunt, W. H. Freeman and Company. Copyright © 1979)

required. The destruction of the organic matter is generally far less marked in reducing anaerobic conditions. This can be summarized by a concentration of planktonic fossils and an absence of benthonic forms being required. Although anaerobic bacteria do exist and these can destroy organic matter, what will stop these bacteria operating is not clear. Various suggestions as to why anaerobic bacteria stop degrading organic matter have been made, including lack of other nutrients, toxicity of their by-products, and extremes of temperature and pressure. In spite of the existence of anaerobic bacteria, reducing anaerobic conditions do seem too be the most conducive to the preservation of organic matter.

High rates of production of organic matter do not guarantee that large amounts of that organic matter will be preserved in the sediments. The amount that is preserved will depend upon the oxic–anoxic conditions and the rate of sedimentation. Certainly in oxic conditions the organic carbon content of sediments is primarily a function of the sedimentation rate. The faster the rate of sedimentation, the faster the organic material will be covered by the inorganic matter and hidden and buried away from the scavenging organisms and bacteria, and thus greater will be the chance for the survival of the organic matter which is liable to be destroyed by aerobic bacteria and animals. However, it should be noted that in oxic waters the organic matter tends to be hydrogen-deficient and this is due to bioturbation which accelerates the organic matter degradation processes and occurs at, or just below, the water–sediment interface. Thus any kerogens that are produced from such organic matter will tend to be gas prone.

In anoxic waters the organic matter types shew a wider variation and are present in greater quantities. However, the most important point about organic matter incorporated into sediments in anoxic environments is that the organic material is more commonly reduced. It is not attacked by aerobic bacteria and it becomes hydrogen- and lipid-rich, and thus any kerogen which is formed under such conditions will tend to be oil prone.

It should be realized that in a totally anoxic water (i.e. anoxic from the surface to the sediment) it will be almost impossible for any life form to exist in order to produce the primary organic matter. In such cases of totally anoxic water there will be very little, if any, life and the most likely way for organic material to be deposited is by debris brought in by rivers or winds or which floats in from other parts of the system. The vast majority of such organic material will be dead material. Any living thing that swims into this anoxic water will die and sink into the sediment. There are cases of extremely well preserved fossil fish which are believed to have died by swimming into such toxic waters and fallen into the anoxic muds below, where they were preserved. Thus in a truly anoxic water the amount of organic matter will be very limited because there will not be the continuing life forms to create a life cycle and produce organic matter which can die and be deposited in the sediment.

Hence, when, in the petroleum context, anoxic waters are discussed, the only system of value is one where there is an oxic layer at the surface where the organic matter lives, reproduces, and dies as a continuing life system. When this organic matter dies it drops into an anoxic mud where it is preserved. The actual position of the oxic–anoxic boundary may vary a few centimetres above or below the water–sediment interface. There may be a layer of anoxic water above the sediment or the top of the sediment may be oxic (Figure 6.19). When organic matter which has been produced in the oxic photic zone dies, it will quickly be incorporated into an anoxic mud where it can be preserved. To this may be added the various types of detritus and debris brought in by river systems, winds, etc. The most prolific generator of

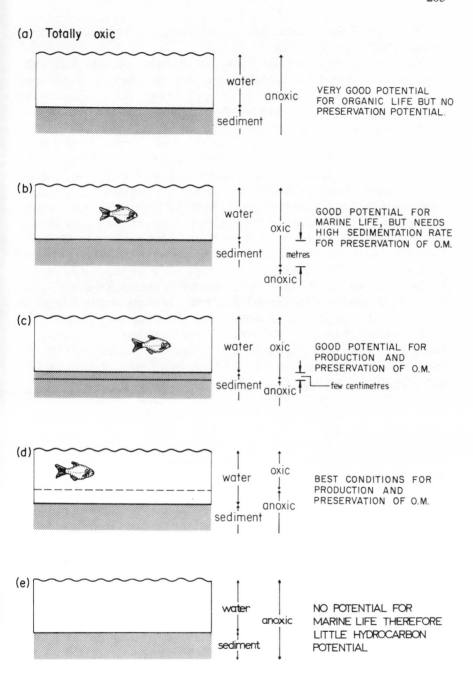

Figure 6.19 Oxic–anoxic boundaries and preservation of source material

oil source rocks is the anoxic mud overlain by anoxic water which is underneath an oxic, productive photic zone.

Lakes are rarely anoxic at depth, but if warm climates with stable temperatures and moderate rainfall are prevalent then lakes which exist under such conditions may be anoxic at depth and sediments rich in organic matter can be deposited in them. Lakes can also become anaerobic if conditions for life are too good and life forms, including those which live upon the debris of other organisms, flourish in too great an abundance. Water can hold all the ingredients necessary for the support of life, but it can hold less oxygen than any other component and once the oxygen is used up it is difficult to replace it. Thus when life flourishes in a closed system such as a lake all the oxygen can be used up and the water become stagnant and the life forms in that water will die. In many modern lakes and rivers the inflow of food and nutrients is making, or has made them anoxic. Such eutrophication has occurred in the Baltic, Lake Erie and the River Thames. The most well known anoxic lake system is the Eocene Green River Formation in the western United States of America. It has been shown that permanent anoxic conditions existed at the lake bottom during the deposition of the Green River Shales, and thus conditions deadly to life forms led to an absence of bioturbation and a high preservation of the organic content. Crude oils which come from such anoxic lake systems are highly paraffinic, have high pour points, and low sulphur content. The non-marine lower Cretaceous oils of China, Brazil, and some of the West African coastal basins were sourced from such anoxic lakes.

Another potential system for the deposition of organic-rich sediment is silled basins. However, the majority of these are not anoxic because of circulation of water within them. If more water flows out than in, the result will be a stagnant anoxic water body. The Black Sea and Mediterranean Sea are contrasting silled basins. In the Black Sea more water flows out than in and there is an anoxic sea with a 200 m oxic layer permanently on top which supports abundant life. When organisms die they can be preserved in the sediment (Figures 6.19 (d) and 6.20 (a)). The Black Sea only became anoxic 5,000–7,000 years ago, at which time the organic carbon in the sediments increased tenfold as did the lipid content of the organic matter in the anoxic sediments.

The Mediterranean is a different system as it has a net inflow of water which evaporates, making the water enriched in salt. These hypersaline waters are denser than normal sea water and sink, taking their dissolved oxygen with them, thus ensuring that the sea is oxygenated throughout. This process is assisted by the outflow of the lower heavier water which ensures that the water circulates, is oxygenated (except where polluted by man) and depleted in nutrients. Thus although life is abundant, organic matter is not incorporated and preserved in the sediments (Figure 6.20 (b)).

Schmalz (1969) suggests that the Mediterranean is at the formative stage of deepwater evaporite formation and if evaporation increases, progressively more dense brines will be produced at the surface. Each denser brine will sink and, in turn, replace the less dense bottom water of the basin until the bottom

(a) Anoxic e.g. Black Sea (positive balance basin)

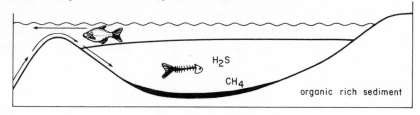

WATER OUTFLOW > INFLOW

(b) Oxic e.g. Mediterranean Sea
(negative balance basin)

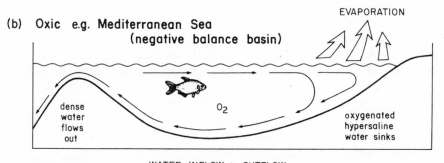

WATER INFLOW > OUTFLOW

Figure 6.20 Silled basins (after Demaison & Moore, 1980. Reproduced by permission of The American Association of Petroleum Geologists)

water will be so dense that it cannot be replaced by oxygenated brines. Thus these bottom waters will become stagnant and aerobic benthos will die, and the decay of organic matter will produce strongly reducing conditions. However, nektonic and planktonic organisms will continue to live in the upper part of the basin and their remains will contribute to the organic content of the bottom sediments. Subsequent infilling of such a basin can occur and the organic rich layers of sediment could be a potential source rock. If Schmalz is correct, basins such as the present Mediterranean could eventually have anoxic bottom waters and generate and preserve sufficient organic matter to form source rocks.

Areas of large-scale upwelling are also potential source bed generators. An example of this is the Peruvian shelf, where there are a lot of phytoplankton and a complex counter-current system forms a nutrient trap so that the organisms and nutrients cannot be dispersed in the open ocean. Sediments deposited in anoxic muds below such conditions are enriched in sapropelic material and are rich in hydrogen. However, not all areas of upwelling and high primary productivity are associated with anoxic layers lower down the water column. Thus even if there is upwelling and high primary productivity, the lack of an anoxic water and anoxic muds will mean that the dead organic detritus will not be preserved and organic-rich sediments will not occur. Examples where there is upwelling but poor preservation due to lack of anoxic water are around Antarctica and the Grand Banks of Newfoundland. Offshore Peru has the correct combination of high organic productivity and upwelling above anoxic water layers and muds which allows for the deposition of organic-rich sediments.

Parrish (1982) has compared Paleozoic petroleum source beds and the distribution of upwelling zones predicted by palaeogeographical reconstructions and shewn a close correspondence between them. Abell and Margolis (1982) note that the organic content of arid region sediments are miniscule compared to those of temperate climes.

There are also anoxic open ocean areas. Normally the difference in water temperature between the poles and the equator is a very effective heat engine which causes the circulation of oxygenated water in the oceans. This is especially true if there is an open corridor between the poles. The Atlantic is a perfect example of open communication and well circulated and oxygenated water. However, in ocean areas such as Northern Pacific and the Indian Ocean there are not openings to the North Pole, with the result that these oceans tend to be anoxic in their northern regions. Because of the spin of the earth this anoxicity tends to be concentrated on the western side of these oceans.

Deltas do not tend to be anoxic and this, combined with the fact that the organic matter may be predominantly terrestrial in origin, mean that the resulting kerogens can be gas prone. However, there are various circumstances under which deltaic kerogens can be oil prone. The river which supplies the delta will bring with it quantities of inorganic matter. If this inorganic material is supplied at the correct rate then the organic matter can be preserved and taken quickly down into the anoxic zone. In addition, if anoxic layers of water intersect wedges of the delta, a good oil kerogen can be formed and thus transgressions and regressions are favourable for the deposition of sediments rich in organic matter. The resulting kerogens will be of a mixed variety and will differ both vertically and laterally. The more marine kerogens will represent a transgression with the palaeoshore line distant from the place of deposition, whereas a predominantly humic kerogen will indicate a regression and closeness of the palaeoshore line to the point of deposition. Such kerogens will have a lower overall convertibility to hydrocarbons than pure sapropelic marine kerogens and the resulting oil will tend to be waxy.

In situations where the seas are anoxic, silled basins will also be anoxic and this is where the majority of the world's large oil basins occur (Demaison & Moore, 1980).

Dow (1978) considers that continental slopes are good petroleum source beds because they are sites of high marine organic production combined with reducing bottom conditions, quiet waters, and intermediate sedimentation rates. All these conditions are ideal for the production of rich source beds and because the sites are aquatic the resulting organic matter will be a rich oil source.

The above categories of marine environment which may be suitable for the production of organic-rich beds have been suggested using conditions in present-day oceans as a guide. However, the origins of some ancient organic-rich sediments are still open to argument. Gallois (1976) has suggested that the Kimmeridge Clay oil shales (and therefore much of the oil in the North Sea) were formed from algal blooms, in an environment between open ocean and enclosed marine basin. He suggests that conditions favoured the production of such blooms, which by deoxygenating and poisoning the water could temporarily create bottom conditions suitable for the formation of organic-rich sediments. Coccolith blooms are characteristic of seas rich in land-derived nutrients, and still occur in the North Sea. Tyson *et al.* (1979) suggest that the variations in organic content of the Kimmeridge clays, shales, and limestones may be related to the variations in the oxic–anoxic boundary. With this boundary well below the water–sediment interface, the organic debris will not survive and an organic clay will result (< 10% organic matter). With the oxic–anoxic boundary a few centimetres either side of the water–sediment interface, more organic matter will be preserved to give a bituminous shale (10–30% organic matter). When the anoxic layer begins well up in the aphotic zone (but not near the euphotic zone), much more organic matter is preserved giving rise to oil shale (up to 70% organic matter). When the oxic-anoxic interface is near the boundary of the euphotic and aphotic zones, storms can cause mixing with the oxygen of the upper layer killing anaerobic bacteria in the lower zone, while nutrients from the aphotic zone cause coccolith blooms at the water surface which results in organic-lean coccolith limestones (Figure 6.21).

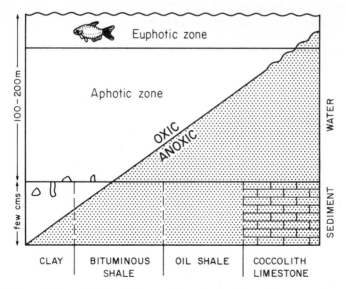

Figure 6.21 Origin of Kimmeridge shales (after Tyson *et al.*, 1979. Reproduced by permission of R. C. L. Wilson)

210

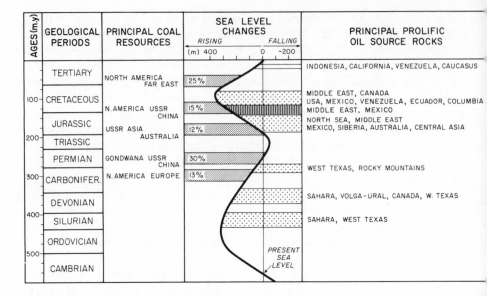

AGES(m.y)	GEOLOGICAL PERIODS	PRINCIPAL COAL RESOURCES	SEA LEVEL CHANGES (m) 400 ... 0 -200		PRINCIPAL PROLIFIC OIL SOURCE ROCKS
			RISING	FALLING	
	TERTIARY	NORTH AMERICA FAR EAST	25%		INDONESIA, CALIFORNIA, VENEZUELA, CAUCASUS
100-	CRETACEOUS				MIDDLE EAST, CANADA USA, MEXICO, VENEZUELA, ECUADOR, COLUMBIA
		N.AMERICA USSR CHINA	15%		MIDDLE EAST, MEXICO
	JURASSIC	USSR ASIA	12%		NORTH SEA, MIDDLE EAST
200-		AUSTRALIA			MEXICO, SIBERIA, AUSTRALIA, CENTRAL ASIA
	TRIASSIC				
	PERMIAN	GONDWANA USSR CHINA	30%		
300-	CARBONIFER.	N.AMERICA EUROPE	13%		WEST TEXAS, ROCKY MOUNTAINS
	DEVONIAN				SAHARA, VOLGA-URAL, CANADA, W. TEXAS
400-	SILURIAN				SAHARA, WEST TEXAS
	ORDOVICIAN				
500-				PRESENT SEA LEVEL	
	CAMBRIAN				

Figure 6.22 Correlation of sea level and organic matter type

Tissot (1979) considers that changes in sea level affect the amount of oil and coal, produced and preserved. The productivity and composition of marine phytoplankton seem to be greatly influenced by the changes in sea level. The increases in phytoplankton and epibenthos are directly related to the global rises in sea levels (Figure 6.22). Likewise, shallow epicontinental seas, transgressive over continental depressions surrounded by emergent land provide mineral nutrients and encourage algal blooms. These cause eutrophic conditions and temporary anoxic waters which are ideal for the preservation of sedimentary organic matter.

In Western Europe the successive Jurassic transgressions over the Triassic continent deposited the source rocks of the North Sea and the Toarcian and Kimmeridgian oil shales. However, the most important coal occurrences are formed by accumulations of terrestrial higher plants in coastal or paralic basins. The most typical environment is provided by coastal swamps which develop after regression of the sea. The comparison of the major coal deposits with the sea level curve (Figure 6.22) shews a frequent correlation with worldwide regression periods.

This relationship between sea level petroleum source rocks is especially noticeable in the Cretaceous. There are dark clay-rich sediments with up to 30% organic carbon in Cretaceous sequences 85–115 My ago. At the peak of this transgression the ratio of shelf seas to oceans was 0.16 : 1, very much higher than today (0.03 : 1), and the Cretaceous was a time of great increase in the growth of land plants, especially flowering plants. Thus the large amount of

211

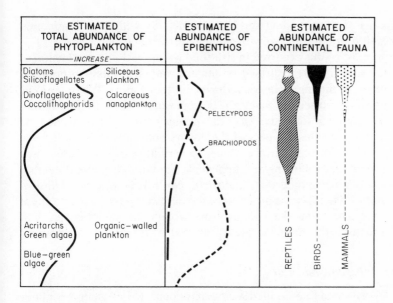

(after Tissot, 1979). (Reprinted by permission from *Nature*, 277, 463. Macmillan Journals Limited)

vegetation exported to the shelf seas would help use up the sea's oxygen content and cause thick oxygen-deficient layers throughout the world's oceans. These layers have been estimated to have extended from a few hundred metres below the surface to several thousand metres. When the oxygen-deficient layers met the sea bed, anoxic sediments were deposited. Thus on continental shelves the sediments are rich in organic matter which contains much terrestrial matter, whereas on oceanic rises the sediments are still organic rich but with very little, if any, terrestrial matter. In the deeper basins the anoxic layer may not have extended to the sea floor, but if their lower waters were not saturated with oxygen that small amount of oxygen could be used up in the degradation of organic matter brought in by turbidity currents and anoxic conditions produced.

Summerhayes (1981) reports that the middle Cretaceous black shales of the deep North Atlantic consist of alternating layers of organic-rich and poor sediments. The organic-rich sediments contain abundant amorphous organic matter derived from marine plankton with a small contribution from higher land plants. The marine material is abundant because of upwelling currents bringing nutrients and essential minerals from the deep waters. This produced highly productive surface waters and where the land climate was arid with little continental runoff (i.e. no terrestrial input), marine kerogen predominates. Where the climate was more humid with increased continental runoff, the resulting kerogen is diluted on deep sea fans by the terrestrially derived organic matter. In areas of no upwelling and low productivity with high continental runoff the result is organic matter with a highly terrestrial facies.

In addition, most of this occurred in stratified waters with the production of organic matter occurring in the surface oxygenated waters, and its incorporation into the sediment occurring in anoxic reducing bottom waters. When periodic climatic changes overturned the water, the bottom waters were oxygenated and benthic faunal activity destroyed most of the organic matter which reached the bottom. Alternatively, the contents of the anoxic bottom waters could have poisoned the life in the oxic surface waters so very little organic matter was produced for some time following such upheavals. Summerhayes also notes that eustatic sea level changes could participate in the control of the accumulation of the organic matter. He reports a progressive increase in organic matter content as the sea level rose from the Hauterivian through the Cenomanian with interruptions to this in the middle to late Aptian and in the middle Cenomanian.

Deposition of inorganic material

The active deposition of fine-grained sediments appears to be essential for later substantial petroleum genesis. Conditions that favour the sedimentation of fine-grained inorganic particles also favour the deposition of organic matter, and the concentration of the latter in the former will be inversely related to the rate of deposition of the inorganic material. The sediment covers the organic matter and protects it from attack by bottom-dwelling organisms. The faster the organic matter is buried, the greater is the degree of its preservation. The system is a fine balance between the rates of deposition of organic and inorganic particles. Too little inorganic matter and the organic debris will be destroyed by scavengers.

In some areas of the present-day oceans high organic productivities occur. These are due to the upwelling currents bringing nitrate and phosphate nutrients from the deeper waters. However, where these currents are seaward of deserts no source rocks will form because the deserts will be adding little of no inorganic material to the sea from which the sediments could be made. Too much inorganic material will decrease the source potential of the rock and its carbon content will be lowered. This matrix, consisting of inorganic particles with organic material within it, will also contain a large amount of water which may be available to remove the generated petroleum from the source bed on the first stage of its journey to the reservoir.

The size of the individual grains of the inorganic matter is of equal, if not greater, importance than the amount of the material. There is a proven relationship between the organic content and the grain size in sediments, ancient and modern. The relationship is that the smaller the size of the inorganic particles, the greater is the organic content of the sediment.

This relationship between fine-grained particles and high organic content comes about in the following way. The action of waves and tides tends to separate the fine-grained particles from the larger sand-sized particles. The fine-grained material, which separates from the larger pieces, settles through the water, deposits in quieter waters, and it will contain the clay-sized particles and the small low-density pieces of organic matter. This connection between the clay and the organic matter is accentuated by the fact that the settling clay particles, which often have very large surface areas, may absorb some of the organic matter, removing it from solution and transferring it into the sediment almost before that sediment has been formed.

The large particles, the sand-sized grains which tend to settle out in a high-energy environment, are not associated with high organic contents, for the inverse reasons that the clay particles and the organic matter are so associated. Thus there is no tendency of the sand to absorb the organic material onto its surface and the density difference between the two mean that they will not sink under the influence of gravity under the same conditions of water turbulence. In addition, the environments in which sand grains are deposited are high-energy environments which will be well oxygenated, which, as has already been demonstrated, are not conducive to the survival of the organic matter.

Ibach (1982) states that for any given sedimentation rate the total organic carbon varies with lithology, increasing along the series calcareous, calcareous--siliceous, siliceous sediments, to black shales. This observation is in agreement with the results of the earlier discussion concerning the conditions required for the preservation of sedimentary organic matter. Ibach also comments upon the fact that for any lithology, the total organic carbon varies with the sedimentation rate, but in a more complex fashion. Low rates of sedimentation mean that the organic matter will spend a long time in the oxic zone and thus be very susceptible to degradation. The result is a low organic carbon content in the ensuing sediment. As the sedimentation rate increases, the organic matter has a quicker passage through the near surface zone of intense organic degradation, and the resulting sediments have higher total organic carbon values. Thus increased rates of inorganic sedimentation should increase the oil-generation potential of oil-prone organic matter in those sediments. However, Ibach points out that this will only be true if the primary productivity of the organic matter also increases at a corresponding rate. If the primary productivity of the overlying water does not increase along with the sedimentation rate then the total organic carbon content will fall.

Presence of suitable reservoir and trap

The conditions for the production, deposition, and preservation of organic matter have been discussed and the conversion of organic matter into petroleum will be discussed in the next chapter. There is, however, another condition which needs to be fulfilled so that generated oil is preserved. This condition is the availability, at the correct time, of a suitable reservoir.

It was earlier demonstrated that the inorganic particles, which are deposited at the same time as the organic matter, will give a mud. This mud will later be compacted by burial and the organic matter it contains converted to petroleum, which will have three possible futures. Firstly, it can stay trapped within the shale if, for instance, the burial was only just sufficient to cause conversion of the organic matter to fluid hydrocarbons but not enough to cause the expulsion of those hydrocarbons. Secondly, when the generated oil is expelled from its source rock it rises until it is oxidized. Thirdly, the oil may migrate to a suitable reservoir where it can be trapped. Thus a porous and permeable reservoir is needed reasonably near to the place of generation together with a feasible migration pathway between the two.

The question of the ratio of sand to shale now becomes important. If there is too much shale in relation to sandstone, there may not be the migration pathways available for the oil to escape from its place of generation and, conversely, too little shale will mean that not enough oil can be generated either to provide pressure for migration or even to fill the reservoir. Thick source beds may not release all their generated oil because of difficulties of migration through the relatively impervious shale. Since only the oil from the shale next to a sand is likely to migrate a series of laminated sands and shales is more likely to fill a trap than equal thicknesses of sand and shale but in two individual units. Once again, the importance of a minimum organic content of a potential source rock, this time to produce the necessary amount of oil in order for migration to occur, is seen.

Kirkland & Evans (1981) consider that the evaporitic environment can be very productive of organic matter. Few species survive in brines, but those that do are commonly present in great profusion, and in a marine evaporitic embayment the inflow of surface currents is persistently to the regions of higher salinity, thus ensuring a continuous supply of nutrients. Only carbonates precipitate in the mesosaline (4–10% salinity) parts of such an evaporitic environment and no great dilution of organic matter by clastic or biogenic sediments occurs.

Because stratification of the brine may occur and reducing conditions may be associated with bottom waters, much of the organic matter will be preserved and upon maturation the result will be a rich carbonate source rock, frequently unrecognized in the geologic column. In the Middle East, mesosaline conditions were known to have occurred many times from the Triassic to the Cretaceous and they may be responsible for the vast reserves of petroleum in that area. Thus carbonates can fulfil all the conditions required for the generation of oil, i.e. an area with a rich marine life, and an anoxic sediment to preserve the dead organic matter.

Basins which provide the correct conditions were discussed earlier and there is plenty of evidence of abundant organic life in mesosaline waters.

Many of the situations which favour the production of organic-rich shales, and which were described earlier in this chapter, are restricted circulation systems. Besides favouring the production and preservation of organic matter, basins with restricted circulation will contain evaporites which can be the cover rocks or seals to petroleum accumulations.

MARINE AND NON-MARINE ENVIRONMENTS

It is now necessary to consider the relative influence of marine and non-marine environments in the formation of petroleum and its precursors. For the reasons given in the preceding paragraphs and because the majority of oil does come from marine environments, it has been agreed by most people that a marine environment is necessary for the formation of oil, and those oil accumulations which are in continental strata were considered to have been the result of the migration of hydrocarbons generated elsewhere. Some gas pools, however, have been found in situations that indicate that the gas was generated in either fresh or brackish water conditions, and gas pools such as these are almost certainly of local origin.

Reducing conditions, which it is known are essential to the deposition and preservation of organic matter and its accompanying inorganic sediments, can exist at the bottom of inland lakes (e.g. Lake Tanganyika, or where the Green River shales were deposited) and therefore it is not inconceivable that petroleum could be formed in these circumstances. The significant differences between this sort of environment and a genuine marine system are the type of organic matter involved and the absence of salt from the former conditions. This latter factor is an unknown in the genesis of petroleum and, of course, in its primary migration.

Laplante (1974) and Hunt (1979) consider there to be different kerogens for oil and gas with a 'coaly' type of kerogen, which yields gas upon maturation, being derived mainly from the nitrogeneous and humic residues of continental and marine plants, whereas the 'oily' type of kerogen, which is the amorphous remains of plankton and the high-lipid residues of spores and pollens, yields oil upon maturation. Vandenbrouche *et al.* (1976) compared, by geochemical means, the organic matter from the Doula basin in the Cameroons and the Toarcian shales from the Paris basin. The organic matter in the Doula basin originates from land-derived material, whereas that from the Toarcian shales comes from autochthonous marine organisms. The kerogens which are isolated from each area are different; that from the Paris basin being highly saturated, as compared with the aromatic kerogen from Doula. The saturated Paris basin kerogen was shewn to give naphtheno-aromatics followed by alkanes, whereas the aromatic Doula basin kerogen only gave alkanes, and these in much smaller quantities. Both Laplante's (Figure 6.17) and Hunt's (Figure 6.18) hypotheses of the different structures of oil-prone and gas-prone kerogens would be supported by this evidence.

Philippi (1974) notes that oils which originate from the fat-rich marine organic matter have normal paraffin concentrations which taper off after the C_{20} alkanes, whereas the molecular weight range of the fatty acids of plant waxes is much higher than the molecular weight range of those found in the lipids of marine organisms. Thus, if the plant waxes do contribute strongly to the oil source then the molecular weight distribution of the *n*-alkanes will be abnormal. If the source has significant plant wax then the ratio of $(C_{21} + C_{22})/(C_{28} + C_{29})$ will be in the range 0.6–1,2, whereas if the source is purely marine this ratio will lie between 1.5 and 2.0. Thus, the composition of the petroleum paraffins will depend upon the ratio of oils and fats to waxes, and the more plant material that has contributed to the source, the higher will be the average carbon chain length of the n-alkanes in the resulting oil.

Hence the effect of terrestrial matter on the kerogen, and upon the oil generated from it, will vary from increasing the wax content of the generated oil, by adding the wax compounds from the waterproofing of plant leaves, to altering the structure of the kerogen and making it more gas prone. This process can vary in extent depending upon the relative amounts of terrestrial and marine organic matter that forms the kerogen. The terrestrial input matter not only adds wax-rich components but also affects the structure of the kerogen so that it is more aromatic and less aliphatic than marine kerogens, and when a terrestrial kerogen matures it produces less hydrocarbons (more of which will be gas) as compared with a marine kerogen. The liquid hydrocarbons which are produced will be rich in high-molecular alkanes and thus very waxy and viscous. Thus it is not contradictory to say that a terrestrial influence will make the kerogen more gas prone but the oil more paraffinic and waxy, as terrestrial kerogens can produce some oil, even if it is as a minor product.

The reason that terrestrial organic matter makes kerogen more aromatic and gas prone, and less aliphatic and oil prone, can be found in the discussion earlier in this chapter on the types of sedimentary organic matter. Although lipids and proteins occur in both the marine (animal and plant) life and in the higher land plants, the latter contains much smaller quantities of these materials which can eventually become oil. In addition, the higher land plants contribute to the kerogen large amounts of lignin which is an aromatic material, whereas marine life does not synthesize such aromatic compounds. The aromatic character of marine kerogens is solely due to condensation and cyclization reactions and is thus small in amount, especially in comparison to the amount of aliphatic material. In land-derived kerogens the aromatic character is due to the components as well as to the condensation and cyclization reactions. This, in addition to the fact that land plants contribute very little lipid material, means a highly aromatic gas-prone kerogen.

Because of the great variation in depositional conditions, kerogens can vary from exclusively marine to completely terrestrial. As the terrestrial influence on the kerogen increases, the absolute convertibility to hydrocarbons reduces and the relative potential for the production of liquids, as opposed to gases, also decreases. At the same time the oil produced will contain a greater amount of wax.

This picture of the types of products produced by the kerogens will be complicated by other factors. The subsequent history of the kerogen may be more important in deciding the types of products of kerogen maturation than its actual composition. For instance, if an oil-prone kerogen is subjected to enough heat for long enough it will produce gas. Even if the kerogen produces only oil, that oil may be significantly altered during and after migration, i.e. a high-wax paraffinic oil may be cracked to a lighter oil by the combined effects of temperature, pressure, and time. Both these aspects will be dealt with in subsequent chapters.

In addition to the above factors there is the question of the uneven distribution of terrestrial and marine matter in a source rock. As the distance between the sediment in which organic matter is being deposited and any source of terrestrial organic matter increases, the resulting kerogen will be more marine. Hedges & Parker (1975) and Shultz & Calder (1976) have studied the sedimentary organic matter in the Gulf of Mexico and have shown that although terrestrial organic matter is carried into the Gulf by rivers and then deposited, the amount of such terrestrial matter decreases with the distance from the river mouth. In addition, the rate of flow of the river will decide the distance that such terrestrial organic matter is carried before settling into the sediment. Crisp *et al.* (1979), studying the organic constituents of recent sediments, have shewn that the concentration of organic matter is greatest nearest to the shore.

Hedberg (1968) considered that the high wax content of some crude oils appears to be a characteristic connected with their genesis or with the type of organic matter from which they are derived, and he noted that the high-wax crudes were restricted to certain stratigraphic sequences and to certain regions. These sequences have the following in common: (a) a shale–sandstone lithology; (b) a non-marine origin or an origin in waters of lower than normal salinity; (c) a common association with coal or other carbonaceous strata; (d) a stratigraphic range from Devonian to Pliocene; (e) a continental or near-continental environment; and (f) an association with low-sulphur crudes. He thus considered it not unreasonable to suggest that such high-wax crudes might be reflecting the contribution to their genesis of terrigenous organic matter, which although derived from aquatic sources, was associated with waters of lower than average salinity.

FORMATION OF KEROGEN

Welte (1965) has studied the relationships between petroleum and source rocks and explained some of the differences between crude oils and the organic matter in recent sediments. He sets out the basic parameters for the generation of oil and lists the sequences of events from the deposition of the organic matter through to the formation of kerogen and its subsequent breakdown to form petroleum. These are illustrated in Figures 6.23 and 6.24.

Let us now consider the first part of this sequence of events, that is the generation of the kerogen, and how the composition of the kerogen will depend upon the original organic matter input and its depositional environment. The latter parts of the sequence will be discussed in later chapters.

The formation of petroleum from deposited organic matter can be conveniently split into three phases. Diagenesis is a process of biological, chemical, and physical alteration during which the deposited organic matter is

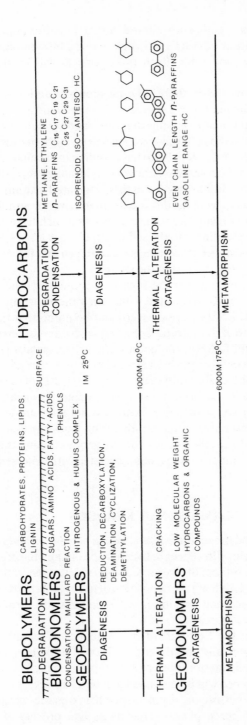

Chapter 6.23 Schematic representation of the transformation of organic matter into petroleum (after Hunt. Reproduced by permission of The American Association of Petroleum Geologists)

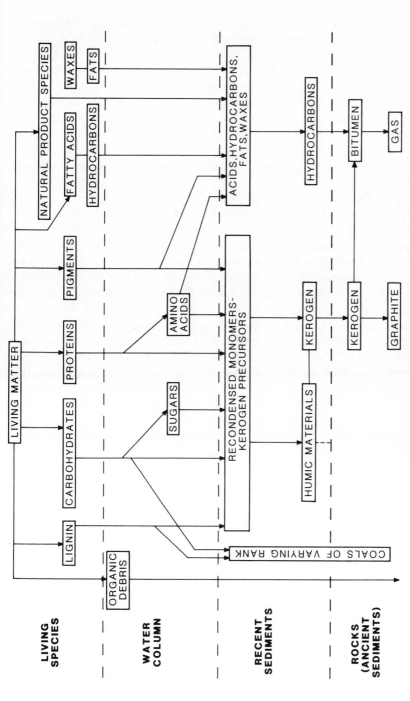

Figure 6.24 Diagrammatic illustration of the alteration of sedimentary organic matter (after Barker, 1979. Reproduced by permission of The American Association of Petroleum Geologists)

broken down into smaller units. Many of these smaller units subsequently recombine to form kerogen. Catagenesis the second phase when the kerogen, under moderate conditions of temperature and pressure, forms hydrocarbons which are mostly liquid. However, towards the end of the catagenetic stage mostly gaseous and fewer liquid hydrocarbons are formed. In order for these fluid products to be mainly saturated the kerogen nuclei become hydrogen-deficient. The third stage is metagenesis when the kerogen is completely converted to gaseous hydrocarbons and a graphite-like skeleton which is almost devoid of hydrogen. Diagenesis will be discussed in the remainder of this chapter, while catagenesis and metagenesis are described in a later chapter.

Much of the initial breakdown of the freshly deposited organic matter is carried out by micro-organisms, which are very small bacteria able to utilize a vast number of types of chemical reaction in order to live. These bacteria consume organic matter and, with the aid of oxygen, produce energy with which to live and reproduce. They can also incorporate parts of their food into their body construction. They require free water, various essential minerals which will be in solution, nitrogen in a usable form, and their energy source (the organic matter), whilst aerobic bacteria also require oxygen. Such bacteria have been reported to exist within the temperature range -10 to $105\ °C$; up to pressures of 1,000 atm.; in pH ranges 1–11; and in both fresh and salt water (Zobell, 1964). When they attack organic matter they use enzymes to decompose the proteins, lipids, carbohydrates, etc., into the simple units from which these are built.

Some of these molecules are used as an energy source while others, or simple polymers of them, are used to make new cell walls, etc. The components that have been identified in sedimentary organic matter and in crude oils can mostly be consumed by bacteria. These compounds include the fats, and their component alcohols and acids, cellulose, and proteins. The simple low molecular weight aromatic compounds are toxic to bacteria and are not consumed by them and the more highly branched hydrocarbons, such as the isoprenoids, are not readily used as substrates and tend to concentrate in body fats, e.g. copepods, shark's livers. The only hydrocarbon to be produced by bacteria is methane and this is the result of anaerobic bacteria reducing carbon dioxide.

As was discussed earlier, organic matter can be deposited in a wide range of oxic environments. The range is from systems that are completely oxic, including both the water and the underlying sediment, to systems that are completely anoxic, both the sea and the mud being devoid of oxygen. In the former case the conditions favour the production but not the preservation of organic matter, while in the latter the preservation but not production is favoured, and neither will give rise to organic-rich sediments. The most prolific organic-rich sediment-gathering situation is where the sea is oxic near the surface in the euphotic zone while the bottom waters and the mud are strongly reducing. This sort of situation favours the greatest concentrations of sedimentary organic matter. However, the most normal condition is where all the sea, as well as the top few centimetres of the mud, are oxic but the rest of the sediment is anoxic. This where the rapid deposition of fine-grained sediments is important so that the organic matter is quickly taken through the oxic zone and preserved in an anoxic environment. In such cases the most resistant organic matter, including the humic material, resins, waxes, as well as the lipids, can be preserved.

During this stage, in which temperatures and pressures are fairly low, the large polymeric molecules are broken down by the microbes into their monomers or simpler and shorter polymers, which, once out of the oxic zone, can be recombined to the precursors of kerogen (the fulvic and humic acids and humins), then the geopolymers, and finally the kerogen. If the contribution of land material is great, especially in relation to the amount of mineral matter being deposited, the products of the terrestrial plant lignin will be peat and eventually brown coals (lignite).

There are during this stage various chemical reactions and alterations taking place, including disproportionation, decarboxylation, elimination, reduction, dimerization, and aromatization. These, although minor in quantity, can produce some of the simpler hydrocarbons found in recent sediments.

Disproportionation is a reaction in which two similar molecules exchange hydrogen so that one becomes saturated and will be an alkane or cycloalkane, while the other becomes hydrogen deficient to become olefinic or aromatic. Decarboxylation is the removal of organic acid groups from a fatty acid to produce carbon dioxide and an alkane. Elimination is a reaction in which a small molecule such as water or ammonia is produced as a by-product during the reaction of two large molecules to form a larger one. Reduction is the taking up of hydrogen by an unsaturated species such as an olefin. The reaction of the phytol side chain of chlorophyll illustrates these reactions (Figure 6.9). If the phytol is oxidized it will produce phytonic acid, the decarboxylation and reduction of which will produce pristane, whereas the hydrogenation of phytol give the alcohol dihydrophytol and the elimination of water from this, followed by reduction will give phytane. Dimerization will involve the linking of two identical molecules together to form a dimer and if this is continued it will become polymerization (but note that not all polymers are made of a single monomer). Aromatization will involve the reactions of disproportionation and Diels Alder (Figure 6.16) to form simple aromatic compounds.

STRUCTURE AND COMPOSITION OF KEROGEN

Having seen the way in which kerogen is formed and the relative importance of its constituent components, let us now examine the structure and composition of the different kerogen types.

Kerogen is a three-dimensional macromolecule consisting of a series of aromatic nuclei cross-linked by various bridges which are hydrocarbon chains, esters, ketones, or sulphides. Quite often within this three-dimensional matrix will be trapped other molecules and these can either be purely physically trapped or held to the kerogen by weak absorption forces. The nuclei may contain linked sheets of polyaromatics with 7–16 fused rings, occasionally with heterocyclic elements, normally nitrogen but sometimes sulphur and oxygen, incorporated. On these nuclei there are side chains which are normally alkyl but occasionally naphthenic rings as well as functional groups such as hydroxide, amide, acid, ketone, or etherial linkages. In addition there will be many compounds which are loosely attached to or physically trapped within the kerogen. These compounds will be the geochemical fossil type of material. Burnham *et al.* (1982) consider that most of the isoprenoid hydrocarbon structures in the Green River shale are chemically bonded to the kerogen. Because such isoprenoid hydrocarbons are released concomitantly with carbon dioxide before the generation of oil it is likely that they are bound to the kerogen via acid linkages. The links between the nuclei are either more aliphatic chains, either normal or branched chain, or other groups with heteroatoms, such as dialkyl sulphide, ether, ester, or amine. The aliphatic content of the kerogens will vary with source and type but it must constitute a major part of the immature kerogens, especially the more marine sapropelic types. Immature kerogens have the layers well separated and not planar but as maturity increases the sheets become closer together, more planar, and more aromatic.

The different types of kerogen, especially when they are immature and in the principal zone of oil generation (i.e. diagenesis and catagenesis) can be identified and characterized on the basis of microscopic examination, elemental analysis, infrared spectroscopy, pyrolysis, etc. Such detailed examinations have indicated three main types of hydrocarbon-producing kerogen, plus some sub-varieties and inert kerogens.

CLASSIFICATION OF KEROGENS

There is no single accepted method of classifying kerogens and no accepted figure for the number of different kerogen types. The most profilic oil-generating kerogens are of the hydrogen rich amorphous types which are either algal or sapropelic in nature. The algal amorphous kerogen has a lower density than the sapropelic amorphous kerogen, which has a slightly lower oil potential than the algal material. Such kerogens are highly oil prone because the algae contribute lipid material (polymethylene chains) to the kerogens, whereas the cell-wall material which is the basis of terrestrial kerogens does not contain such lipids. (Ishiwatari & Machihara, 1982). There is also a hydrogen lean amorphous kerogen which has a structured component of about 10%. This kerogen is gas prone even though it is believed to be derived from algae. Whether this kerogen originates from the same algae as oil-prone kerogens but undergoes a different conversion, possibly with a less reducing environment, or if different algae produce gas- and oil-prone kerogens, is not known.

With the exception of the gas-prone algal kerogen described above, the open marine environment favours the production of oil-prone kerogens with high convertibilities. In contrast, kerogens produced from terrestrial organic matter are highly gas prone and have low convertibility. Any oil from these terrestrial kerogens will be very waxy. Whether the transition between these two extremes is continuous or in a series of steps is unknown.

The open marine situation considered here does not imply open ocean conditions, as the open ocean is a relatively poor source of organic matter. 'Open marine' means an environment where the conditions for abundant life are favourable, but there little terrestrial organic matter can be incorporated into the resulting kerogen. Such areas include continental shelves and near-shore areas with little organic terrestrial input.

Tissot *et al.* (1974) describe three kerogen types which they say represent the three basic kerogen types, principally based upon elemental analyses. Their type III kerogen is a terrestrially derived kerogen which can be considered as one end member. Their type I kerogen is either algal or the remains of disseminated organic matter which has been extensively reworked by organisms. The algal type I kerogen can be considered as the other end member of this series of kerogens with the other sub-type of the Tissot type I as the next kerogen in the series. The type II kerogen described by Tissot *et al.* is another marine kerogen nearer to the type I than to the III.

Microscopic examination has shewn that many kerogen types can be identified. Palynologists use the terms algal, amorphous, herbaceous, woody, and coaly to describe kerogens. Other workers use the terms sapropelic and humic or marine and terrestrial, although these latter terms are much broader classifications. Probably no one classification scheme can satisfactorily describe all kerogens and often several terms are used. For instance, many kerogens are amorphous but they can be algal or sapropelic in nature.

The main advantage of the Tissot classification is that it simplifies naming and discussions. From here on the names type I, II, and III will be used under the following conditions. Type I is used to describe the open-marine algal kerogen and type III is a terrestrially derived kerogen. In between these two extremes there will be many kerogens of which type II is one example which is predominantly marine in origin and is oil prone. The Tissot *et al.* classification and kerogens are described below.

Type I (alginite) is a kerogen which, in its immature state, has a high hydrogen content (i.e. atomic H/C ratio > 1.5) but a low oxygen content (atomic O/C ratio < 0.1). It comprises lipid material with a lot of aliphatic chains but a low content of polyaromatic nuclei and heteroatomic bonds (Figures 6.25 and 6.26). Its source is either from algal material, which can be marine or lacustrine, e.g. torbanite or algal-rich oolitic limestones, or it can originate from the remains of disseminated organic matter which has been extensively reworked by organisms. This kerogen, which has the greatest potential for oil generation with often up to 80% convertibility, is the least commonly found (Figure 6.26(a)).

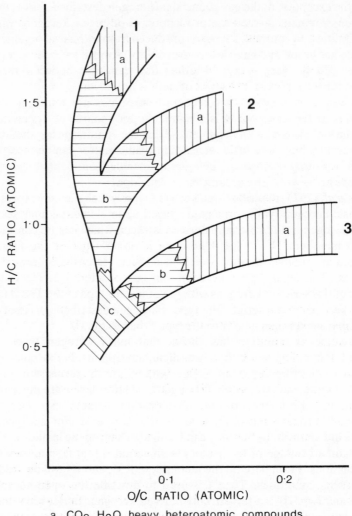

a. CO_2, H_2O, heavy heteroatomic compounds

b. OIL

c. GAS

Figure 6.25 Kerogen types and maturation paths (after Tissot & Welte, 1978. Reproduced by permission of Springer-Verlag, Heidelberg)

Type II (sapropelite) is the most commonly found oil-source kerogen and has a hydrogen content not quite as high as type I (H/C; 1.2–1.45) and its oxygen content is slightly higher than type I (O/C; 0.1–0.20). There are more of the polyaromatic nuclei and heteroatoms present plus more ketone and carboxylic acid groups. However, there are still abundant aliphatic chains of reasonable length as well as naphthenic rings and other cyclic structures. Sulphur can be present in substantial amounts. This type of kerogen is usually

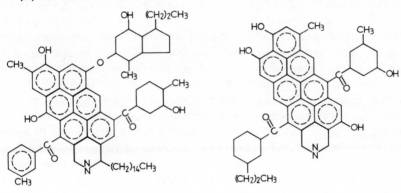

(a) ALGAL KEROGEN

	(a)	(b)	(c)
ATOMIC H/C	1.65	1.28	0.84
ATOMIC O/C	0.06	0.10	0.13
HYDROCARBONS	OIL	OIL+GAS	GAS

(b) LIPTINITIC KEROGEN

(c) HUMIC KEROGEN

Figure 6.26 Basic kerogen types showing some of the organic structures believed to be present when kerogen is just approaching the oil window (after Dow, 1977. Reproduced by permission of Elsevier Scientific Publishing Co.)

deposited in marine sediments in reducing conditions from autochthonous organic matter which is derived from a mixture of phyto- and zooplankton, micro-organisms, and the lipid-rich residues of terrestrial spores and pollens. If this type of organic matter is deposited in oxidizing conditions it will not be preserved. Even if it is preserved, future geological events will, as with all kerogen types, determine its future and whether or not it produces oil and/or gas. Later uplift and the effect of turbidity may well oxygenate the sediment and allow the material to be oxidized whereas insufficent burial will not allow its conversion to hydrocarbons (Figure 6.26(b)).

Type III has a relatively low initial hydrogen content (H/C ratio < 1.0) and a high oxygen content (O/C > 0.20). In this kerogen there is a much smaller amount of aliphatic chain material, mostly methyl and other short-chain species plus a few longer-chain alkyl groups originating from the waxes of the higher terrestrial plants. Much of this wax is, however, physically rather than chemically bound to this type of kerogen. The proportion of polyaromatic nuclei and heteroatomic bonds is much higher than in types I and II. This kerogen is the least likely of all the three kerogens to yield oil but it does have a much greater gas potential than any other kerogen type. Although this kerogen is deposited in aquatic conditions, it is composed of terrestrial material mainly coming from the higher plants and it is often incorporated into the sediment with the minimum of microbial attack. Where such microbial oxidation has occurred, the small oil potential of this kerogen will be destroyed (Figure 6.26(c)).

In summary, the effect of marine and non-marine origins on the organic matter which has to be deposited in the sediment is as follows. Terrestrial matter derived from land plants incorporated into sediments under oxidizing conditions will, upon subsequent breakdown of any kerogen formed from it, yield mostly gas. When, however, the same terrestrial organic matter is deposited under more reducing conditions, the eventual end product will be paraffinic crude oil and wet gas in the oil window (catagenesis zone), and dry gas will not be formed until the metagenesis stage is reached. Marine organic matter, when deposited under reducing conditions will, subsequent to the formation and breakdown of the kerogen, yield both oil and gas at the catagenesis and metagenesis stages of maturation, respectively (Harwood, 1977). The convertibility of this kerogen, at about 60%, is higher than the terrestrial kerogen (40%). The algal kerogen will be the most prolific producer of paraffinic oils with a potential convertibility of about 80%.

Hence the 'coaly' kerogen comes from the carbohydrates, the lignin, the humic acids, the nitrogeneous and humic materials of higher terrestrial plants, whereas the 'oily' kerogen originates from the lipid derivatives and the amorphus remains of plankton with the high-lipid residues of spores and pollens (Hunt, 1979; McIver, 1967). While these are the commonest kerogens, two other types have been identified. One is mostly composed of structured algal material and/or highly removed marine material and it will yield very paraffinic crude oils (Tissot et al., 1974). The other kerogen is 'dead' carbon, inertinite, which consists of recycled material, often of a terrestrial origin that has been oxidized prior to its incorporation into the sediment (Erdman, 1975; Tissot & Welte, 1978). The majority of kerogens are varying mixtures of the first two types, II and III (or marine and terrestrial), together with some inertinite. Because the inertinite is rich in carbon but almost totally devoid of hydrogen and oxygen, it has no potential for the generation of hydrocarbons. When mixed with other kerogens it will increase the carbon content of the sediment without increasing that sediment's genetic potential. Thus purely chemical assays to ascertain genetic potential of sediments can be upset when significant amounts of inertinite (type IV kerogen) are present.

REFERENCES AND FURTHER READING

Abell, P. I., & Margolis, M. J. (1982) n-Paraffins in the sediments and in situ fossils of the Lake Turkana Basin, Kenya. Geochim. & Cosmochim. Acta, 46, 1505.

Albrecht, P., & Ourrison, G. (1969) Diagenesis of hydrocarbons in a series of sediments. Geochim. Cosmochim. Acta, 33, 138.

Albrecht, A., Vandenbrouche, M., & Mandengue, M. (1976) Geochemical studies on the organic matter from the Doula Basin (Cameroon), I, Evolution of the extractable organic matter and the formation of petroleum. *Geochim. Cosmochim. Acta*, **40**, 791.

Baker, E. G. (1959) Origin and migration of oil. *Science*, **129**, 871.

Baker, E. G. (1960) A hypothesis concerning the accumulation of sediment hydrocarbons to form crude oil. *Geochim. Cosmochim. Acta*, **19**, 309.

Barghoorn, E. S., Meinschein, W. G., & Schopf, J. W. (1964) Biological remains in a precambrian sediment. *Science*, **145**, 262.

Barghoorn, E. S., Oro, J., Nooner, D. W., Wikstron, S. A., & Zlatkis, A. (1965) Biological matter in a 2 million year old precambrian sediment. *Science*, **148**, 77.

Barker, C. (1979) *Organic Geochemistry in Petroleum Exploration*. AAPG Continuing Education Course Note Series No. 10.

Barton, D. H. R., Carruthers, W., & Overton, K. H. (1956) A triterpenoid lactone in petroleum. *J. Chem. Soc.*, 788.

Bayliss, G. S. (1968) The formation of pristane, phytane and other related isoprenoid hydrocarbons by the the termal degradation of chlorophyll. *ISSM Am. Chem. Soc. Nat. Meet. Div. Petv. Chem. Reprint*, F117–131.

Blumer, M., Mullin, M. M., & Thomas, D. W. (1963) Pristane in zooplankton. *Science*, **140**, 974.

Blumer, M., & W. D. Snyder, (1965) Isoprenoid hydrocarbons in recent sediments. *Science*, **150**, 1588.

Blumer, M., & Thomas, D. W., (1964) Porphyrin pigments of a Triassic sediment. *Geochim. Cosmochim. Acta*, **28**, 1147.

Blumer, M., & Thomas, D. W. (1965) Phytadienes in zooplankton. *Science*, **149**, 1148.

Blumer, M., Guillard, R. R. L., & Chase, T. (1971) Hydrocarbons of marine phytoplankton. *Mar. Biol.*, **8**, 183.

Bray, E. E., & Cooper, J. E. (1968) A postulated role of fatty acids in petroleum formation. *Geochim. Cosmochim. Acta*, **27**, 1113.

Bray, E. E., & Evans, E. D. (1961) Distribution of normal paraffins as a clue to the recognition of source beds. *Geochim. Cosmochim. Acta*, **22**, 2.

Bray, E. E., & Evans, E. D. (1965) Hydrocarbons in non-reservoir rock source beds. *BAAPG*, **49**, 248.

Bray, E. E., Evans, E. D., & Stevens, N. P. (1956) Hydrocarbons in soils and muds of the Gulf of Mexico. *BAAPG*, **40**, 975.

Breger, I. A. (1960) Possible sources of hydrocarbons in soils and muds of the Gulf of Mexico. *BAAPG*, **40**, 975.

Breger, I. A., & Brown, A. (1962) Kerogen in the Chattanooga Shale. *Science*, **137**, 221.

Brooks, J. D., & Smith, J. W. (1967) Diagenesis of plant lipids during the formation of coal and petroleum. *Geochim. Cosmochim. Acta*, **31**, 2389.

Brooks, J. D., Gould, K., & Smith, J. W. (1969) Isoprenoid hydrocarbons in coal and petroleum. *Nature*, **222**, 257.

Buckley, S. E., Hocott, C. R., & Taggart, Jr., M. S. (1958) Distribution of dissolved hydrocarbons in subsurface waters. In: L. G. Weekes (Ed.), *Habitat of Oil: A Symposium*, Tulsa, Okla., AAPG, p. 850.

Green River kerogen decomposition. *Geochim. & Cosmochim. Acta*, **46**, 1243.

Cantarow, A., & Schepartz, B. (1954) *Biochemistry*. W. B. Saunders & Co., Philadelphia & London.

Clark, R. C., & Blumer, M. (1967) Distribution of *n*-paraffins in marine organisms and sediment. *Limnol. Oceanogr.*, **12**, 79.

Cox, B. B. (1946) The transformation of organic matter to petroleum under geologica conditions. *BAAPG*, **30**, 645.

Craig, H. (1953) Geochemistry of stable carbon isotopes. *Geochim. Cosmochim. Acta*, **3**, 53.

228

Crisp, P. T., Brenner, S., Venketesan, M. I., Ruth, E., & Kaplan, I. R. (1979) Organic chemical characterization of sediment trap particles from San Nicolas, Santa Barbara, Santa Monica and San Pedro basins, California. *Geochim. Cosmochim. Acta*, **43**(11), 1791.

Day, W. C., & Erdman, J. G., (1963) Ionene—A thermal degradation product of β-carotene. *Science*, **141**, 808.

Demaison, G. J., & Moore, G. T., (1980) Anoxic environments and oil source bed genesis. *BAAPG*, **64**(8), 1179.

Dembicki, H., Meinschein, W. G., & Hattin, D. E. (1976) The significance of the predominance of even numbered n-alkanes (C_{22}–C_{30}). *Geochim. Cosmochim. Acta.*, **40**, 203.

Douglas, A. G., & Grantham, P. T. (1973) Docosane in rock extracts. *Chem. Geol.*, **12**, 249.

Douglas, A. G., Eglington, G., Maxwell, J. R., Ramsey, J. N., & Stallberg-Stenhagen, S. (1966) The occurrence of isoprenoid fatty acids in Green River Shale. *Science*, **153**, 113.

Dow, W. G. (1977) Kerogen studies and geological interpretations. *J. Geochem. Expl.*, **7**, 79.

Dow, W. G. (1978) Petroleum source beds on continental slopes and rises. *BAAPG*, **62**, 1584.

Draffen, G. H., Eglington, G., & Rhead, M. M. (1971) Hydrocarbons from the alteration of cholesterol. *Chem. Geol.*, **8**, 277.

Eckleman, W. R., Broechker, W. S., Whitlock, D. W., & Allsup, J. R. (1962) Implications of carbon isotopic compositions of total organic carbon of some recent sediments and ancient oils. *BAAPG*, **46**, 699.

Eisma, E., & Jung, J. W. (1964) Petroleum hydrocarbon generation from fatty acids. *Science*, **144**, 1451.

Erdman, J. G. (1975) Time and temperature relations affecting the origin, explusion and preservation of oil and gas. *9th World Petroleum Cong. Proc.*, **2**, 139.

Erdman, J. G., & Schwendinger, R. (1964) Sterols in recent aquatic sediments. *Science*, **144**, 1575.

Feux, A. N. (1977) The use of stable carbon isotopes in hydrocarbon exploration. *J. Geochem. Expl.*, **7**, 155.

Gallois, R. W. (1976) Coccolith blooms in the Kimmeridge clay and origin of North Sea oil. *Nature*, **259**, 473.

Gehman, H. M. (1962) Organic matter in limestones. *Geochim. Cosmochim. Acta*, **26**, 885.

Gohring, K. E. H., Schenck, P. A., & Engelhardt, E. D. (1967) A new series of isoprenoid isoalkanes in crude oils and cretaceous bituminous shales. *Nature*, **215**, 503.

Groennings, A. (1953) Determination of porphyrins in petroleum. *Anal. Chem.*, **25**, 938.

Han, J., & Calvin, M. (1969) Occurrence of C_{22}–C_{25} isoprenoids in Bell Creek crude oil. *Geochim. Cosmochim. Acta*, **33**, 7333.

Han, J., & Calvin, M. (1970) Branched alkanes from blue-green algae. *J. C. S. Chem. Comm.*, 1490.

Harwood, R. J. (1977) Oil and gas generation by laboratory pyrolysis of kerogen. *BAAPG*, **61**, 2082.

Hedberg, H. D. (1964) Aspects of the origin of petroleum. *BAAPG*, **48**, 1755.

Hedberg, H. D. (1968) The significance of high wax oils with respect to the genesis of petroleum. *BAAPG*, **52**, 736.

Hedges, J. I., & Parker, D. L. (1976) Land derived organic matter in surface sediments in the Gulf of Mexico. *Geochim. Cosmochim. Acta*, **40**, 1019.

Hoan, M. E.., & Meinschein, W. G. (1977) Fatty acids in fossil fruits. *Geochim. Cosmochim. Acta*, **41**, 189.

Hunt, J. M. (1961) Distribution of hydrocarbons in sedimentary rocks. *Geochim. Cosmochim. Acta*, **22**, 37.

Hunt, J. M. (1972) Distribution of carbon in the earth's crust. *BAAPG*, **56**, 2273.

Hunt, J. M. (1979) *Petroleum Geochemistry and Geology*. W. H. Freeman & Co., San Francisco.

Ibach, L. E. J. (1982) Relationship between sedimentation rate and total organic carbon content in ancient marine sediments. *BAAPG*, **66**(2), 170.

Ishiwatari, R., Ishiwatari, M., Kaplan, I. R., Rohrback, B. G. (1977) Thermal alteration experiments on organic matter from recent marine sediments in relation to petroleum genesis. *Geochim. Cosmochim. Acta*, **41**, 815.

Ishiwatari, R., & Machihara, T. (1982) Algal lipids as a possible contributor to the polymethylene chains in kerogen. *Geochim. & Cosmochim. Acta.*, **46**, 1459.

Ishiwatari, R., Rohrback, B. G., & Kaplin, I. R. (1978) Hydrocarbon generation by thermal alteration of kerogen from different sediments. *BAAPG*, **62**, 687.

Kaplin, I. R., Schopf, J. W., & Smith, J. W. (1970) Extractable organic matter in the precambrian. *Geochim. Cosmochim. Acta*, **34**, 659.

Kidwell, A. L., & Hunt, J. M. (1958) Migration of oil in Recent Sediments of Pedernales, Venezuela. In L. G. Weekes (Ed.), Habitat of oil; a symposium. Tulsa, Okla. *AAPG*, 790.

Kirkland, D. W., & Evans, R. (1981) Source rock potential of evaporitic environment. *BAAPG*, **65**(2), 181.

Knenvolden, K. A. (1966) Molecular distribution of fatty acids and paraffins in lower cretaceous sediments. *Nature*, **209**, 573.

Kolatturkurdy, P. E. (1980) Biopolyester membranes of plants. *Science*, **208**, 990.

Koons, C. B., Jamieson, G. W., & Ciereszko, L. S. (1965) Normal alkane distributions in marine organisms. Possible significance to petroleum origin. *BAAPG*, **49**, 301.

Laplante, R. E. (1974) Hydrocarbon generation in Gulf Coast tertiary sediments. *BAAPG*, **58**, 1281.

McIver, R. D. (1967) Composition of kerogen—Clue to its role in the origin of petroleum. In *7th World Petroleum Congress Proc., Mexico City*, Vol. 2, London, Elsevier, p. 26.

Mair, B. J. (1964) Terpenoids, fatty acids and alcohols as source materials for hydrocarbons. *Geochim. Cosmochim. Acta*, **28**, 1303.

Martin, R. L., Winters, J. C., & Williams, J. A. (1963) Distribution of *n*-paraffins in crude oils and their implications to origin of petroleum. *Nature*, **199**, 110.

Meinschein, W. G. (1961a) Significance of hydrocarbons in sediments and petroleum. *Geochim. Cosmochim. Acta*, **22**, 58.

Meinschein, W. G. (1961b) Origins of petroleum. *BAAPG*, **43**, 925.

Mitchell, P. H. (1950) *A Textbook of Biochemistry*. McGraw-Hill, New York.

Mold, J. D., Stevens, R. K., Means, R. E., & Ruth, H. M. (1963) 2,6,10,14-tetramethyl pentadecane (pristane) from wool wax. *Nature*, **199**, 283.

Parrish, J. T. (1982) Upwelling and Petroleum Source Beds, with reference to Paleozoic. *BAAPG*, **66**, 750.

Perry, G. J., Volkman, J. K., Johns, R. B., & Bavor, H. J. (1979) Fatty acids of bacterial origin in contemporary marine sediments. *Geochim. Cosmochim. Acta*, **43**(11), 1715.

Philp, R. P., Brown, S., & Calvin, M. (1978b) Isoprenoid hydrocarbons produced by thermal alteration of *Nostoc mascorum* and *Rhodopseudomonas* spherides.

Philp, R. P., Calvin, M., Brown, S., & Young, E. (1978a) The organic geochemical studies on kerogen precursors in recently deposited algal mats and oozes. *Chem. Geol.*, **22**, 207.

Philippi, G. T. (1965) Depth, time and mechanisms of petroleum generation. *Geochim. Cosmochim. Acta*, **29**, 1021.

Philippi, G. T. (1974) The influence of marine and terrestrial source material on the composition of petroleum. *Geochim. Cosmochim. Acta*, **38**, 947.

230

Powell, T. G. (1975) Geochemical studies related to the occurrence of oil-gas in the Dampier sub-basin of W. Australia. *J. Geochem. Expl.*, **4**, 441.

Powell, T. G., & McKirdy, D. M. (1973) Effects of source material, rock type & diagenesis on the *n*-alkane content of sediments. *Geochim. Cosmochim. Acta*, **37**, 623.

Powell, T. G., & McKirdy, D. M. (1975) Geological factors controlling crude oil composition in Australia and Papua-New Guinea. *BAAPG*, **59**, 1176.

Raschid, M. A. (1979) Pristane–phytane ratios in relation to source and diagenesis of ancient sediments from the Labrador shelf. *Chem. Geol.*, **25**, 109.

Scalan, R. S., & Smith, J. E. (1970) An improved measure of the odd even predominance in the normal alkanes of sediment extracts and petroleum. *Geochim. Cosmochim. Acta*, **34**, 611.

Schmalz, R. F. (1969) Deep water evaporite deposition, a genetic model, *BAAPG*, **53**, 798.

Schultz, D. J., & Calder, J. A. (1976) Organic carbon $^{13}C/^{12}C$ variations in esturine sediments. *Geochim. Cosmochim. Acta*, **40**, 381.

Smith, P. V. (1952) The occurrence of hydrocarbons in recent sediments from the Gulf of Mexico. *Science*, **116**, 437.

Stevens, N. P. (1956) Origins of petroleum. *BAAPG*, **40**, 51.

Stevenson, F. J. (1960) Some aspects of the distribution of biochemicals in the geological environment. *Geochim. Cosmochim. Acta*, **19**, 261.

Summerhayes, C. B. (1981) Organic facies of Middle Cretaceous black shales in Deep North Atlantic. *BAAPG*, **65**(11), 2364.

Tissot, B. (1979) Effects on prolific petroleum source rocks and major coal deposits caused by sea-level changes. *Nature*, **277**, 463.

Tissot, B., Durrant, B., Espitalie, J., & Combaz, A. (1974) Nature and diagenesis of organic matter. *BAAPG*, **58**, 499.

Tissot, B. P., & Welte, D. H. (1978) *Petroleum Formation and Occurrence: A New Approach to Oil and Gas Exploration*. Springer-Verlag, Berlin-Heidelberg-New York.

Trask, P. D., & Wu, C. C. (1930) Does petroleum form in sediments at the time of deposition? *BAAPG*, **14**, 1451.

Tyson, R. V., Wilson, R. C. L., & Downie, C. (1979) A stratified water column environmental model for the type Kimmeridge clay. *Nature*, **277**, 377.

Vandenbrouche, M., Albrecht, P., & Durand, B. (1976) Geochemical studies on organic matter from Doula Basin (Cameroon) III. *Geochim. Cosmochim. Acta*, **40**, 1241.

Waples, D. W., Haug, P., & Welte, D. H. (1974) Occurrence of regular C_{25} isoprenoid hydrocarbons in tertiary sediments. *Geochim. Cosmochim. Acta*, **38**, 381.

Weekes, L. G. (1958) *Habitat of Oil: A Symposium*. Tulsa, Okla, AAPG.

Welte, D. H. (1965) Relation between petroleum and source rock. *BAAPG*, **49**, 2246.

Welte, D. H., & Waples, D. W. (1973) Uberdie, Bevorzugung geradzahliger *n*-alkane in sedimentgesteinen. *Naturwissen-Schaften*, **60**, 516.

Williams, B., (1982) Synthetic fuels. *Oil & Gas J.*, **80**, 26, 71.

Zobell, C. E. (1964) Geochemical aspects of the microbial modification of carbon compounds. In U. Columbo and G. D. Hobson (Eds), *Advances in Organic Geochemistry*, New York, MacMillan, pp. 339–356.

Chapter 7

Factors and Processes in Petroleum Formation

INTRODUCTION

In the last chapter the various types of organic matter that eventually form petroleum were discussed, as were the relative influences of marine and terrestrial organic material upon kerogen formation. The structures of the different immature kerogens were also described. It is now proposed to examine the various factors which convert that organic matter into petroleum; to describe the changes in kerogens with maturation; and to see if any other processes can contribute to the final generated oil. The study of the migration and secondary alteration of that petroleum will be dealt with in the following chapter.

Geologic time

Changes in the composition of crude oils, their specific gravity, optical rotation, colour, etc., with age have all been advanced as evidence to prove that time, on a geologic scale, is necessary for the formation of oil. However, it is probably the simplest fact that provides the most convincing argument for the necessity of long periods of time for petroleum formation. That fact is that true petroleum is absent from recent sediments. There will be a minimum time after which, if Connan (1974) is correct, the formation of oil begins and from then on the longer the organic matter has to mature, the greater will be the changes within it. This simple picture will be complicated by the effects of depth of burial, pressure, and temperature and as the sediments containing the organic matter get older, the oil generation threshold will get shallower. For instance, after 10 million years the oil generation threshold may be in the region of 3,000 metres, but after 350 million years it may only be 1000 metres. This is illustrated in Figure 7.1 where the maturation of sedimentary organic matter of different ages, as measured by vitrinite reflectance, is plotted against depth. This substitution of time for temperature will be discussed later in this chapter.

Timing

It is quite probable that a lot, if not even the majority, of oil formed during the life of sediments has not been able to accumulate either because no migration has occurred due to the conditions of temperature, solubilizers, pressure (all

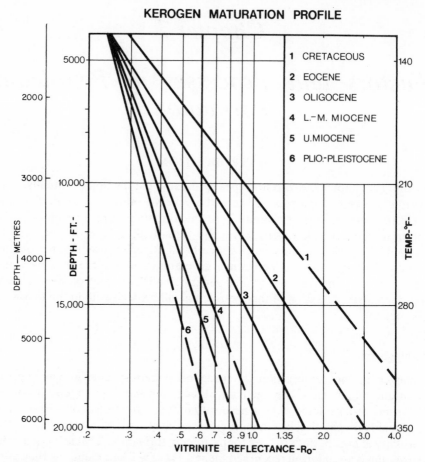

Figure 7.1 Kerogen maturatio profile (Dow, 1978. Reproduced by permission of The American Association of Petroleum Geologists)

factors to be discussed later) not being correct; or because a suitable reservoir rock was not available at the time that migration was occurring; or that the reservoir rock that was available during migration did not have an adequate seal or cap and the oil was unpreserved; or because at the time migration occurred, taking the oil through a suitable reservoir complete with an effective seal, the correct conditions required for the deposition of that oil were not available. Magara (1977) points out that the peak of oil generation must occur before the peak of oil accumulation, which itself precedes the peak of gas generation. Thus if, through studies, it is possible to predict the time at which maximum oil generation was occurring, it must be assumed that the peak of oil accumulation was later.

In situ formation

This is unlikely to occur for several reasons. Firstly, because the organic matter would have had to have been deposited within a reservoir rock and it has already been shown that the conditions for the deposition of clastic source beds and reservoir rocks are different. A source rock needs a low-energy reducing environment whereas a reservoir rock tends to be deposited under high-energy oxidizing conditions. The one exception to this is the deposition of organic-rich carbonates. Secondly, because a cap rock would have been needed to have been deposited fairly quickly, as would the occurrence of any deformations of those rocks to form traps from which the oil could not leak out. Thirdly, the lack of any organic residues could only be accounted for by the complete conversion of the deposited organic matter into mobile hydrocarbons. As will later be shewn, in other source rocks, even in the rare case of a kerogen reaching its maximum convertibility and producing as much fluid hydrocarbon as it can, there will always be the insoluble graphitic residue left.

Figure 7.2 provides a summary of the genesis of petroleum. In brief, the best situation is when organic substances originating from living matter are laid down with mineral matter during sedimentation in a reducing environment. This organic material is a mixture of biopolymers consisting of carbohydrates, proteins, lipids, etc., which are split up by bacterial action into biomonomers when they can lose their functional groups. At this stage, and with only very moderate burial, there are a few hydrocarbons mostly consisting of isoprenoid compounds and a limited range of normal alkanes which have an odd carbon number preference. It is soon after this that the biomonomers unite to form geopolymers, via such reactions as were discussed in the last chapter (reduction, deamination, decarboxyation, cyclization, disproportionation, etc.) These geopolymers are the precursors of kerogen which is formed by the uniting of the geopolymers. As the sediment is slowly buried, oil genesis proper begins, which is shown by the n-alkanes, where not only does the quantity of alkanes increase dramatically, but also the gaps in the range are filled in and the carbon preference index falls towards unity. Likewise, the oil becomes richer in aromatic and heterocyclic compounds, which are absent in recent sediments. At this stage migration will be starting to take place.

These processes continue with further burial and compaction with the result that the hydrocarbons get lighter, contain less nitrogen, oxygen, and sulphur, and there are equal amounts of odd- and even-numbered carbon chains, until eventually oil generation ceases and further burial causes the oil to be cracked into gas, mostly methane, which is the only final labile product at that stage. At the same time, the remains of the kerogen will have become more and more hydrogen-deficient until it eventually resembles graphite. The great generation of hydrocarbons swamps those few compounds that were originally present as shewn by the alkanes or the isoprenoid compounds.

The names given to these stages of the conversion of organic matter into petroleum are as follows. Diagenesis is the process up to and including the formation of kerogen; catagenesis follows diagenesis and describes the

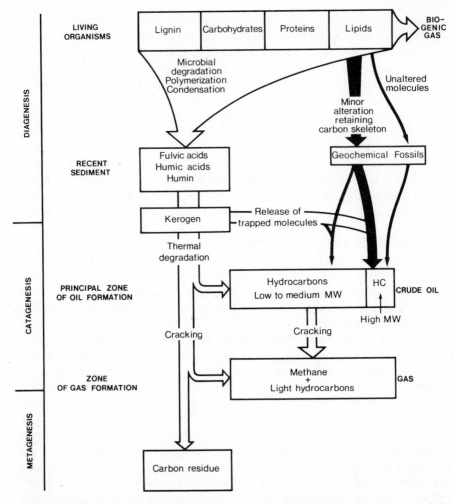

Figure 7.2 Kerogen formation (after Tissot & Welte, 1978. Reproduced by permission of Springer-Verlag, Heidelberg)

process up to, and including, the principal zone of oil generation; whilst metagenesis is the final breakdown of the kerogen and the cracking of the already generated material.

VARIATIONS IN PETROLEUM

It has already been noted that all petroleums contains the elements carbon, hydrogen, nitrogen, oxygen, and sulphur, together with some trace metals, but because of the infinite number of combinations in which these elements can be arranged into compounds and the varying amounts of each compound in different crude oils, no two crude oils are the same either in physical or chemical properties. The differences in the relative amounts of each compound can either be due to different genesis or to different secondary alteration. Different genesis will include different input material or a different subsequent history either in terms of temperature or catalysts, and alteration could include differences in migration, water washing, bacterial attack, deasphalting, temperature, etc. Different genesis will be the predominant factor although secondary alteration will play a not unimportant role, and in a sequence of closely related oil sands in one field separated by relatively impervious shales or water barriers, the systematic variations in oil density, viscosity, refractive index, colour, odour, wax content, and chemical composition suggest that local differences in either the source material or conditions of organic change are responsible.

In certain broad characteristics, oils from different areas of more or less the same age, and regardless of local conditions, may be indicative of differences in source material or conditions of origin. Consider, for example, various Cretaceous–Tertiary oils; those from Indonesia are high in aromatics; those from the Middle East have high sulphur; those from West Africa are, however, low in sulphur; Maracaibo gives oils with a high vanadium content; while the Po valley gives only gas. Brooks (1948) suggested that these varied oil characteristics reflect differences in the catalytic activity of the host rocks and minerals, and certainly such a hypothesis could explain the inconsistencies in the relationship of petroleum to age, depth, temperature, etc., just as well as the idea of different genesis.

Baker (1959) suggests that these selfsame differences are due to differences in the collodial solutions that transported the oil during its migration from source bed to eventual reservoir. Crude oils from the older saline waters will tend to be more paraffinic as compared with those from fresher waters which will tend to contain more cyclic compounds. Baker (1962), having studied the Cherokee group, states that each lithologic group has its own characteristic organic matter composition and therefore this organic matter composition is a property of the rock which will cause some rocks to develop more oil than others and will cause each to produce a different type of oil depending upon this 'rock property'. He further considered that migration would hardly affect the primary hydrocarbon distribution because only a small amount of the hydrocarbons are removed by the inefficient process of solution in water. He also stated that there exists a quantitative relationship between hydrocarbons and the organic matter in the Cherokee group. Barton (1934) shewed a decrease in density and an increase in the paraffin content with depth, and this he attributed to the effects of temperature and depth.

It has been noted that Tertiary and Jurassic crudes are in general lighter than Cretaceous oil, however the reverse of this, i.e. heavier oils with age and depth, is known in Venezuela, Burma, and Russia; but in these cases it could be in part due to the loss of gas or contact with water and the subsequent changes such a contact could involve (see Chapter 8).

Weekes (1958) finds that oil gets heavier as the chloride salinity decreases, the carbonate content increases within the associated connate waters, and that the heavier oils near the margins of the basin and the lighter oils at the centre could be due to one of several related causes. Firstly, the heavier oils with a greater number of heteroatoms which have been inherited from the original organic matter may well have been formed first and thus may have had to travel further to the edge of the basin which will be shallower, thus allowing differential secondary alteration. Secondly, the differences in the clays between the centre and the edges of the basin could affect the degree of catalysis involved in the alteration of the kerogen. Thirdly, there will be differences between the near shore and open sea sources of organic matter. Fourthly, bacterial action, upon which the formation of petroleum is dependent, and which can also be involved in its destination, may vary from area to area.

This decrease in oil density towards the centre of a basin has often been noticed and, in general, oils indigenous to a fresh or brackish water environment tend to be heavier while asphaltic oils are often associated with calcareous rocks and have higher sulphur contents than oils derived from clayey sources. There are, however, exceptions to this general rule.

OIL VERSUS GAS

The vast majority of oil has some gas associated with it either in solution and/or as free gas in the form of a gas cap. This gas will, in all probability, have the same source as the oil with which it is associated. Gas separate from oil can occur in three guises. The first is as marsh gas which is solely biogenic in origin and owes nothing to the thermogenic 'conventional' hydrocarbon-forming processes. Because methane is the only hydrocarbon which can be produced biogenically, such gas accumulations will contain almost nothing but methane. Thus any gas samples that do not contain the other low molecular weight alkanes, must be considered as being products of the diagenesis stage of organic matter alteration.

However, some biogenic gas deposits do contain very small quantities of higher molecular weight compounds which result from low-temperature thermogenic reactions and Snowdon & Powell (1982) have shewn that condensate range hydrocarbons can be produced from resinite at vitrinite reflectance levels of less than 0.6% and these can become associated with biogenic methane. In these cases isotropic studies will give indications whether the gas is biogenic or thermogenic.

The second form of the occurrence of gas alone will be as condensate and wet gas where the majority of the gas will be methane, but with some of the lighter gaseous hydrocarbons with significant amounts of the normally liquid hydrocarbons coexisting in the vapour state with the gaseous hydrocarbons. These normally liquid hydrocarbons can be extracted from the gas by cooling. The third form will be dry gas which contains a very high percentage of methane but almost always with small amounts of other gaseous hydrocarbons such as ethane, propane, and their unsaturated derivatives.

The question that has to be answered is how to account for these different occurrences of oil and/or gas. Is it due to differences in origin, differences in maturation process or level, or to differences in migration or accumulation?

Various theories of differential migration from a common source have been advanced and these involve differences in the porosity and permeability of the rock through which the fluid flows together with viscosity differences of the various components of the fluid. Leythaeuser et al. (1982) have measured the effective diffusion coefficients for the normal alkanes C_1 to C_{10}. They have shewn that gas diffusion can occur on a large enough scale to account for gas accumulations, that the process will not contribute significantly to the migration of the petroleum-range hydrocarbons, and that diffusion can explain accumulations with high gas-to-oil ratios and condensates in sediments of low maturity.

Movement through opening and closing faults has been suggested as explanations for the different occurrences. These faults would open so as only to transmit one part of the contents of a hydrocarbon pool depending upon where that fault occurs. For instance, if the fault occurs at the top the structure which contains a pool of reservoired hydrocarbons the gas, and possibly some of the oil, will escape, whereas if the fault is at the edge then perhaps the oil may depart to leave only the gas. While theories such as these can explain one or two individual cases, they cannot account for all situations, and in order to discuss the oil/gas problem any further it is necessary to carry on with the consideration of the formation of petroleum to see how the different factors involved in petroleum formation will effect the end-product.

The starting point for the generation of petroleum is the organic matter that is laid down in a sedimentary environment, often in a shallow deltaic situation, and this organic matter can be a mixture of marine and terrestrial material. Some of this organic matter will be directly converted to hydrocarbons while some of it will already be in the form in which it will finally occur in crude oils. However, the majority of the deposited organic matter will be converted to humic acids, a little of which may be altered to form hydrocarbons, but the bulk of which will first of all produce kerogen, which upon further maturation will give the vast majority of the oil and gas.

Cox (1946) gives the essentials for the conversion of organic matter to petroleum as: (a) a marine environment; (b) a maximum temperature of 100 °C; (c) a pressure between 2,000 and 5,000 psi; (d) sufficient time, since no oil has been formed since the Pliocene; (e) bacteria; and (f) catalysts. These will be dealt with in the following paragraphs.

DIRECT FORMATION OF HYDROCARBONS

The hydrocarbons that are generated directly from the lipid constituents of the original organic matter probably constitute a small percentage of the total hydrocarbons in crude oil. Sensitive analytical methods capable of detecting 2 parts in 10^9 have failed to shew any hydrocarbons in the range C_4–C_8 in recent sediments, and yet it is these compounds that are not only present in ancient sediments but form a major part of crude oils. It has been seen that the carbon preference index of those alkanes that are found in recent sediments differs considerably from that of alkanes in ancient sediments and crude oils

(Kvenvolden, 1966). While it is possible to produce all the low molecular weight hydrocarbons by the thermal decomposition of the unreacted organic products (lipids, etc.) left following the diagenesis stage, there is some doubt as to whether there would be sufficient lipid material available since the fatty acid lipid concentration of recent sediments seems far too low to account for all the hydrocarbons in petroleum.

Recent investigations on contemporary marine sediments have shewn that many marine sedimentary fatty acids are produced by bacteria (Perry et al. 1979). This could allow for a continuing production of fatty acids which in turn could yield paraffins. The role of bacteria in sediments has been investigated by Oppenheimer (1960). However, there is abundant evidence that bacteria do consume alkanes and other hydrocarbons (Button, 1976; Grant et al., 1943) and thus the overall role of bacteria in hydrocarbon generation is far from clear. It is possible that the deposited organic matter is quickly attacked by bacteria and then undergoes chemical reactions to become humic acids. In sea water it has been shewn that the concentration of saturated fatty acids in a vertical column falls slowly with depth whereas the concentration of unsaturated acids decreases rapidly, and this latter case is probably due to the dissolved oxygen in the water reacting with the multiple bonds to give peroxides which can then more readily cross-link to form kerogen. However, there are other chemical pathways by which the fatty acids can be removed from the environment.

Most low molecular weight alkanes and arenes can be produced by the decarboxylation and deamination of aminoacids and the decarboxylation of fatty acids. However, these reactions rarely produce compounds above C_5. Philippi (1977) has investigated the proteins as a possible source of petroleum hydrocarbons. He states that 5 of the 21 most common aminoacids in proteins contain alkyl side chains. He therefore heated egg white at 130–150 °C for 40 hours and analysis of the product showed various amounts of methane, ethane, ethylene, propane, isobutane, n-butane, isopentane, and toluene.

For the higher hydrocarbons the difference between recent sediments, ancient sediments, and crude oils is not only in the quantity, but the quality, particularly as expressed by the CPI. It has been noted that with increasing burial of the sediment not only are the missing hydrocarbons formed, but the total alkane abundance increases. Differences between the organic matter in recent sediments and that in ancient rocks and crude oils may be summarized as: (a) the light hydrocarbons, which constitute up to 50% of an average crude oil, are absent in recent sediments; (b) hydrocarbons are in general more abundant in ancient sediments than in recent ones of the same facies; (c) crudes from ancient sediments are often chemically similar regardless of age; (d) the odd carbon number preference of recent sediments disappears in ancient sediments and crude oils; (e) the enrichment of ^{13}C in kerogen and crude oils compared with living organisms (Bray & Evans, 1965). There are exceptions to the odd carbon preference (Dembicki et al., 1976), but these are normally the result of special circumstances. Alkanes, as generated, are not those found in crude oil, so it is necessary to consider some of the ways that these differences may be overcome. It is worth pointing out that some of these changes will add alkanes, some will remove them, whilst some of the changes will alter the primarily produced alkanes. Some of these processes will apply to other components as well as the alkanes (Koons, *et al.*, 1965; Kvenvolden, 1966).

It has been seen that the fatty acids can be the source of alkanes (Bray & Cooper, 1968), but that the natural fatty acids have carbon numbers 14, 16, 18, and 20 which upon their decarboxylation will yield alkanes with carbon chains of 13, 15, 17, and 19 atoms. Whilst this might not be unreasonable for recent sediments, it is most unhelpful for crude oils and therefore other processes must be involved. One such process would be production of new alkanes from the side chains of chlorophyll and similar compounds. Because many of these chains are isoprenoid derivatives and the normal side chains are of limited molecular weight range, additional processes would need to be involved even if these side chains played a part.

Philippi (1974) has suggested a free radical reaction mechanism for the decarboxylation of fatty acids which would explain not only the smooth alkane distribution found in crude oils, but also the higher molecular weight alkanes for which there are no corresponding fatty acids. In the first stage the fatty acid is split into two free radicals one of which is an acid and the other an alkane:

$$C_nH_{2n+1}COOH \rightarrow {}^*C_pH_{2p+1} + {}^*C_qH_{2q}COOH \tag{1}$$

where $n = p + q$

The free radical acid is then free to react with a hydrogen donor (H_2D) as in reaction (2a) below, which results in a new lower molecular weight fatty acid which can start the process again.

$$^*C_qH_{2q}COOH + H_2D \rightarrow C_qH_{2q+1}COOH + {}^*HD \tag{2a}$$

At the same time the alkane free radical may also react with a hydrogen donor to give a new alkane (reaction (2b)) which will be of lower molecular weight to the original fatty acid.

$$^*C_pH_{2p+1} + H_2D \rightarrow C_pH_{2p+2} + {}^*HD \tag{2b}$$

The alkane free radical can also react with a hydrogen acceptor (D), as in reaction (2c), which will result in the production of an olefin of lower molecular weight than the original fatty acid.

$$^*C_kH_{2k+1} + D \rightarrow C_kH_{2k} + {}^*HD \tag{2c}$$

If the point at which the original fatty acid splits varies, i.e. various values of p and q, a whole range of lower molecular weight alkanes with a smooth distribution can be produced from the original naturally occurring fatty acids by decarboxylation involving this free radical mechanism. Of equal importance is the ability of the free radical alkane to produce an olefin, because this olefin may then react with more free radical as in reaction (3a), to give a larger alkane free radical, which by reacting with a hydrogen donor will give an alkane (reaction (3b)).

$$*C_pH_{2p+1} + C_kH_{2k} \rightarrow *C_{p+k}H_{2(p+k)+1} \tag{3a}$$

$$*C_{p+k}H_{2(p+k)+1} + H_2D \rightarrow C_{p+k}H_{2(p+k+1)} + *HD \tag{3b}$$

Now while both p and k are going to be less than n, since they were originally formed from a fatty acid of chain length n, their sum may be greater than n; and thus a C_{20} fatty acid could yield a C_{19} alkane free radical (together with an acetic acid free radical). If that C_{19} radical were then to react as in reaction (2c) to form the olefin and that olefin subsequently reacted with more alkane free radical and then a hydrogen donor as in reactions (3a) and (3b), the longest alkane that could be produced would be $C_{38}H_{78}$. This alkane would have been produced from a C_{20} fatty acid. It so happens that the C_{38} alkane is about the longest alkane that is found in ancient sediments and thus these high molecular weight alkanes could have originated from the naturally occurring shorter fatty acids. Higher alkanes ($>C_{38}$) are normally only observed in terrestrially derived oils and sediments are derived from the plant waxes.

The only question is whether there would have been enough fatty acid available? Harrison (1976) has studied the thermally induced diagenesis of aliphatic acids and shewn that the following sequence of events takes place. Firstly, the free fatty acids and those that are very slightly bound to the kerogen are decarboxylated. Secondly, the fatty acids which are attached to the kerogen matrix, either by weak chemical bonds or by physical trapping, are released. Lastly, those acids which were liberated in the second stage are decarboxylated. It was however pointed out that the acids are liberated much faster than they can be decarboxylated, and thus the extra acids may migrate to the reservoirs and be degraded there. These extra acids may also play an important part in the transport of the primary produced oil from the source to the reservoir. Perry *et al.* (1979) also considered that bacteria could be a continuous source of fatty acids.

Johns & Shimoyama (1971) conducted experiments with the $C_{21}H_{43}$ COOH fatty acid by heating it in the presence of a montmorillonite for between 50 and 500 hours at temperatures varying between 200 and 250 °C. Analysis by gas liquid chromatography of the reaction mixtures subsequent to these experiments shewed that the C_{21} paraffin was predominant after 50 hours, but after 300 hours only compounds lower than C_{21} were found. This so far could be explained by catalytic cracking reactions. However, after 500 hours compounds up to C_{28} were detected. The normal paraffins that were detected in these experiments accounted for only 1% of the acid used and no other acids were detected. The most likely explanation of this is that the rest of the acid was converted to kerogen which at higher temperatures would become a hydrocarbon source. The CPI values for the produced paraffins started at 5.6 after 50 hours, and dropped to between 1.5 and 1.7 after 500 hours, i.e. they progressed in a similar manner to the CPI of alkanes in sediments.

Eisma & Jung (1964) also heated behenic acid ($C_{21}H_{43}COOH$) but this time at 200 °C in a sealed tube with bentonite for varying times. At the conclusion of the experiments the residues were extracted with n-pentane and analysed. Whilst the major product was the C_{21} alkane, there were some other alkanes, the majority of which were in the range C_{14}–C_{36}. The alkanes between C_3 and C_5 were noted to increase in quantity with time. When water was added to the reaction mixtures the most significant change in the product was an increase in the ratio of branched to normal alkanes.

These experiments shew, as Philippi (1974) postulated, that the fatty acids that occur naturally in nature, even though they are of limited molecular weight range, can be a source of a whole range of alkanes which, given sufficient time and temperature, will have a CPI similar to that of the alkanes in crude oils.

So far the reactions of fatty acids that could produce the hydrocarbon pattern of crude oils have been discussed. It is necessary to see if there are any other types of mechanism that could produce the same result. Waples & Tomheim (1978) have developed various mathematical models which they used to simulate both random and non-random thermal cracking of branched and normal hydrocarbons. The results of these studies indicate that for normal paraffins thermal cracking alone is not enough to give the present-day paraffin distribution. Even allowing for the fact that thermal cracking cannot produce the higher molecular weight compounds, the low molecular weight compound distribution is not correct. However, there is some evidence in the case of the isoprenoid compounds that non-random cracking could play some part in achieving the distributions seen in crude oils.

It has been shewn that in the early stages of the alteration of sedimentary organic matter, bacteria and microbes play an important part. Is there therefore any part they could play in the production of the whole range of hydrocarbons? The answer to this question seems to be in the negative if one is considering the primary production of liquid hydrocarbons. Methane is the only hydrocarbon that can be produced by microbial action; however, it will be later demonstrated that bacterial action can play a very important part in the alteration of already produced hydrocarbon mixtures.

The question of the effect of natural radiation on the sedimentary organic matter also needs to be answered. Leverson (1954) considered that although alpha-particles can produce hydrocarbons from organic matter and that while there are plenty of radioactive elements in sediments, it is unlikely that much of the oil was formed by such processes. The reasons that he advanced for this were, firstly, that these reactions would produce heavy oil and hydrogen, a combination which is not encountered, and secondly, in highly radioactive shales (e.g. the Cambrian in Sweden or the Devonian in the USA) there is not the great excess of oil that would be expected if radioactivity were important in this transformation. However, it is not inconceivable that the effect of radiation could be to alter already generated hydrocarbons.

Other compounds besides the n-paraffins are produced in the sediment by the breakdown of other molecules, often of a biological origin. Brooks *et al.*

(1969) considered that phytane is a post-depositional geochemical product as it does not occur in recent sediments but is found in oils and ancient sediments. Philp *et al.* (1978) have shewn how such isoprenoid hydrocarbons can be obtained by the thermal alteration of plant material. Other natural product species have been shewn to break down in the sediment to give hydrocarbons. β-Carotene has been shown to be capable of producing under simulated sedimentary conditions toluene, *m*-xylene, 2,6-dimethyl naphthalene, and 1,1,6-trimethyl, 1,2,3,4-tetrahydronaphthalene (Erdman, 1961; Day & Erdman, 1963). Draffen *et al.* (1971) have shewn how cholesterol, when heated with clays and shales, will yield various hydrocarbons.

ALTERATION AND MATURATION OF KEROGEN

It has so far been demonstrated that the full range of hydrocarbons that are found in ancient sediments and in crude oils cannot have come via the individual reactions mentioned above and that the bulk of these compounds must be derived from the breakdown of the kerogen. In the last chapter the first stage in the alteration of the deposited organic matter was discussed. This stage is called diagenesis and it occurs at shallow burial where temperature and pressure are low, with most of the alteration being due to microbial activity. The biopolymers are destroyed and their components are then incorporated into geopolymers which are the precursors of kerogen, and methane is the only hydrocarbon generated at this stage. If plant matter is the predominant component of the organic material then the final products will be peat and eventually brown coal (Brooks & Smith, 1967). Whatever the input at the end of this stage, there is a very limited amount of solvent-soluble extract which Snowden & Powell (1982) consider has been produced by the low temperature alteration of resinite. The reflectance of the polished vitrinite grains will be about 0.5%. The carbon dioxide which is generated during the diagenesis (and early catagenesis) stages may alter the acidity of the pore waters. An increase in the carbon dioxide contents of the pore water will allow calcite (calium carbonate) to dissolve

$$CaCO_3 + H_2O + CO_2 = Ca(HCO_3)_2$$

Also in the last chapter were discussed the different types of kerogen; how they depend very much upon the type of organic matter from which they are made and how their structure and composition vary. In this chapter the next stages of the evolution of kerogen will be described together with their products and the factors that can affect the further degradation of the kerogen.

Catagenesis is the name given to the next stage of kerogen alteration which involves burial down to several kilometres and temperatures in the range 50–150 °C. The products of this stage are liquid petroleum, wet gas, condensate, and gas, in that order, as depth increases. The relative amounts of each will depend upon the kerogen type. During the later stages of this phase significant amounts of methane are also generated due to thermal cracking.

The catagenesis stage contains the 'principal zone of oil formation' and once it is over no more liquid hydrocarbons will be generated. The vitrinite reflectance at the end of the catagenesis will be about 2%.

The final stage is metagenesis which is the last stage of organic maturation before metamorphism, and it occurs at great depth, or next to local hotspots. The labile product is mainly methane which is formed both from the cracking of already generated hydrocarbons and from the final breakdown of the kerogen. Some structural rearrangement of the kerogen occurs at this stage to leave carbon residues which are graphitic in nature. This stage of the transformation of organic matter is entirely in the dry gas zone and at it the vitrinite reflectance is about 4%.

The experiments that have been discussed so far have only dealt with changes in the alkane spectrum as the progression from recent to ancient sediments to oil takes place. There are, however, other groups of compounds in sediments and oil that vary, and while some of them can be accounted for by the direct reaction of the original organic matter, e.g. the porphyrins from chlorophyll, or the aromatics from β-carotene (Figures 6.9, 6.10, and 6.16), many of these non-alkyl compounds (and some alkanes) must be derived from the breakdown of kerogen.

Effect of kerogen type

It has been shown that kerogen is the fraction of sedimentary organic matter which is insoluble in organic solvents and has the potential to generate hydrocarbons. It has no universal formula and is generally considered to vary in type depending upon the conditions of its formation. The more coaly type of kerogen is derived from the nitrogeneous and humic residues of continental and marine plants and consists of high molecular weight molecules containing condensed aromatic rings with sulphur, nitrogen, and oxygen atoms in the ring. There are some methyl groups attached to the rings but very few long aliphatic chains. It will contain only 3–5% hydrogen and upon heating yields mostly gas, possibly plus a small quantity of liquid hydrocarbons. The oily type of kerogen is the amorphous remains of plankton and the high-lipid residues of spores and pollens. It contains many fewer condensed rings than the coaly type and has long aliphatic side chains and/or naphthenic rings. It contains 6–10% hydrogen and when heated will yield first liquid hydrocarbons and then gaseous ones. Besides these two types, there are also the algal kerogen which is a variation of the oily type but an even more prolific producer of hydrocarbons; and the graphitic type which contains less than 3% hydrogen, will yield neither oil nor gas, is often found in metamorphic rocks, and is virtually graphite.

Thus the type of hydrocarbon that is generated by the alteration of this sedimentary organic matter will depend upon the original organic debris from which that particular kerogen was produced. It has been established that the organic matter is laid down in aquatic systems with both marine and terrestrial material being involved. Quite often these situations are in deltas with river systems bringing in the terrestrial matter. Obviously the physical conditions of the estuary will determine how much land matter is taken how far out to sea, with the faster flowing river into the deeper estuary tending to deposit the terrestrial organic matter over a wider area (Hedges & Parker, 1976). Thus, under such a system the kerogen will be more uniform with a little land influence all over, as compared with the slow, shallow river system. In this latter case there will be a change from highly terrestrial kerogen near the river mouth to almost exclusively marine kerogen further out. Obviously in any system, however fast the river flow, there will be a limit to the distance the land material travels. Also, the extent of the influence will depend upon the relative overall amounts of marine and terrestrial organic matter as well as how their deposition is distributed (Hedges & Parker, 1976; Shultz & Calder, 1976).

Philippi (1974b) discusses the relative influence of marine and terrestrial organic matter, and he noted that land-derived matter is rich in the five-ring naphthenes and the higher alkanes (C_{30}, C_{32}, and C_{34}) whereas marine material is rich in C_{16} and C_{18} compounds. He therefore considered that for oils of terrestrial origin the ratio $(C_{21} + C_{22})/(C_{28} + C_{29})$ should lie between 1.6 and 1.2, whereas for oils of marine origin that ratio is in the range 1.5–3.0.

There is other evidence that oils derived from terrestrial sources can be recognized by the investigation of biologically produced hydrocarbons. These come from material derived from the terrestrial environment and contain waxes (normal or iso-alkanes of high molecular weight) which are often the waterproofing on the leaves of plants, whereas the marine alkanes are of lower molecular weight. Petroleums derived from source rocks deposited in non-marine marginal facies are frequently rich in waxes, indicating that these products survive and are incorporated in the oil (Hedberg, 1968).

Abell & Margolis (1982) have shewn that alkanes of different origins can be identified in sediments. The alkanes in the C_{15} to C_{19} range are derived from algae whilst those in the C_{21}–C_{35} range are from plant waxes. The latter have a high CPI value which differentiates them from alkanes in the C_{20}–C_{30} range which have a low CPI (approximately unity) and are the result of bacteria acting upon organic matter in soils.

Brooks & Smith (1967) agree with this theory and suggest that fragmentation could even up the carbon preference index, while Hedberg considered the wax content of oils to be related to their genesis or to the kind of organic matter from which they were derived. He noted that these oils were restricted to certain stratigraphic sequences and regions. These sequences had the following in common: (a) shale–sandstone lithology; (b) non-marine origin, or origin in water of lower than average salinity; (c) common association with coals or other highly carbonaceous strata; (d) a stratigraphic range Devonian to Pliocene; and (e) an association with low sulphur crudes. All of these point to the waxes being of terrestrial origin.

Thus the origin of the starting material needed for the generation of petroleum can well affect the amount and type of those compounds that are going to occur unaltered in the final petroleum. Thus the type and origin of the original starting material will almost certainly affect the compounds in petroleum which are secondary, i.e. those that are produced by simple reactions such as decarboxylation, as well as those which are produced via the intermediate stage of kerogen. It is these variations in kerogen which are produced from the different sources that must be now considered. Much of the evidence for the different products produced by the thermal alteration of various kerogens has come from laboratory experiments on immature kerogens. Thus it is fitting to consider these experiments and the results obtained from them before the breakdown of kerogens within their natural environment are discussed.

Laboratory experiments

Harwood (1977) has studied the oil and gas produced by the laboratory pyrolysis of kerogen where thermally immature kerogens were heated under air-free conditions at temperatures between 250 and 450 °C for several weeks, and significant amounts of bitumens, hydrocarbon gases, water, and carbon dioxide were formed. He noted that the immature kerogens, whose atomic H/C ratio was greater than unity, generated liquid hydrocarbons upon pyrolysis, whereas the lower the atomic H/C ratio of the immature kerogen, the more gaseous and fewer liquid products were obtained. The principal liquid hydrocarbon generation occurred when the carbon weight percentage of the kerogen was between 77 and 85% and the H/C ratio was greater than 0.8. Similarly, the principal gas generation occurred when the kerogen carbon weight percentage was between 85 and 89% and the atomic H/C ratio lay between 0.4 and 0.8, thus shewing that the principal gas generation occurs after the principal oil generation. When the carbon content of the kerogen reached 90% and the atomic H/C ratio was below 0.4, most kerogens had completed their hydrocarbon formation. He noted, however, that if the kerogen had a very high initial hydrogen content the oil generation would last until a carbon weight percentage of 87% was reached and the subsequent gas generation would continue until the carbon content was as high as 92%. However, in both these latter cases the atomic H/C ratio remained at the values previously given for the two generation zones. In other words, it is impossible to generate hydrocarbons if both ingredients, i.e. carbon and hydrogen, are not present, and a high carbon content is worthless unless there is the hydrogen to go with it to produce saturated hydrocarbons.

Harwood also showed that the kerogen derived from the lower plants and animals with high lipid and high hydrogen contents tend to be oil sources whereas kerogens from the remains of vascular plants have low lipid and low hydrogen content and are, in general, gas sources. He pointed out that even the richest oil-type kerogens will ultimately generate gas and the quantity of gas that these kerogens will generate can possibly be even greater than the gas which will be generated from kerogens which had initially lower hydrogen contents. This is because the convertibility of the kerogens varies and in general the higher the initial hydrogen content, especially as expressed in the atomic H/C ratio, the greater will be the convertibility of that kerogen and the greater will be its potential for generating all hydrocarbons, both liquid and gas. Shibaoka et al. (1978) have shewn how organic matter with a high exinite content yields oil whilst a high vitrinite content gives gas.

The gas that is generated from the oily type kerogens will come from two sources. Firstly, from the final stages of the conversion of the kerogen, and secondly, from the thermal cracking of the previously generated bitumens. Up to 30% by weight of the rock can be extractable bitumen if the kerogen is of the very high hydrogen type and these same kerogens can generate up to 13% gas and are often known as wet gas type kerogens.

Ishiwatari et al. (1978) has also studied the generation of hydrocarbons by the thermal alteration of kerogens from different sediments. In these studies they heated samples for 24 hours at temperatures between 150 and 410 °C and analysed both the liquid and solid products. They found that the greatest liquid product came from the mainly marine sources whereas the smallest liquid product came from the terrestrial source even though it may have had a higher initial carbon content.

Thus the products of kerogen maturation will depend upon the type of kerogen, which will in turn depend upon the organic matter from which that kerogen originated, but as Harwood pointed out the kerogen's subsequent history will also determine the final products of kerogen maturation.

Thus the term 'kerogen' is applied to a wide variety of mixtures of organic matter and the differences are due to variations in the source material, and in the environment of deposition and maturation. Kerogen is essentially a diagenetic alteration product of the original sedimentary organic matter and is characterized by a complex chemical composition. It is the main source of commercial hydrocarbon reserves and its structure is that of a macromolecule of condensed cyclic nuclei linked by heteroatomic bonds or aliphatic chains with varying amounts of side chain material which can vary in length.

Kerogen types

It is essential to determine the type of kerogen because the hydrocarbon-generating potential of any kerogen will be related to its chemical composition. In the past, kerogens have been divided into sapropelic and humic types according to the physical appearance of their constituents as seen under the microscope. The humic kerogen has structured and angular particles derived from land plants. If such material, especially if derived from the larger plants, were deposited in swampy, oxygenated conditions, the result of alteration would be peat and coal. Major components are the constituents of the plant cells, e.g. the celluloses, tannins and lignins (see Chapter 6). The sapropelic material is normally finely divided into amorphous particles and is derived from the lipid-rich components of such marine life as planktonic algae (Ishiwatari & Machihara, 1982) and terrestrial spores and pollens. The sapropelic material is thus mainly marine derived and in general oil-prone, as compared with the terrestrially originating gas-prone humic material. The pure algal amorphous sapropelic material can be considered as one extreme, with the pure humic material as the other. However, because a material is amorphous it cannot be assumed that it will produce oil: some amorphous kerogens have been shown to be gas-prone. Such amorphous kerogens are more dense and have some structure.

On the basis of microscopic studies the subdivisions of the coal macerals have been attributed to sedimentary kerogens which have similar optical properties. These macerals are firstly liptinite (alginite, sporinite, and exinite), secondly vitrinite (huminite), and thirdly inertinite. The first two groups are capable of generating petroleum but it is the liptinite kerogens that have the greater potential for liquid production. Of the components of liptinite, the alginite has the greatest generating capacity, but is the least common. Inertinite is composed of recycled organic matter, often oxidized, is unaffected by heat, and has no genetic potential (Figure 7.3).

Petroleum palynologists have classified kerogens as algal, amorphous, herbaceous, woody, and coaly. The algal obviously corresponds to the alginite, or Tissot's type I, while the amorphous is finely divided sapropelic matter. The herbaceous kerogen is composed of the recognizable terrestrial plant material such as spores, pollens, and cuticles. The amorphous and herbaceous types together more or less correspond to the sporinite and exinite macerals and to Tissot's type II. The woody kerogen is fibrous material, mostly terrestrial and equivalent to the vitrinite macerals or type III. The coaly kerogen is the dead carbon or exinite, sometimes called type IV (see Figure 7.4).

Tissot et al. (1974) have differentiated three types of kerogen according to their evolution paths and based upon their elemental composition. They have compared their classification with that proposed by Van Krevelen (1961) for coal macerals. They pointed out that the evolution paths of their kerogen types I, II, and III, as shewn upon an atomic H/C vs. O/C diagram, resembled the carbonization paths of alginite, exinite, and vitrinite respectively as reported by Van Krevelen (Figure 6.25).

Type I kerogen (alginite) is mainly composed of structured algal material and organic matter enriched by bacterial activity. It is rich in aliphatic chains but has few polyaromatic rings and heteroatomic bonds, and its initial hydrogen content is high, but its oxygen content is low. Upon pyrolysis between 550 and 600 °C it gives the largest yield of volatile hydrocarbons of any type of kerogen, with a maximum convertibility upwards of 80%.

Type II kerogen (sapropelite) is the kerogen most commonly found in petroleum source rocks and shales and is composed of autochthonous organic matter of marine origin which has been deposited in a reducing environment. It consists of sapropelic organic matter as well as some algal particles and planktonic remains together with some terrigeneous material. The hydrogen content (H/C) is high and the oxygen content, whilst higher than for type I, is still low. There are more polyaromatic nuclei and heteroatomic bonds than in type I, but fewer saturated chains. The yield of volatile hydrocarbons upon pyrolysis is lower than for type I, but this type of kerogen, which has a convertibility of up to 60%, forms the major source of commercial oil and gas occurrences.

Type III (huminite) is derived from the allochthonous terrestrial organic material containing higher plant and coaly material with a high proportion of condensed polyaromatic nuclei and heteroatomic bonds, especially oxygen. Aliphatic chains are much less in evidence, while the initial hydrogen content (H/C) is low and the oxygen content (O/C) is high. This type of kerogen is normally deposited in deltaic environments on continental margins where a high rate of sedimentation restricts the amount of bacterial activity that can occur and thus the organic matter is reasonably preserved. This type of kerogen is the main source of gas reserves but its potential for the generation of liquid hydrocarbons, as shewn by pyrolysis, is low with an overall maximum convertibility of about 35–40%. If this type of organic matter is subjected to less reducing conditions during the early stages of burial and diagenesis, the liquid hydrocarbon potential will be reduced to almost nil.

Type IV (inertinite) consists mostly of the reworked organic residues and coal particles from older sediments, or highly oxidized contemporaneous organic matter, and is considered as 'dead' carbon as it shews no convertibility when heated (Dow, 1978). Its initial hydrogen content is low whereas its oxygen content can be high. When viewed in transmitted light the material is black and opaque.

Two points should be stressed at this stage, the first of which is that these classification schemes are mutually complementary and although based on different analytical techniques are equally valid. Secondly, one does not get, except possibly in very special circumstances, pure kerogens of one type or another. In practice they are mixtures of two or three types, the most common combinations being I and II; II and III; II, III and IV. In addition to this the relative amounts of each type in a source bed will vary as the conditions of deposition varied. It has already been seen that the further from the palaeoshore line the sediment occurs, the smaller will be the terrestrial influence. However besides these lateral variations there will be vertical differences in kerogen type quite often due to the varying of the shore line at the times of deposition or the position of the oxic-anoxic boundary (Tyson et al., 1979).

While a direct comparison between the alginite and vitrinite macerals, the Tissot types I and III, and the palynologist's algal and woody organic matter can be made, such a comparison is more difficult in the case of type II kerogen, liptinite, and the herbaceous and amorphous organic matter. In the former cases all the components in both the macerals and the kerogens originate from the same sources, i.e. algal remains and terrestrial plant material respectively. As discussed in Chapter 6, these two kerogen types can be considered as the end members of an unknown number of kerogens. The type II kerogen of Tissot is an oil-prone marine kerogen with some land-derived material which reduces its convertibility. It does not directly compare with the herbaceous and amorphous kerogen as described by palynologists and it contains more than spores, pollen, exine, etc., as suggested by the macerals liptinite, sporinite, and exinite. It is one of many possible kerogens between the two extremes of open sea and terrestrially derived kerogens.

When kerogens undergo degradation, certain changes take place at different stages of the process. To all kerogens certain general things happen as they mature. These are a decrease in both the oxygen content and the hydrogen content with increasing maturation; all kerogens tend to become graphite-like; and the material, as seen in transmitted light, gets darker and less transparent, with the colour changing from pale yellow through yellow and shades of brown to black (Figures 7.5 and 7.6). The material looks more worn and damaged, with spores being squashed and split open (Figure 7.6) and other particles losing their sharp outlines (Figure 7.7); its reflectivity increases; its infrared spectrum changes with the aliphatic, carbonyl, methyl absorption bands disappearing with maturation while the aromatic and olefinic absorption bands increase. The degree to which these happen depends upon the degree of maturation and also on the kerogen type.

Diagenesis

Diagenesis is the first stage with a maximum vitrinite reflectance level of 0.5% at its conclusion. During this stage there is a slight decrease in the hydrogen content but there is a much larger and more important decrease in the oxygen content (O/C), mainly due to the elimination of carbonyl groups. This is shown well by infrared spectroscopy. Little fluid hydrocarbon is produced at this stage except biogenic methane (marsh gas) and some condensate range hydrocarbons; however, carbon dioxide, water, and hydrogen sulphide are all produced during this stage, which lasts down to depths of 600–800 metres. The principal result of the diagenesis stage is the production of kerogen.

Biogenic gas

Before considering the catagenesis and metagenesis of the kerogen, let us look in more detail at the one major hydrocarbon product of the diagenesis stage, which is biogenic gas. Rice & Claypool (1981) and Claypool et al. (1980) estimate that not only are more than 20% of the world's discovered gas reserves of biogenic origin, but that a higher percentage of gases of predominantly biogenic origin will be discovered in future. They are important targets for exploration because they occur in geologically predictable circumstances and are only widespread in large quantities at shallow depths. They also state that accumulations of biogenic gas have been discovered in Canada, Germany, Italy, Japan, Trinidad, the USA, and the USSR in Cretaceous and younger rocks, often at depths of less than 3,350 m, and that these accumulations occur in both marine and non-marine rocks. These

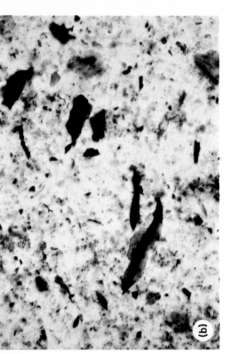

(b)

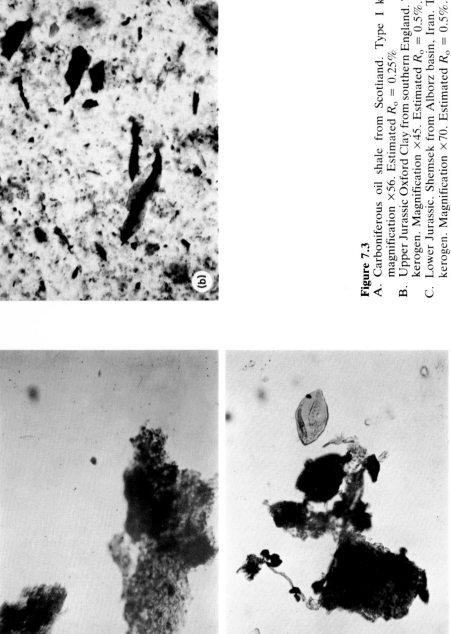

(a)

(c)

Figure 7.3

A. Carboniferous oil shale from Scotland. Type I kerogen magnification ×56. Estimated $R_o = 0.25\%$

B. Upper Jurassic Oxford Clay from southern England. Type II kerogen. Magnification ×45. Estimated $R_o = 0.5\%$.

C. Lower Jurassic. Shemsek from Alborz basin, Iran. Type III kerogen. Magnification ×70. Estimated $R_o = 0.5\%$.

A comparison of kerogen classifications — table:

	SAPROPELIC			HUMIC	
	Algal	Amorphous	Herbaceous	Woody	Coaly
KEROGEN (as viewed in transmitted light)	Algal	Amorphous	Herbaceous	Woody	Coaly
KEROGEN (as viewed in reflected light)	Liptinite		Sporinite Cutinite	Vitrinite Telinite Collinite	Inertinite Fusinite Micrinite
KEROGEN (chemical analysis)	I		II	III	IV
Organic source	Marine or lacustrine			Terrestrial	Recycled O.M.
Initial Chemical Composition C	77	83	82	83	89
H	10	11	8	5	3.5
O	9	6	10	12	7.5
Change in atomic ratios upon maturation H/C	1.7–0.3			1.0–0.3	0.45–0.3
O/C	0.1–0.02			0.4–0.02	0.3–0.02
Hydrocarbon products	Oil, oil shale, boghead and cannel coals		Oil and gas	Little oil, mostly gas humic coals	No hydrocarbon products
CH_4		high	intermediate		
C_2+C_{14}		high	low		
C_{15}^+		high	low		
Biological markers alkane CPI $C_{15}-C_{21}$		odd	none		
$C_{37}-C_{31}$		even or none	odd		
pristane/phytane pristane/n-C		low (<1) low	high (>3) high		
terpenoids $C_{12}-C_{18}$		high	low		
fatty acids $C_{24}-C_{36}$		low	high		

Figure 7.4 A comparison of kerogen classifications. (From *Petroleum Geochemistry and Geology* by John M. Hunt, W. H. Freeman and Company. Copyright © 1979)

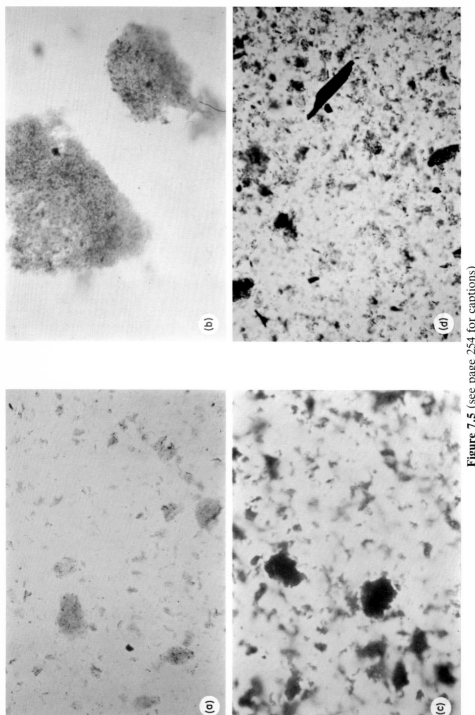

Figure 7.5 (see page 254 for captions)

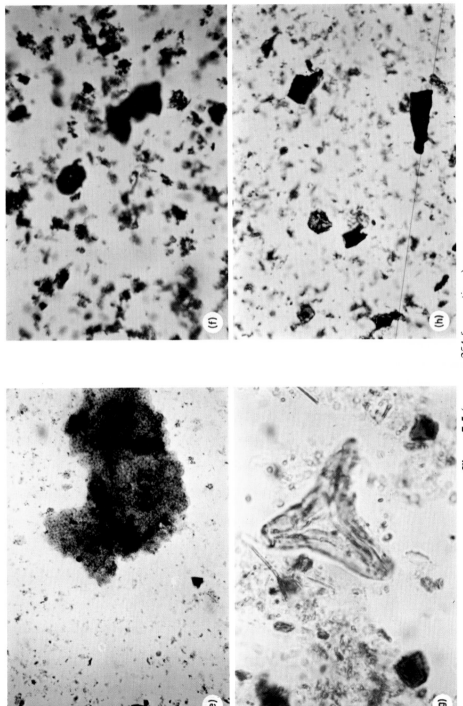

Figure 7.5 (see page 254 for captions)

Figure 7.5

A. Palaeocene Gurpi formation from the Zagros, Iran. Type I kerogen in the diagenesis stage. Magnification ×28. Estimated $R_o = 0.45\%$.

B. Albian Kazhdumi formation from the Zagros, Iran. Type I kerogen in the catagenesis stage. Magnification ×88. Estimated $R_o = 0.7\%$.

C. Lower Cretaceous Garan formation from the Zagros, Iran. Type I kerogen in the metagenesis stage. Magnification ×28. Estimated $R_o = 1.7\%$.

D. Upper Jurassic Kimmeridge Clay from southern England. Type II kerogen in the diagenesis stage. Magnification ×35. Estimated $R_o = 0.5\%$

E. Lower Jurassic Lower Lias from southern England. Type II kerogen in the catagenesis stage. Magnification ×112. Estimated $R_o = 0.55\%$.

F. Cretaceous Sembar formation from Pakistan. Type II kerogen in the metagenesis zone. Magnification ×56. Estimated $R_o = 2.0\%$.

G. Lower Jurassic Shemsek formation from the Alborz basin, Iran. Type III kerogen in the diagenetic stage. Magnification ×112. Estimated $R_o = 0.4\%$.

H. Carboniferous coal measures from South Wales. Type III kerogen in the catagenesis stage. Magnification ×35. Estimated $R_o = 1.2\%$.

I. Carboniferous coal measures from South Wales. Type III kerogen in the metagenesis stage. Magnification ×70. Estimated $R_o = 3.0\%$.

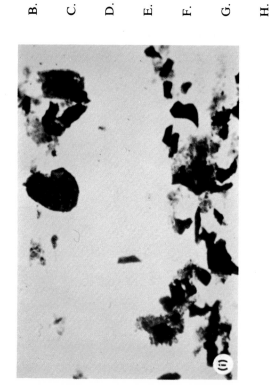
(i)

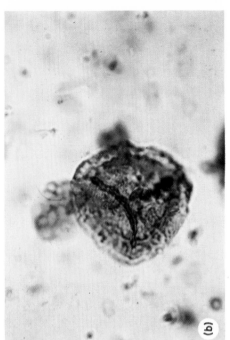

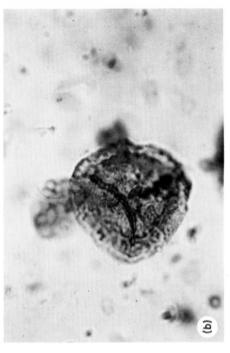

(b)

(a)

(c)

Figure 7.6

A. Albian Gault Clay from southern England. Spores in the diagenesis stage. Magnification ×448. Estiomated R_o = 0.4%.

B. Carboniferous coal measures from South Wales. Spores in the catagenesis stage. Magnification ×280. Estimated R_o = 1.2%.

C. Carboniferous coal measures from South Wales. Spores in the metagenesis stage. Magnification ×280. Estimated R_o = 2.0%.

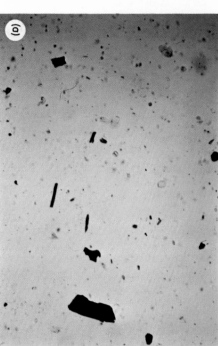

(a)

(c)

(b)

Figure 7.7
A. Upper Jurassic oolitic limestone from Yorkshire. Vitrinite grains in the diagenetic stage. Magnification ×56. Estimated $R_o = 0.5\%$.
B. Carboniferous coal measures from South Wales. Vitrinites in the catagenesis stage. Magnification ×35. Estimated $R_o = 1.2\%$.
C. Carboniferous coal measures from South Wales. Vitrinites in the metagenesis stage. Magnification ×56. Estimated $R_o = 3.0\%$.

occurrences are thus an important contribution to the world's reserves of fossil fuels.

Biogenic gas is predominantly methane, but there will be small amounts of heavier hydrocarbons which can be attributed to minor thermogenic products of the low-temperature degradation of the organic matter which Snowdon & Powell (1982) consider are derived from resinites. They suggest that the proportions of resinite, liptinite and vitrinite in the organic matter of terrestrial source rocks controls the ultimate product (i.e. condensate, oil or gas respectively) and the level of thermal alteration required to generate hydrocarbons. Increasing levels of maturation are required for the macerals (in the order listed above) to generate. The amounts of the hydrocarbon components other than methane are generally proportional to the temperature, age, and organic matter constituents of the sediment, but are still very minor.

The special conditions that need to be met for the generation of biogenic methane are firstly, strictly anaerobic conditions, because the methane-producing bacteria cannot tolerate even traces of oxygen, and secondly, an absence of dissolved sulphate. Typical environments where these conditions are met are dung heaps, anaerobic sewage digesters (where the methane is used as a fuel source), swamps, bogs, and marshes (where the methane can often be seen bubbling to the surface).

In the last chapter the conditions of oxicity needed to preserve deposited organic matter were discussed and the most prolific and common hydrocarbon-generating situation is where there is an aerobic water column often with a small aerobic layer of sediments all of which is underlain by anaerobic sediments. The anaerobic sediment can fall into two parts. On top is the sulphate-reducing zone where there is a high concentration of sulphate ions and sulphate-reducing bacteria predominate, producing hydrogen sulphide as an end product of sulphate reduction. Thus its production is evidenced in the bogs and marshes where the foul smell of H_2S is often recognized. Below the sulphate-reducing zone is a carbonate-reducing zone in which carbon dioxide reduction becomes the dominant process of anaerobic respiration, and this results in the formation of methane. The carbon dioxide, which is to be reduced, can either be the product of metabolic activity by bacteria or the thermal decarboxylation of organic matter.

The two conditions mentioned so far, that is the need for strictly anaerobic conditions and the elimination of sulphate from the waters, are special conditions for this biogenic methane generation. Other conditions that are required include the correct temperature. However, this condition is slightly different to that normally expected for hydrocarbon generation. For normal hydrocarbon generation a sufficiently high temperature is required; however, in biogenic methane generation a sufficiently low temperature is required, and it is normally in the range of 0–75 °C (e.g. *methanobacterium thermoautotrophicus* has an optimum temperature range of 60–70 °C.

Other conditions, such as the amount of organic matter required, are similar to the conditions required for the generation of hydrocarbons through the breakdown of kerogen. For instance, the reservoir with its seal must be present prior to the release of the biogenically generated gas from solution. Thus there is a certain minimum depth to which the sediment must be taken for these conditions to have been met. The other way in which biogenic methane may be trapped is by the formation of early diagenetic carbonates cements, either as layers or concretions forming effective traps. Durand-Souron et al. (1982) have shewn that methane is the last product of the thermal breakdown of kerogen and thus thermal alteration cannot explain the formation of biogenic methane.

It is only fair to point out that some geologists and geochemists believe that most of the gas deposits which have been described as biogenic in origin are actually of thermogenic origin. The main reasons for these views involve the conditions of generation and migration. Although the bacteria involved can survive and produce methane at significant depths and temperatures, the number of bacteria per gram of sediment falls quickly with depth. In addition to this, the pore diameters of even poorly compacted shales are too small to accommodate the required bacteria. Thus the production of biogenic gas is unlikely to occur in consolidated sediments and it is much more likely to be favoured in the shallow, unconsolidated sediments where the bacteria are abundant and the pore spaces are large. Certainly gas is produced in such circumstances, as evidenced by gas bubbling to the surface in marshes and swamps. However, such evidence of the possibility of biogenic gas generation immediately raises a problem in regard to the accumulation of such gas, and that is the availability of suitable reservoirs. Under the conditions at which biogenic gas generation is most favoured, i.e. in shallow, unconsolidated sediments, there are unlikely to be suitable reservoirs available. The only possibility for such shallow generation to source large accumulations is if the gas is held in the sediment until the sediment is more deeply buried and a suitable reservoir is available.

As an alternative it has been suggested that these shallow gas deposits are the result of the accumulation of gas which has been produced thermogenically at great depths and has diffused through the cap rocks of the accumulations whence it came. Because the rates of diffusion of gases are proportional to their molecular weights, the light gases will diffuse fastest. This would explain not only the virtual absence of higher paraffins in such shallow gas deposits but also the fact that such gases are composed of methane which is isotopically light. The fractionation due to diffusion depends upon the square root of the ratio of the molecular weights of the species involved. For $^{13}CH_4$ and $^{12}CH_4$ this fractionation coefficient is 1.031 and thus periods of time on a geological scale would be required to cause such fractionation. Leythaeuser et al. (1982) have shown the feasibility of the required quantities of methane diffusing over even small intervals on the geologic time scale, i.e. half a million years. If such a theory is true and these shallow deposits of isotopically light, almost pure methane have diffused up from great depths, then there could be great volumes of oil and gas awaiting discovery below these gas reservoirs.

Catagenesis

It is during the catagenesis stage, which is the second step of kerogen degradation, that the majority of hydrocarbons are formed. The first part of this stage is the principal zone of oil formation where, as temperature rises, more bonds are broken. Firstly, the ester bonds and then the carbon–carbon bonds are broken, giving rise to more aliphatic molecules being liberated. This stage is marked by an important decrease in the hydrogen content, with the H/C ratio falling. There is a progressive reduction in the aliphatic bonds as shown in the infrared spectrum, matched by an increase in aromatic character due to aromatization of naphthenic rings as hydrogen is used to saturate labile hydrocarbon molecules. At the same time, the rest of the carbonyl groups are eliminated and thus the oxygen content will decrease. The vitrinite reflectance increases from 0.5% at the start of this stage, to 2% at its conclusion.

During this stage some biogenically produced molecules which were loosely associated with the kerogen break away and these molecules are often within the C_{15}–C_{30} range. However, most of the hydrocarbon products of this stage are in the low molecular weight range, being both gas and liquid. The second phase of the catagenesis stage is the production of wet gas and condensate, which occurs as the organic matter is buried to even greater depths and even more carbon–carbon bonds are broken, affecting not only the remaining kerogen, but also the already generated hydrocarbons with the result that gases are the eventual dominant products.

The alkanes produced during this late catagenesis stage have a lower odd/even preference and therefore the CPI of the mobile hydrocarbons falls due to dilution because the quantity of newly produced alkanes is vastly larger than the original hydrocarbons. The molecular weight range of the hydrocarbons falls because those alkanes produced from the kerogen have lower molecular weights than those incorporated in the sediments during diagenesis.

The isoprenoid compounds and other geochemical fossils become a smaller proportion of the extractable organic matter because no more such geochemical fossil molecules are formed during this stage and their numbers can only increase due to the liberation of the few molecules that were physically trapped in the kerogen, whereas the supply of normal alkanes increases greatly due to their production from the kerogen. Thus the geochemical fossils are diluted and the ratio of isoprenoid to normal alkanes decreases with increasing maturity. Likewise during the principal stage of oil formation the steroid concentration falls, either by dilution or by degradation, and the ratio of aromatics to total organic carbon rises with depth but at a slower rate than the corresponding saturated hydrocarbon to total organic carbon ratio, and thus the aromatic to saturates ratio will also fall with increasing maturation. The small increase in the number of aromatic molecules can be accounted for by the thermal degradation of steroid compounds, as well as compounds being thermally cracked off the kerogen nucleus. These reasons for aromatic hydrocarbon production explain why the higher homologues are more common than the parent compound, i.e. most crudes contain more toluene, ethyl benzene, xylene, etc., than benzene.

During the catagenesis stage the atomic H/C rates of the kerogens decrease due to the production of hydrocarbons and the kerogen becomes more aromatic, but the actual changes in H/C ratio will depend upon the kerogen type (Figure 6.25). The end of this stage is often at depths of 3000 m or more.

Metagenesis

The last stage is metagenesis or the dry gas zone where the kerogen is restructured and no hydrocarbons are formed from the kerogen, with the notable exception of methane. The remainder of the liquid hydrocarbons are also cracked to methane and the H/C ratios of the kerogens decrease very slowly (Figure 7.2) with carbon content rising to around 90% and the vitrinite reflectance reaching 4%.

The stacks of the aromatic layers in the kerogen, which had a random distribution, now tend to get a preferential orientation and become parallel and closer, becoming more akin to the structure of graphite. This is shewn in the infrared spectrum by the aromatic bands becoming more intense while the aliphatic absorption bands have almost totally disappeared.

During these changes the hydrocarbons that are generated have isotopically lighter carbon than the kerogen (1–4°/oo). The methane that is the final product is even lighter isotopically and therefore the residual graphite will be even heavier.

All these changes in products, in chemical composition, in spectroscopic character, in reflectance, in colour and shape as viewed under the microscope have all been confirmed by experimental artificial maturation of immature kerogens.

Summary of kerogen alteration

Figure 7.2 shows how Tissot (1974) illustrates his three kerogen types as identified by elemental analysis and the principal products of their alteration. Figure 7.8 shews the changes in structure of an oily kerogen as it matures. The breakdown of the kerogen is essentially a series of disproportionation reactions in which, from one source, two sets of products are formed. Of these two sets of products, one is hydrogen rich and one is hydrogen deficient with respect to carbon. This is well illustrated in Figure 7.8 where the breakdown of an oily-type kerogen is shown. As the hydrocarbons (fluid products) are generated they are, for the most part, saturated or nearly saturated with hydrogen, which has to be supplied from the rest of the source molecule. Hence the solid product of kerogen breakdown becomes leaner and leaner in

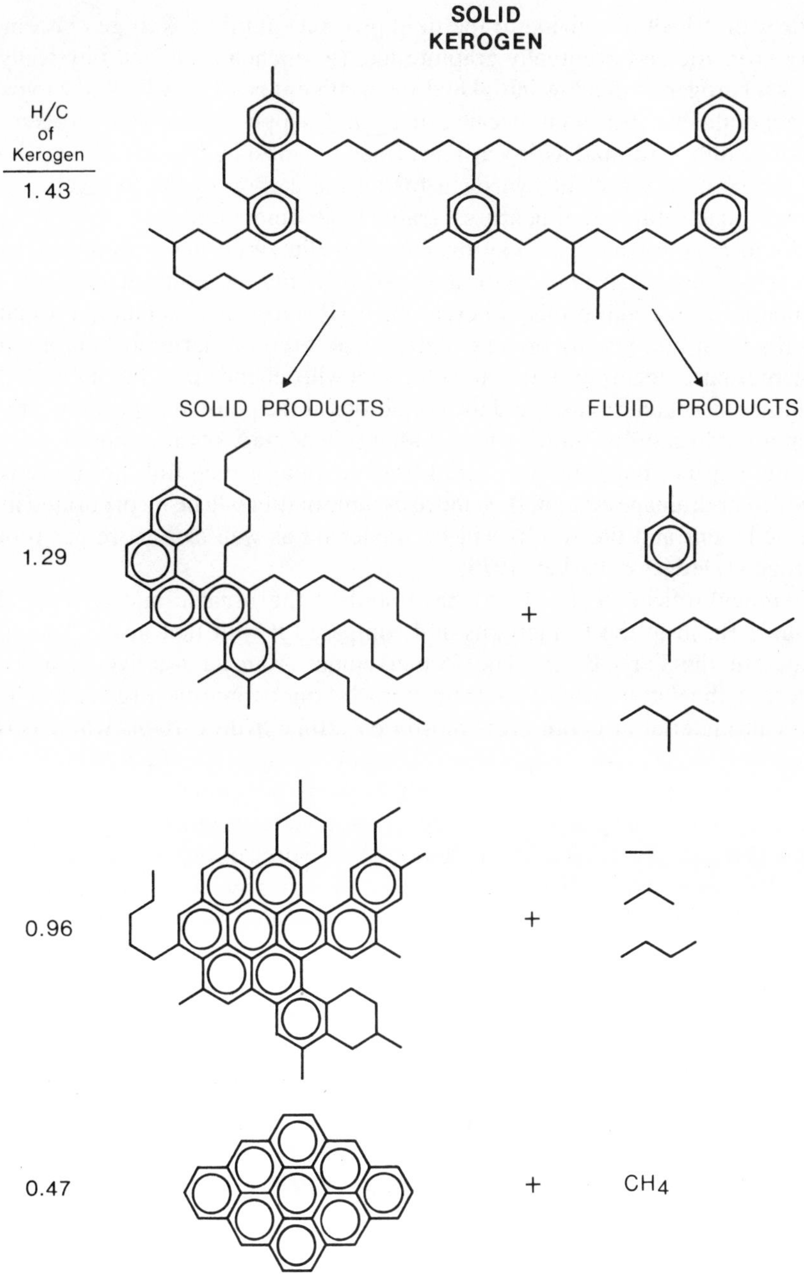

261

Figure 7.8 Structural changes in kerogen consequent to maturation. (From *Petroleum Geochemistry and Geology* by John M. Hunt, W. H. Freeman and Company. Copyright © 1979)

hydrogen, which it is giving to the fluid products, until the kerogen becomes very aromatic and eventually graphite-like (both chemically and physically). Thus a kerogen with a low initial hydrogen content has a very limited capacity to generate hydrocarbons because it cannot supply sufficient hydrogen to produce fully saturated hydrocarbons. Even the most hydrogen-rich kerogens do not contain sufficient hydrogen for all the carbon of the kerogen to be converted to saturated or nearly saturated labile compounds.

We have noted that type II kerogen (sapropelite) is normally produced from a marine source and upon maturation yields either paraffinic–naphthenic or aromatic–intermediate oils, whereas type III kerogen (huminite) normally results from non-marine organic matter. The ratio of marine to non-marine organic matter incorporated in any sediment will depend upon the runoff of the continental organic matter and the conditions at the point of deposition. If the continental runoff is limited the resulting sedimentary organic matter will be mainly marine, but if the terrestrial organic input is plentiful, i.e. there is a large river drainage system, then more organic matter will be incorporated into the sediment and the results will be waxier oil as well as a more gas-prone kerogen (Hedges & Parker, 1976).

Vandenbrouche *et al.* (1976) have studied the organic matter from the Douala Basin in the Cameroons and compared it with that in the Toarcian shales of the Paris Basin. The former comes from land-derived material, whereas the latter originates from autochthonous marine organisms. The African material at depth gives mostly paraffinic hydrocarbons whereas the Toarcian shale gives naphtheno–aromatic compounds first. Only with increasing depth are paraffinic hydrocarbons generated, and in greater quantity than from the land-derived organic material. In fact the Toarcian shale generates twice as much medium-range hydrocarbons as compared with the African sediment. The two kerogens are different, with the alginite/exinite making the Toarcian kerogens highly saturated with a high H/C ratio while the vitrinite give the Douala kerogen richness in aromatics with few aliphatic side chains that eventually give a low yield of paraffins.

It should be remembered the maturation of the generated oil, which will be discussed later, will also be occurring at the same time as the maturation of the kerogen is taking place. This alteration of the already generated material is also going to be, in part at least, heat dependent.

OTHER NON-HYDROCARBON GASES

The origins of the hydrocarbon gases have already been discussed and it has been shewn that these occur either as biogenic gas early in the organic alteration process or as the last stage of kerogen maturation and by the cracking of heavier hydrocarbons which have already been generated. Most accumulations of hydrocarbon gases also contain a variety of inorganic gases. These other gases include helium (plus minor amounts of other inert gases), nitrogen, hydrogen sulphide, and carbon dioxide.

Helium is the second lightest element, containing two protons and two neutrons in its nucleus (cf. hydrogen which has a single proton). Helium is chemically inert, colourless, and odourless. Some natural gases contain up to 8% of helium. The other inert gases are neon, argon, krypton, xenon, and radon, of which neon and radon have been found in small amounts in a few natural gases.

The higher atomic weight elements have more neutrons than protons in their nuclei and this leads to instability and radioactive decay. With atomic numbers above 84 there are no stable nuclei, but the rate of decay can vary. The rate of decay is expressed as half-life of the nucleus and is the time after which only half the original amount of that element remains. The naturally occurring unstable elements have long half-lives but some of the elements produced by radioactive decay have short half-lives.

One method by which unstable nuclei decay is alpha emission. This is the ejection of a helium nucleus (i.e. 2 protons and 2 neutrons) from the nucleus of the decaying atom. Thus alpha emission is a source of helium. Often there are chains of decay whereby one element changes into another with the emission of a helium nucleus. One such chain is:

$$_{92}U^{234} \xrightarrow{\alpha} {}_{90}Th^{230} \xrightarrow{\alpha} {}_{88}Ra^{226} \xrightarrow{\alpha} {}_{86}Rn^{222} \xrightarrow{\alpha} {}_{84}Po^{218} \xrightarrow{\alpha} {}_{82}Pb^{214}$$

(The subscript prefix denotes the atomic number or number of protons and the superscript suffix denotes the atomic weight or number of protons and neutrons. α denotes an alpha particle or helium nucleus.)

The rates of these reactions may vary. The reaction

$$_{92}U^{234} \rightarrow {}_{90}Th^{230} + {}_2He^4$$

has a half-life of 4.5×10^9 years whereas the reaction

$$_{90}Th^{230} \rightarrow {}_{88}Ra^{226} + {}_2He^4$$

has a half-life of only 24.1 days. ($_2He^4$ is a helium nucleus.) The other inert gases are accounted for by the radioactive decay of other elements. Radon is produced by the radioactive decay of radium, with an equal amount of helium formed at the same time:

$$_{88}Ra^{226} \rightarrow {}_{86}Rn^{222} + {}_2He^4$$

Thus the most likely origin for helium in natural gas is from the radioactive decay of the transuranic minerals in deep-seated basement rocks such as granite. Wells (1929) has shown that silica-rich igneous rocks at moderate temperatures (200–500 °C) have much higher permeabilities to helium than to other gases, which would allow the helium to rise to the reservoir beds.

Nitrogen is a colourless, odourless gas which is the principal component of air. It is chemically fairly inert although in high-temperature flames it will be oxidized to produce poisonous oxides of nitrogen. The source of nitrogen in natural gas is believed to be threefold. A large amount of air will be trapped in sediment either as the gas or dissolved in water (solubility of nitrogen in water is 2.3 ml per 100 ml at 0 °C). A second source of elemental nitrogen is from nitrogen-containing organic compounds, possibly by bacterial action. A third possible source of nitrogen has been suggested because of the relationship of nitrogen and helium in many natural gases. If the helium is derived from basement igneous rocks then the nitrogen could have the same source.

Carbon dioxide is a colourless, odourless gas which is poisonous at concentrations greater than 8%. Interestingly, carbon dioxide has a great physiological effect on the lungs of animals and humans. It is the lungs' tendency to remove carbon dioxide, rather than to inhale oxygen, which causes our lungs to operate. Thus people in restricted places with no changing of the air tend to pant because of the high concentration of carbon dioxide rather than the low concentration of oxygen. If the carbon dioxide is removed but no oxygen added, one would peacefully fall asleep in such conditions.

Some natural gases can contain up to 40% of carbon dioxide and it is believed to have both an organic and inorganic origin. Carbon dioxide is a product of the diagenetic stage of the alteration of organic matter (Figure 1.13). Fatty acids can be decarboxylated to give alkanes,

$$RCOOH \rightarrow RH + CO_2$$

and many young immature oils contain such fatty acids. Subsequent maturation of these and other carbonyl-containing compounds would add carbon dioxide to the oil and gas. Bacterial degradation of organic matter will also produce carbon dioxide.

$$2C_nH_{2n+2} + (3n + 1)O_2 \rightarrow 2nCO_2 + (2n + 2)H_2O$$

Fermentation is also a source of carbon dioxide. The fermentation which is best known is conversion of sugars to alcohol and carbon dioxide, as in reaction (A) below:

$$C_6H_{12}O_6 \xrightarrow[\text{enzyme}]{\text{yeast}} 2C_2H_5OH + 2CO_2 \tag{A}$$

but in intestines it is known that bacteria can convert sugars to carbon dioxide and methane:

$$C_6H_{12}O_6 \rightarrow 3CO_2 + 3CH_4 \tag{B}$$

There is no reason to suppose that sedimentary bacteria are incapable of the same process.

The inorganic sources of carbon dioxide are the carbonate sediments. The action of igneous intrusions (C) or acid waters will liberate carbon dioxide (D):

$$CaCO_3 \xrightarrow[\text{550 °C}]{\text{heat}} CaO + CO_2 \tag{C}$$

$$CaCO_3 + 2HCl \rightarrow CaCl_2 + H_2O + CO_2 \tag{D}$$

Hydrogen sulphide is a colourless gas with unmistakeable odour of bad eggs. It is extremely poisonous with a fatal concentration of 1 part in 1000. It is often present, at very low concentrations, in spa and mineral waters. Besides being poisonous, hydrogen sulphide is also corrosive and in the warm will react with metals to give metallic sulphides and hydrogen, e.g.:

$$M + H_2S \rightarrow MS + H_2$$

where M denotes a metal. Thus this gas is generally unwelcome as it needs removal to prevent corrosion and to render the gas and oil suitable for public combustion. If there are high concentrations of hydrogen sulphide its extraction and the resale of the sulphur may be financially adavantageous. As with carbon dioxide, there can be both organic and inorganic sources of hydrogen sulphide.

Hydrogen sulphide is produced by bacterial reduction of proteins and their hydrolysis products. A by-product of this reaction is nitrogen. An overall equation for such a reaction might be:

$$CH_2N_2S \xrightarrow[\text{reduction}(+2H_2)]{\text{bacterial}} CH_4 + N_2 + H_2S$$

Metal sulphates may be converted into carbonates with the release of hydrogen sulphide. At low temperatures in the oozes at the bottom of lakes, bacteria are known to take part in the conversion:

$$C_6H_{12}O_6 + 3MSO_4 \rightarrow 3MCO_3 + 3CO_2 + 3H_2S + 3H_2O$$

It is worth noting that natural gas containing hydrogen sulphide often occurs where large amounts of evaporites are present. Thus the conversion of anhydrite to calcite may take place deeper in the sediment with the production of hydrogen sulphide provided sufficient organic matter is present.

EFFECTS OF DEPTH, TEMPERATURE, AND PRESSURE UPON ORGANIC MATTER MATURATION

Depth

It has often been suggested that a minimum depth of burial is required for hydrocarbon generation; however, for the overburden pressures which are needed to expel the generated fluid, great depths of burial are not necessarily required. Teas & Miller (1933) suggest that oil at Racoon Bend occurred at only 450 ft of burial, while Kidwell & Hunt (1958) found petroleum-like substances in a sand lens at 110 ft. Weber & Maximov (1976) have reported the occurrence of up to 1% gaseous hydrocarbons (C_4–C_7) in recent sediments and there have been Russian reports of commercial gas fields which are at shallow depths; they are believed to consist of hydrocarbons that have formed during the early stages of diagenesis and these cases will involve biogenic gas (Rice & Claypool, (1981).

However, it is generally accepted that as the depth of burial increases the oil matures and increases in quality up to a certain point, after which cracking of the generated hydrocarbons to gas occurs. The only exception to this rule of greater burial meaning greater oil quality is when shallow organic matter is matured by the action of igneous intrusions.

Sikander & Pittion (1978) used reflectance measurements to shew that increases in organic metamorphism are related to the thickness of the sequence, except where there has been igneous activity. Albrecht & Ourrison (1969) have shewn how the n- and iso-alkanes from different levels in a thick sedimentary series change as the result of evolution by diagenesis after burial, since environmental conditions and the original organic matter are the same along the series, and these changes are depth related. Perregaard & Schiener (1979) have shewn how the organic matter in sediments is altered by igneous intrusions. This thermal alteration was recognizable up to one-half of a dyke width from the intrusion. There is the possibility that this 'depth rule' could be a secondary alteration effect with hydrocarbon formation occurring in the early stages of diagenesis, and then the temperature and pressure would play little part in hydrocarbon generation, but only in any subsequent alteration. It has been demonstrated that changes in kerogen and the production of hydrocarbons occur with increasing thermal maturation. This and other work still to be described, indicate that heat will play a part in generation as well as alteration.

Albrecht et al. (1976) examined the materials from cores obtained from depths between 700 and 4000 metres with special reference to organic material contained within them (Figure 7.9). Down to depths of 1200 m the soluble material was a low percentage of the total organic matter, and equivalent to that found in recent sediments. However, humic acids, which were extracted with sodium hydroxide, were present in large quantities. Between 1,500 and 2,200 m there was a large increase in both the total solvent-soluble extractable compounds and also in the saturated and aromatic compounds within that extract. Further burial brought about a decrease in the quantity of extractable hydrocarbons (Figure 7.9(a)). More detailed examination shewed that the generation of the saturated compounds actually increased at a much greater rate than the aromatics, but both eventually were destroyed (Figure 7.9(b)). This increase in soluble organic matter was considered to be due to the loss of functional groups from the kerogen as it broke down and altered under the influence of depth and pressure. Albrecht et al. shew that the preference for odd-numbered carbon atom chains decreased with depth, and that after 2,200 m cracking of the generated hydrocarbons began and the amount of solvent-soluble material markedly decreased.

As the depth and temperature from which the samples were obtained increased, the ratio of C_{15+} solubles to the insoluble kerogen first increased and then subsequently fell. The increase reflected the generation of C_{15+} molecules from the kerogen and the decrease was due to the thermal cracking

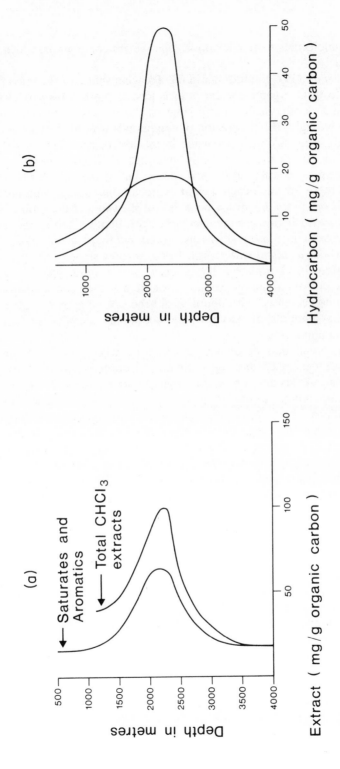

Figure 7.9 Effect of depth upon hydrocarbon generation (after Albrecht *et al.*, 1976. Reproduced by permission of Pergamon Press Ltd)

of these molecules to give low molecular weight compounds, which were not analysed.

Tissot *et al.* (1971) noted that in the Toarcian shales of the Paris basin, the extract to total organic carbon ratio is low at depths down to 1500 m, but below that depth the ratio rises due to the effects of thermal alteration. Correspondingly, the proportion of compounds with high molecular weights (resins or asphalts) increases while in the hydrocarbons the light members increase at a faster rate than the heavier ones. Hydrocarbons lighter than C_{15} were shewn to be scarce at shallow depths, but by the time 2,500 m had been reached these low molecular weight hydrocarbons were accounting for up to one-third of the total hydrocarbons. Tissot shewed that the ratio of saturated to unsaturated hydrocarbons decreased with depth whereas the amount of carbon dioxide increased. It was also noted that the structure of the saturated hydrocarbons changed with depth. In the shallow samples the percentage of two-, three-, and four-ring naphthenes, which were derived from living matter with a minimum of alteration, was high but the concentration of these compounds decreased with depth as did the optical activity of the samples, whilst the amounts of alkanes and monocyclic naphthenes underwent a corresponding increase.

As the total amount of alkane increased with depth, the odd-number preference decreased and the number of isoprenoid compounds also decreased due to the dilution of the original compounds with those generated from the breakdown of the kerogen.

Cummings *et al.* (1965) studied the Green River oil shale and found that for alkanes within the range C_{25}–C_{37} the odd–even preference decreased from 3.6 to 1.2 with increasing depth, and at the same time the C_{20} isoprenoid compounds decreased but the C_{15}, C_{16}, C_{17}, C_{18}, and C_{19} isoprenoid compounds increased in concentration and this was attributed to the decomposition of source material generating new compounds rather than the cracking of hydrocarbons already present. Nixon (1973) also noted that as depth increases hydrocarbons become a greater percentage of the crude oil as compared with the asphaltic material.

Vandenbrouche *et al.* (1976) investigated changes in depth on the generation of hydrocarbons from marine and non-marine kerogens and shewed that whilst the non-marine kerogens have paraffins at depth, the marine kerogen gave naphtheno-aromatics first followed, at greater depths, by paraffinic hydrocarbons. Also reported was the fact that the marine source material generated twice as much medium-range hydrocarbons as did the non-marine kerogens.

Petroleum has been found from the surface to depths of 25,000 ft, under pressures up to 20,000 psi and temperatures of up to 150 °C, but in general, and up to a certain point, oil quality improves with depth. However, excessive depth has been shewn to destroy the oil. Bailey *et al.* (1971) state that in western Canada oil quality increases with temperature, i.e. the oil becomes more paraffinic and contains less sulphur, but if sufficiently high temperatures are encountered the liquid hydrocarbons will be destroyed to leave only gas.

Temperature and pressure

Up to a point temperature will improve oil quality but if the temperature becomes too high the liquids are destroyed and the only product will be gas (Figure 7.10).

It is now possible to start to consider in more detail the relationships between depth, pressure, and temperature. The effects of depth on the alteration of sedimentary organic matter are really manifest in the effects of temperature and pressure. Depth *per se* will have very little effect other than that it is associated with increases in temperature and pressure and these will be of assistance in the migration of any generated hydrocarbons. It is now necessary to decide which of temperature and pressure is the most important in the generation and subsequent alteration of petroleum.

Philippi (1975) examined the gasolene-range (85–180°) hydrocarbon generation (this he considered to be concentrations of 80 ppm or more) in the Los Angeles and the Ventura basins. In the Los Angeles basin this generation occurred just after 8,000 ft, whereas in the Ventura basin generation did not occur until after 12,500 ft. The pressure at which generation occurred in the two basins was considerably different as this is directly related to depth of burial. However, the generation of the hydrocarbons occurred in both basins at the same temperature. Thus it would seem that it is the subsurface temperature and not the pressure which is the vital factor in kerogen breakdown; for if pressure had been the important factor, generation in these two basins would have occurred at the same depth but at different temperatures.

Gasolene-range hydrocarbons mature with depth, temperature, and age of the sediment and higher subsurface temperatures are required for the generation of gasolene-range hydrocarbons than for the heavy hydrocarbon generation; while wet and then dry gas is formed even later in the process and at successively higher temperatures. Brooks (1949) did not believe that present bottom hole temperatures would be high enough to form petroleum and thus he inferred that catalysts are an essential part of oil generation, and Dickey (1975) believes that the majority of petroleum is formed at temperatures between 60 and 150 °C. It is certainly very unlikely that many oils have been exposed to temperatures in excess of 120–150 °C. The moderate temperatures to which petroleum has been subjected will have aided its generation and migration.

The relationship between temperature and the rate of chemical reactions is stated by the Arrhenius equation:

$$k = A \exp(-E_a/RT)$$

where k is the reaction rate, A is a constant, E_a is the activation energy, k is the universal gas constant, and T is the absolute temperature. Thus reaction rates

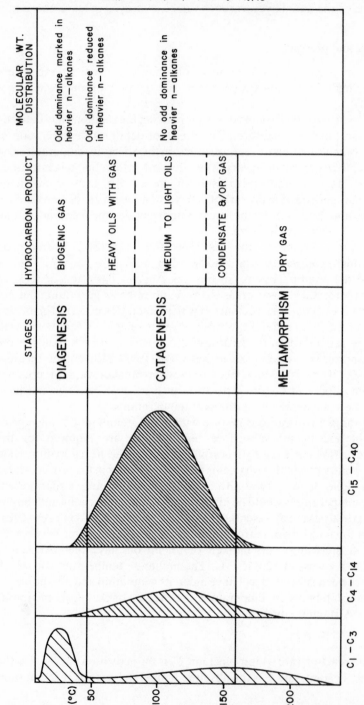

Figure 7.10 Principal products of the main maturation stages

increase exponentially with temperature and for every 10 °C rise in temperature, the reaction rate will double. Hence relatively small increases in temperature will cause large increases in the rate of generation. As far as migration is concerned, the increase in temperature will mean that the oil becomes more soluble and its viscosity will be lowered and thus an increase in temperature will aid migration.

The existence of porphyrins in petroleum has been cited as an indication of the low-temperature history of oil; however, certain care has to be exercised as it has now been shewn that some porphyrins are able to survive temperatures of up to 360 °C. It is not possible to use porphyrins as a guide to maximum temperature without sophisticated analysis to determine which porphyrins are actually present. However, it is generally accepted that the temperatures to which the organic matter can be subjected without destruction are not high enough to convert it to pretroleum on a large enough scale or in a way to produce the wide range of compounds found in reservoired petroleum. Some of the organic matter in the sediments will be converted at low temperatures. Such products have been seen in recent sediments; however, it has been shewn that these compounds are not sufficient in quality or quantity and it is for reasons of this sort that the involvement of catalysts has been suggested.

THE EFFECTS OF CATALYSTS ON THE GENERATION OF HYDROCARBONS

Catalysts play an important part in the refining of petroleum into its commercial products because they either allow a chemical reaction to proceed at a lower temperature than would be possible otherwise, or they may influence the reaction so that particular products are the result. These products may be considerably different from the types and amounts of products that would be obtained if the reaction had occurred without a catalyst. It is worth noting that because many of the products in the generation and conversion of petroleum are temperature sensitive they would not survive the conditions required to produce them without a catalyst. It is also worth noting that many industrial catalysts are chemically very similar to clay minerals.

Waples & Tomheim (1978) have developed mathematical models to simulate both random and non-random thermal cracking of branched and normal hydrocarbons. For normal paraffins they shewed that thermal cracking alone is not enough to get the present-day crude oil paraffin distributions, but that for the isoprenoid compounds the non-random cracking could be important. Thus for the majority of hydrocarbons heat alone is not sufficient to produce the range of compounds we now see and it is likely that catalysts play an important part.

Evidence for the action of catalysts in the formation of petroleum has been given by Brooks (1948, 1949), and these include: (a) the low-temperature history of petroleum; (b) thermal stability of the paraffins; (c) the complexity of petroleum, its multiplicity of isomers; (d) the evidence of change of composition of generated petroleum with age and depth; (e) the fact that catalytic activity, which is needed for such changes, is available in the shales, sandstones, clays, etc.; and (f) the presence of aromatic hydrocarbons in oil. Brooks suggested that it was most likely that the catalysts that are involved in these reactions might well be acid silicate members of the host rocks and shales. However, this idea has not been universally accepted as it has been considered that the water film might prevent the organic matter and the catalyst meeting. Certainly, industrial catalysts are often inhibited by even trace amounts of water. In spite of this objection, the involvement of catalysts in the generation of petroleum remains the primary means of the conversion of sedimentary organic matter into petroleum and for the subsequent alteration of the generated petroleum. As far as the effect of water is concerned, experiments have been carried out on the breakdown of hydrocarbons, akin to the secondary cracking of petroleum, with varying amounts of water added to the reaction mixture. These experiments have shown that the catalytic activity declines with increasing water concentrations, but does not cease completely, and this may well be due to the fact that organic matter is able to form a film between the water and the active surfaces on the catalyst.

Grim (1947) has studied the relationship between clay mineralogy and the origin and the recovery of petroleum, especially with reference to the absorptive powers of the clays. He concluded from the relationship of clay minerals to the organic matter in argillaceous sediments that the clays are a key factor in the transition of sedimentary organic matter to petroleum. Erdman & Mulik (1963) studied the genesis of the low molecular weight hydrocarbons in organic-rich sediments. In these experiments β-carotene was heated in an aquatic sediment with 60% water at 188 °C for 72 hours and it was found that the products of this reaction were a selection of benzenoid compounds including toluene, meta-xylene, naphthalene, and 2,6-dimethyl naphthalene. In a related experiment using a marine sediment under otherwise identical conditions, the reaction products were benzene, toluene, and meta-xylene, as well as carbon dioxide, methane, hex-2-ene and n-butane.

Espitalié et al. (1980) have shewn that clay minerals can, in pyrolysis experiments, trap heavy hydrocarbons and, at higher temperatures and after longer times, convert them to lighter hydrocarbons and a graphite residue. These authors have suggested that this effect could be important in the generation and migration of petroleum. The situation in which the effect would be minimal is where the organic matter is of type II and at a high concentration in a carbonate rock, whereas for a type II kerogen in low amounts in a high-activity mineral matrix (clastic-clayey), the effect would be maximized. In that latter case most of the heavy hydrocarbon that was generated would be retained and eventually could be cracked to give light hydrocarbons, so the only available product would be light oil and gas.

Laboratory experiments

Eisma & Jung (1964) heated behenic acid ($C_{21}H_{43}COOH$) at 200 °C in a sealed tube with bentonite clay for between 89 and 760 hours. It was discovered that the proportions of propane, butane, and pentane increased with time and that the major component was the n-alkane $C_{21}H_{74}$, but that alkanes ranging from $C_{14}H_{30}$ to $C_{36}H_{74}$ were present in varying amounts. Johns & Shimoyama (1971) heated behenic acid with montmorillonite and analysed the products. The results in this case shewed that after 50 hours at 200 °C only those alkanes below C_{24} had been formed but that after 150 hours some higher alkanes were to be seen; however, the quantity of these decreased with further heating.

Another important observation of this experiment was that the majority of the acid that underwent reaction had been used within the first 50 hours of the experiment and that no more was used even after 150 hours, even though less than 1% of the original acid content had been used. This was attributed partly to possible poisoning of the catalyst by the organic debris and also, in part, to the fact that the acid could be used to produce kerogen. The analysis of the products did not shew any other fatty acids than the parent acid and production of the normal C_{21} alkane occurred mostly between 50 and 150 hours, after which time the concentration of that alkane declined, at the same time the concentrations of the C_{16} and C_{17} n-alkanes increased, and this was interpreted as if the C_{21} alkane was produced from the parent acid and with time was cracked into lighter alkanes. A similar situation was observed for the C_{18} and C_{19} alkanes as had been for the C_{21} alkane, and the inference was that these alkanes had been produced directly from the C_{21} acid and after some time at higher temperatures they had been cracked in to lower alkanes. In the experiments at higher temperatures the results were the same but the reactions occurred at a faster rate.

The lack of higher fatty acids could possibly be a problem in the understanding of the production of the higher alkanes. If the process described by Philippi (1975) is involved then the higher acids must be very quickly involved in further reactions or, if not, the higher alkanes are produced by catalytically engineered condensation reactions. The CPI values for the paraffins produced after 50 hours at 200 °C started at 5.8 and decreased to between 1.6 and 1.7 after 500 hours. The figures for the experiments conducted at 250 °C were slightly lower, probably due to the higher reaction rates. These figures are akin to the changes in CPI that have been seen in alkanes from sediments of increasing age.

Johns & Shimoyama (1972) again discussed the effect of clay and minerals on the petroleum-forming reactions during burial and diagenesis, and they demonstrated that two reactions were taking place. Firstly, there was a decarboxylation reaction to give an alkane with one carbon atom less than the parent acid; and there was a second thermal cracking reaction in which the already generated alkane is broken into shorter-chain compounds. It was shewn that the temperature or depth of burial will affect which reaction occurs. In the shallow depth range of 0–4000 ft, or at a maximum temperature of about 65 °C, the main reaction is that of decarboxylation, but below that depth it is the second, cracking, reaction that becomes the most important.

Henderson *et al.* (1968) pyrolysed *n*-octacosane. When heated alone this compound decreased in quantity and the amount of cyclic, aromatic, and unsaturated compounds rose in proportion to the temperature increase. When octacosane was heated with bentonite, similar products were obtained, but the aromatic yields were especially high.

Harrison (1976) carried out laboratory experiments on a modern organic-rich sediment by heating it and studying the fatty acid changes. The results shewed the following sequence of events: (a) an early, low-temperature decarboxylation of the free and slightly bound fatty acids; (b) the cleavage of the more tightly bound fatty acids from the kerogen matrix; (c) the decarboxylation of the acids liberated in stage (b). Harrison also noted that the liberation of the acids was faster than their decarboxylation and that some acids could travel to the reservoir, possibly carrying generated oil in micelles, and then be decarboxylated in the reservoir. It was also noted that the maximum alkane yields were at the highest temperature and although the odd carbon number preference of the alkanes disappeared with temperature, that of the fatty acids did not.

Horsfield & Douglas (1980) have studied the influence of minerals on the pyrolysis of kerogen and have shewn that the composition of kerogen and coal maceral pyrolysates may change when minerals are present during the pyrolysis. They suggest that this result means that the products of natural kerogen alteration will depend upon the types of minerals present as well as the organic matter type and maturity.

Durand-Souron *et al.* (1982) have studied the formation of hydrocarbons by the pyrolysis of immature kerogens. In some experiments industrial catalysts were added to the kerogen with the result that the temperature at which hydrocarbons were released was reduced by up to 100 °C and the products were altered. The aromatic compounds were unaffected but the normal paraffins decreased and monocyclic paraffins and olefins increased.

The thermal degradation of kerogen has so far been assumed to obey the usual laws of chemical kinetics as described by the Arrhenius equation. Using this equation it has been calculated that the activation energies for the thermal breakdown of kerogen are in the range 11,000–15,000 calories per mole and it has been noted that such values are much lower than for other chemical reactions which involve the breaking of carbon–carbon bonds (approximately 58,000 calories per mole). This result is one of the pieces of evidence for the involvement of catalysts in the formation of fluid hydrocarbons. Waples (1981) states that no catalysts are known which could lower the activations energy of kerogen decomposition by such a large amount. He quotes Juntgen & Klein (1975), who suggest that because the breakdown of the kerogen involves a series of parallel reactions, the overall rate of oil generation will depend upon the sum of the rates of all the individual reactions involved. This summation is a mathematical operation which results in a final figure for the overall reaction activation energy which is lower than the simple sum of its components and thus it is not a true activation energy. If true, this could invalidate some of the evidence for the action of catalysts in oil and gas formation. Tissot & Welte (1978) suggest that activation energy of a reaction increases with the increase in temperature at which that reaction occurs (the differential numerical increase in the value of the activation energy compared with the temperature will mean that the reaction rate will increase and approximately double for every 10 °C rise in temperature). Tissot & Welte suggest that the differences between the calculated activation energies of oil generation and of other chemical reactions could be due to the variations in temperature. At lower temperatures, less energy (and thus presumably a lower activation energy) are required to split hydrocarbons from the kerogen nucleus as compared with the final stages of kerogen alteration. Thus the low values of activation energy could relate to the onset of intense oil generation and the higher activation energy values correspond to the gas-generation stage. Certainly the reactions involved at the start of oil generation will involve many that will have lower energy requirements, e.g. breaking of carbon–oxygen bonds, release of loosely held material, etc.

TIME AND TEMPERATURE

It has been shewn that the generation of petroleum is covered by the laws of thermodynamics, as expressed in the Arrhenius equation. The direct implication of this is that time can be traded for temperature in oil generation. This is illustrated in Figure 7.1 where organic matter in rocks of different ages reach the same level of maturation, as measured by vitrinite reflectance, at different depths (which means at different temperatures) and the older the sediment, the shallower will be the depth (or lower the temperature) required to reach the same level of maturation. Study of the beginning of the 'oil window' (0.6% reflectance) on Figure 7.1 will illustrate this well.

Assuming that the Arrhenius equation is valid in hydrocarbon generation, the rate of hydrocarbon-producing reactions will double for each 10 °C rise in temperature, and thus the same amount of oil will be formed after 80 million years at 100 °C as after 40 million years at 110 °C, or after 20 million years at 120 °C. Obviously this assumes the same kerogen type and this simplistic approach will be modified by the interplay of all other variables of hydrocarbon production, but the basic elements of the truth of the interchangeability of time and temperature are contained in the above statement. Many of the examples that have already been quoted have shewn that although the general trend of increasing oil maturation and quality is related to increasing depth, there is a definite trade-off of time against temperature.

Connan (1974) considered the relationships between time and temperature in oil generation and considered that the threshold of the principal zone of oil generation was defined by the age of the formation in which the oil occurs (t in millions of years) and the corresponding existing temperature at that depth (T °C). He shewed that a graph of the natural logarithm of the age (t) against the reciprocal of the absolute temperature was a linear regression and this was interpreted as showing that the degradation of the kerogen followed the laws of chemical kinetics, i.e. that Arrhenius' equation does apply. From the slope of this line, Connan calculated a pseudo-activation energy for the degradation of the kerogen which was between 11 and 15 kcal per mole. This figure needs to be compared with the energy required to cleave a normal carbon–carbon bond, which is of the order of 58 kcal per mole. Thus Connan perceived that catalysts must play a very important part in the formation of hydrocarbons from kerogen and that, because the graphs which he produced (Figure 7.11) were linear, longer times could compensate for lower temperatures during generation.

Peters et al. (1981) carried out simulated burial maturation experiments on the precursors of kerogen and have shewn that the generation of kerogen during burial of the sediment is the net result of both time and temperature. Thus the whole of the sequence of processes involved in the production of hydrocarbons, from generation of kerogen right through to the breakdown of that kerogen and probably including the secondary thermal alteration of generated hydrocarbons, depends upon temperature and time, and one may be substituted for the other. Organic matter which is heated quickly will need a shorter time to form kerogen and that kerogen will need a shorter time to break down as compared with organic matter that is heated more slowly.

Connan's diagram (Figure 7.11) illustrates the main oil and gas generation zones as functions of time and temperature, again shewing that longer times can be a substitute for lower temperatures in the formation of oil and gas. Thus when evaluating the suitablility of a source rock, due consideration must be taken, not only of the amount and type of organic matter, but of its maturation state in terms of both the age of the sediment and the temperature regime to which that sediment has been subjected. Old sediments subjected to high temperatures for long periods of time will be in the super-mature stage and will be past their peak hydrocarbon production stage, whereas either a young sediment which has experienced the same temperature regime or an old sediment which has had a lower heat input could both be in the oil-producing zone. A discussion of the practical application of this principal is in Chapter 9.

Reference has been made to Figure 7.1 which is based upon the work of Dow (1978) describing the effects of time and temperature (as illustrated by vitrinite reflectance) in a series of wells between the Cretaceous and Pleistocene from the Louisiana coast. A series of reflectance values in these wells gives a kerogen maturation profile for each well and these are expressions of increasing rank with depth (or temperature) and age. Selected wells had very similar geothermal gradients and the variation in maturation slope gradients along the trends is primarily due to exposure time, or average sediment age. The depths and present temperatures start at the beginning of the principal zone of oil generation ($R_{max} = 0.6\%$) and end at the finish of the principal zone of oil formation ($R_{max} = 1.35\%$). In each case the trends are different due to different exposure times but the older the sediment, the shallower is the depth of burial at which the start of principal zone of oil formation is reached. If,

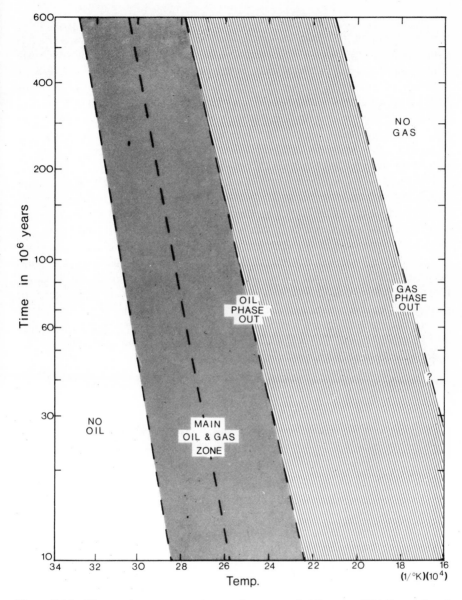

Figure 7.11 Time and temperature in petroleum genesis (Connan, 1974. Reproduced by permission of The American Association of Petroleum Geologists)

however, the exposure time is constant, the depth to the start of the principal zone of oil generation will change as the geothermal gradient changes. Thus if the geothermal gradient doubles, the depth to the principal zone of oil formation will halve for each trend. Thus it is evident that the depth of burial required to reach the principal zone of oil formation will vary according to both exposure time and geothermal gradient.

The Connan time–temperature chart of petroleum genesis (Figure 7.11) was based upon the idea of continually subsiding basins. However, this supposition is not always valid and, while the Connan model is obviously simplified, it still has practical value when used to evaluate the relative effects of time and temperature upon the formation of oil and gas. More work needs to be done to consider the geothermal history of a basin and it will involve analysis of the time intervals during which the sediment is subjected to various temperatures. Such an approach is probably the best way of evaluating the hydrocarbon-generation potential of a basin so long as reasonable palaeotemperatures are available.

Lopatin (1971, 1976) has carried out such studies, originally based upon coals, but extended to oil and gas. He obtained the estimated maturation levels from vitrinite reflectance measurements and calculated a 'time–temperature index of maturation' (τ) which was the product of the coalification in each 10 °C interval and the temperature coefficient of the speed of reactions of maturation within that interval. He considered that this index was valid because it gave a straight line when plotted against vitrinite reflectance (correlation coefficient +0.99). Figure 7.12 illustrates the graph used to calculate this time–temperature index of maturation. $\Sigma\tau$ is calculated by finding the value of for each period of time spent in each temperature regime and summing these. The values of oil and gas generation as determined by Lopatin according to this index were:

70–85	beginning of principal zone of oil generation
160–190	maximum oil generation
380–400	end of principal zone of oil generation
550–650	maximum gas generation
1,500–2,000	end of gas generation

Allowances can be made for varying geothermal gradients and igneous intrusions. A fuller description of the method is given in Chapter 9 (Waples, 1981).

Tissot & Welte (1978) have constructed a mathematical model for kerogen degradation to provide quantitative values of oil and gas produced as a function of time. It involves an assessment of the activation energy of the kerogen breakdown reaction, either by laboratory experiment or by determination of kerogen type and use of previously obtained calibration curves; the burial curve of depth vs. time and the geothermal gradient. This allows the amount of oil and gas per ton of rock to be calculated. It also gives immediately an indication of whether any hydrocarbon has been formed, and if so whether it is oil or gas derived from the oil.

Thus the effects of time and temperature upon the maturation of organic matter can be calculated and used to predict the products of such maturation. The fact that these predictions can be obtained and are accurate is evidence of

AGE (million years)

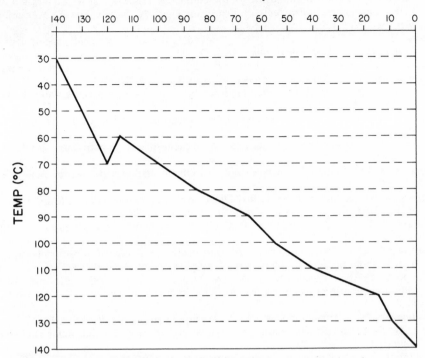

Figure 7.12 Simple subsurface time–temperature grid showing the time–temperature history of a sediment (after Waples, 1981, Lopatin, 1971). (From *Organic Geochemistry fo Exploration Geologists*, 1981, Douglas Waples, Burgess Publishing Company, Minneapolis, Minnesota. Reprinted by permission)

the importance of time and temperature in organic matter maturation and that time and temperature are, to a limited degree, interchangeable as regards the breakdown of kerogen.

REFERENCES AND FURTHER READING

Abell, P. I., & Margolis, M. J. (1982). *n*-Paraffins in the sediments and in situ fossils of the Lake Turkana Basin, Kenya. *Geochim. & Cosmochim. Acta*, **46**, 1505.

Albrecht, P., & Ourrison, G. (1969) Diagenesis of hydrocarbons in a series of sediments. *Geochim. Cosmochim. Acta*, **33**, 138.

Albrecht, P., Vandenbrouche, M., & Mandengue, M. (1976) Geochemical studies on the organic matter from the Doula basin (Cameroons) I. *Geochim. Cosmochim. Acta.*, **40**, 791.

Bailey, N. J. L., Evans, C. R., & Rogers, M. A. (1971) Evolution and alteration of petroleum in W. Canada. *Chem. Geol.*, **8**, 147.

Baker, D. R. (1962) Organic geochemistry of Cherokee group in S. E. Kansas and N. E. Oklahoma. *BAAPG.*, **46**(9), 1261.

Baker, E. G. (1959) A hypothesis concerning the accumulation of sediment hydrocarbons to form crude oil. *Geochim. Cosmochim. Acta*, **19**, 309.

280

Barton, D. C. (1934) Natural history of the Gulf Coast crude oil. In W. E. Wrather & F. H. Lake (Eds.), *Problems of Petroleum Geology*, Sidney Powers Memorial Vol., AAPG p.109.

Bray, E. E., & Cooper, J. E. (1968) A postulated role of fatty acids in petroleum formation. *Geochim. Cosmochim. Acta*, **27**, 1113.

Bray, E. E., & Evans, E. D. (1961) Distribution of *n*-paraffins as a clue to the recognition of source beds. *Geochim. Cosmochim. Acta*, **22**, 2.

Bray, E. E., & Evans, E. D. (1965) Hydrocarbons in non-reservoir rock source beds. *BAAPG*, **49**, 248.

Breger, I. A. (1960) Possible sources of hydrocarbons. *Geochim. Cosmochim. Acta*, **19**, 297.

Brooks, B. T. (1948) Active surface catalysts in the formation of petroleum. *BAAPG*, **32**, 2269.

Brooks, B. T. (1949) Active surface catalysts in the formation of petroleum. *BAAPG*, **33**, 1600.

Brooks, J. D., & Smith, J. W. (1967) Diagenesis of plant lipids during formation of coal and petroleum. *Geochim. Cosmochim. Acta*, **31**, 2389.

Brooks, J. D., Gould, K., & Smith, J. W. (1969) Isoprenoid hydrocarbons in coal and petroleum. *Nature*, **222**, 257.

Button, D. K. (1976) Influence of clay and bacteria on concentration of dissolved hydrocarbons in saline solutions. *Geochim. Cosmochim. Acta*, **40**, 435.

Claypool, G. E., Threlkeld, G. N., & Magoon, L. B. (1980) Biogenic and thermogenic origins of natural gas in Cook Inlet Basin, Alaska. *BAAPG*, **64**(9), 1131.

Connan, J. (1974) Time–temperature relationships in oil genesis. *BAAPG*, **58**, 2516.

Cox, B. B. (1946) Transformation of organic matter to petroleum under geologic conditions. *BAAPG*, **30**, 645.

Cummins, J. J., Robinson, W. E., & Dinneen, G. V. (1965) Changes in Green River shale paraffins with depth. *Geochim. Cosmochim. Acta*, **79**, 249.

Day, W. C., & Erdman, J. G. (1963) Ionene—Thermal degradation product of β-carotene. *Science*, **141**, 808.

Demaison, G. J. (1977) Tar sands and supergiant oilfields. *BAAPG*, **61**, 1950.

Dembicki, H., Meinschein, W. G., & Hatton, D. E. (1976) Significance of the predominance of even numbered alkanes. *Geochim. Cosmochim. Acta*, **40**(2), 203.

Dickey, P. A. (1975) Possible primary migration of oil from source rock in the oil phase. *BAAPG*, **59**, 337.

Dow, W. G. (1978) Petroleum source beds on continental slopes. *BAAPG*, **62**, 1584.

Draffen, G. H., Eglington, G., & Rhead, M. M. (1971) Hydrocarbons from the alteration of cholesterol. *Chem. Geol.*, **8**, 277.

Durand-Souron, C., Boulet, R., & Durand, B. (1982). Formation of methane and hydrocarbons by pyrolysis of immature kerogens. *Geochim. & Cosmochim. Acta*, **46**, 1193.

Eisma, E., & Jung, J. W. (1964) Petroleum hydrocarbon generation from fatty acids. *Science*, **144**, 1451.

Erdman, J. G. (1961) Some chemical aspects of petroleum genesis as related to source bed recognition. *Geochim. Cosmochim. Acta*, **22**, 10.

Erdman, J. G., & Mulik, J. D. (1963) Genesis of low molecular weight hydrocarbons in organic rich sediments. *Science*, **151**, 806.

Espitalié, J., Madec, M., & Tissot, B. (1980) The role of the mineral matrix in kerogen pyrolysis. The influence on petroleum generation and migration. *BAAPG*, **64**(1), 59.

Grant, G. W., Haus, H. F., & Zobell, C. E. (1943) Marine Micro-organisms which oxidize petroleum hydrocarbons. *BAAPG*, **27**, 1175.

Grim, R. E. (1947) The relation of clay mineralogy to origin and recovery of petroleum. *BAAPG*, **31**, 1491.

Harrison, W. E. (1976) Thermally induced diagenesis of aliphatic fatty acids. *BAAPG*, **60**, 452.

Harwood, R. J. (1977) Oil and gas generation by laboratory pyrolysis of kerogen. *BAAPG*, **61**, 2082.

Hedberg, H. D. (1964) Aspects of origin of petroleum. *BAAPG*, **48**, 1755.

Hedberg, H. D. (1968) Significance of high wax oils with respect to genesis of petroleum. *BAAPG*, **52**, 736.

Hedges, J. I., & Parker, D. L. (1976) Land derived organic matter in surface sediments in the Gulf of Mexico. *Geochim. Cosmochim. Acta*, **40**, 1019.

Henderson, W., Eglington, G., Lovelock, J. E., & Simmonds, P. (1968) Thermal alterations as a contribution to petroleum genesis. *Nature*, **219**, 1012.

Horsfield, B., & Douglas, A. G. (1980) The influence of minerals on the pyrolysis of kerogen. *Geochim. Cosmochim. Acta*, **44**(8), 1119.

Hunt, J. M. (1978) *Petroleum Geochemistry and Geology*. W. H. Freeman & Co., San Francisco.

Ishiwatari, R., Ishiwatari, M., Kaplan, I. R., & Rohrback, B. G. (1977) Thermal alteration experiments on organic matter from recent marine sediments in relation to petroleum genesis. *Geochim. Cosmochim. Acta*, **41**, 815.

Ishiwatari, R., & Machihara, T., (1982) Algal lipids as a possible contributor to the polymethylene chains in kerogen. *Geochim. & Cosmochim. Acta*, **46**, 1459.

Ishiwatari, R., Rohrback, B. G., & Kaplin, I. R. (1978) Hydrocarbon generation by thermal alteration of kerogen from different sediments. *BAAPG*, **62**, 687.

Johns, W. D., & Shimoyama, A. (1971) Catalytic conversion of fatty acids to petroleum-like hydrocarbons and then maturation. *Nature (Phy. Sci.)*, **232**, 140.

Johns, W. D., & Shimoyama, A., (1972) Clay minerals and petroleum; forming reactions during burial and diagenesis. *BAAPG*, **56**, 2160.

Kidwell, A. L., & Hunt, J. M. (1958) Migration of oil in recent sediments of Pedernales, Venezuela. In L. G. Weekes (Ed.), *Habitat of Oil: A Symposium*. AAPG, Tulsa, Okla., p.790.

Jungten, H., & Klein, J. (1975) Formation of natural gas from coaly sediments. *Erdol und Kohle, Erdgas, Petrochemie*, **28**, 65.

Koons, C. B., Jamieson, G. W., & Ciereszko, L. S. (1965) Normal distributions in marine organisms with possible significance to petroleum origin. *BAAPG*, **49**, 301.

Kvenvolden, K. A. (1966) Molecular distribution of fatty acids & paraffins in lower cretaceous sediments. *Nature*, **209**, 573.

Leverson, A. I. (1967) *The Geology of Petroleum*. W. H. Freeman & Co., San Francisco.

Leythaeuser, D., Schaefer, R. G., & Yukler, A. (1982) Role of diffusion in primary migration of hydrocarbons. *BAAPG*, **66**(4), 408.

Lopatin, N. V. (1971) Temperature and geological time as factors of carbonification. *Akad. Nauk. SSSR. Ivc. Sev. Geol.*, No.3, p.95.

Lopatin, N. V. (1976) The determination of the influence of temperature and geological time on the catagenic processes of coalification and oil-gas formation. In 'Research on Organic matter of modern and fossil deposits' (*Issledovaniya organicheskogo veshchestva sovremennykh i iskopaemykh osadkov*) Moscow: Akad. Nauk. SSSR, Ixdatel'stvo, 'Nauka', pp.361–366.

Magara, K. (1977) Petroleum migration and accumulation. In Hobson G. D. (Ed.), *Advances in Petroleum Geology*.

Mair, B. J. (1964) Terpenoids, fatty acids and alcohols as source materials for hydrocarbons. *Geochim. Cosmochim. Acta*, **28**, 1303.

Meinschein, W. C. (1961) Origins of petroleum. *BAAPG*, **43**, 925.

Nixon, R. P. (1973) Oil source rocks in Cretaceous Mowry shale of NW interior USA. *BAAPG*, **57**, 136.

282

Oppenheimer, C. H. (1960) Bacterial activity in sediments of shallow marine bays. *Geochim. Cosmochim. Acta*, **19**, 244.

Perregaard, J., & Schiener, E. J. (1979) Thermal alteration of sedimentary organic matter by a basalt intrusive (Kimmeridgian shales, Milne Land, E. Greenland). *Chem. Geol.*, **26**, 331.

Perry, G. J., Volkman, J. K., John, R. B., & Bavor, H. J. (1979) Fatty acids of bacterial origin in contemporary marine sediments. *Geochim. Cosmochim. Acta*, **43**(11), 1715.

Peters, K. E., Rohrback, B. G., & Kaplan, I. R. (1981) Geochemistry of artificially heated humic and sapropelic sediments. 1: Protokerogen. *BAAPG*, **65**(4), 688.

Philippi, G. T. (1974a) Depth of oil origin and primary migration. *BAAPG*, **58**, 149.

Philippi, G. T. (1974b) Influence of marine and terrestrial source material on the composition of petroleum. *Geochim. Cosmochim. Acta*, **38**, 947.

Philippi, G. T. (1975) The deep subsurface temperature controlled origin of gaseous and gasolene range hydrocarbons of petroleum. *Geochim. Cosmochim. Acta*, **39**, 1353.

Philippi, G. T. (1977) Depth, time and mechanism of origin of heavy to naphthenic crude oils. *Geochim. Cosmochim. Acta*, **41**, 33.

Philp, R. P., Brown, S., & Calvin, M. (1978) Isoprenoid hydrocarbons produced by thermal alteration of *Nostoc mascorum and Rhodopseudomonas spherides*. *Geochim. Cosmochim. Acta*, **42**, 63.

Powell, T. G., & McKirdy, D. M. (1973) Relationships between the ratio of pirstane and phytane, crude oil and geological environments in Australia. *Nature*, **243**, 37.

Powell, T. G., & McKirdy, D. M. (1973) Effects of source material, and diagenesis on the *n*-alkane content of sediments. *Geochim. Cosmochim. Acta*, **37**, 623.

Rice, D. D., & Claypool, G. E. (1981) Generation, accumulation and resource potential of biogenic gas. *BAAPG*, **65**, 5–26.

Shibaoka, M., Saxby, J. D., & Taylor, G. H., (1978) Hydrocarbon generation in Gippsland basin, Australia. *BAAPG*, **62**, 1151.

Shultz, D. J., & Calder, J. A. (1976) Organic $^{13}C/^{12}C$ variations in esturine sediments. *Geochim. Cosmochim. Acta*, **40**, 381.

Sikander, A. H., & Pittion, J. L. (1978) Reflectance studies on organic matter in lower Paleozoic sediments of Quebec. *B.Can.Pet.Geol.*, **26**(1), 132.

Snowdon, L. R., & Powell, T. G. (1982) Immature oil and condensate-modification of hydrocarbon generation model for Terrestrial organic matter. *BAAPG*, **66**, 775.

Teas, P., & Miller, C. R. (1933) Raccoon Bend oilfield, Austin County, Texas. *BAAPG*, **17**, 1459.

Tissot, B., Califet-Debyer, Y., Deroo, G., & Oudin, J. L. (1971) Origin and evolution of hydrocarbons in early Toarcian shales, Paris basin. *BAAPG*, **55**, 2177.

Tissot, B., Durand, B., Espitalié, J., & Combaz, A. (1974) Nature and diagenesis of organic matter. *BAAPG*, **58**, 499.

Tissot, B. P., & Welte, D. H. (1978) *Petroleum Formation and Occurrence: A New Approach to Oil and Gas Exploration*. Springer-Verlag, Berlin-Heidelberg-New York.

Tyson, R. V., Wilson, R. C. L., & Downie, C. (1979) A stratified water column environmental model for the type Kimmeridge Clay. *Nature*, **277**, 377.

Vandenbrouche, M., Albrecht, P., & Durand, B. (1976) Geochemical studies on organic matter from Doula basin (Cameroon) III. *Geochim. Cosmochim. Acta*, **40**, 1241.

Van Krevelen, D. W. (1961) *Coal*. Amsterdam, Elsevier Publ. Co.

Waples, D. W. (1981) *Organic Geochemistry for Exploration Geologists*. Burgess Publishing Co., Minneapolis, Minnesota, USA.

Waples, D. W., & Tomheim, I. (1978) Mathematical models for petroleum forming processes. *Geochim. Cosmochim. Acta*, **42**, 457.

Weber, V. V., & Maximov, S. P. (1976) Early diagenetic generation of hydrocarbon gases & dependance on initial organic composition. *BAAPG*, **60**, 293.

Weekes, L. G. (Ed.), (1958) *Habitat of Oil: A Symposium*. Tulsa, Okla., AAPG.

Wells, R. C. (1929) Origins of helium rich natural gas. *Jour. Washington Acad. Sci.*, **19**(15), 321.

Chapter 8

Petroleum Alteration and Migration

ALTERATION PROCESSES

When petroleum is generated it is thermodynamically unstable and is susceptible to alteration by a large number of agencies, which can be chemical, physical, or bacteriological. The oil may be affected in a number of different ways, either making the oil lighter and more valuable or heavier and less valuable (Bailey *et al.* 1971).

Thermal maturation

Thermal maturation is the first process which will be considered. It is a matter of the generated hydrocarbons being subjected to higher temperatures so that there is enough energy available to break down the hydrocarbons into progressively smaller and smaller components with methane as the final fluid product. The result of this is initially to improve the quality of the oil by increasing the percentage of lighter hydrocarbons and its paraffinicity and at the same reducing the percentage of asphaltic compounds which are rich in the heteroatoms of nitrogen, sulphur, and oxygen. However, this process, if carried too far, can result in methane as the only fluid product, together with solid pyrobitumen.

Thermal maturation is controlled by the same factors that regulate the original generation of hydrocarbons from kerogen, namely geothermal gradients, depth of burial, and the time that hydrocarbons spend in a particular temperature regime. With increasing depth and temperature crude oils in reservoirs become lighter with more low molecular weight and fewer high molecular weight compounds and at the same time gases, especially methane, dramatically increase in abundance. The eventual result of all this, given sufficient time and energy (i.e. high enough temperature), will be the cracking of the majority of the oil to wet gas and eventually to methane, or to dry gas (Figure 8.1).

The process is a series of disproportionation reactions, with the transfer of hydrogen from one molecule to another, so that the result is that, as well as the light gaseous products which are hydrogen rich with respect to carbon, there is also a series of compounds with condensed aromatic structures, known as pyrobitumen, which are rich in the heteroatoms and deficient in hydrogen. In many reservoirs within the mature zone only methane and the pyrobitumen, which are respectively the hydrogen-rich and hydrogen-lean products of the cracking of crude oil, are found (Bailey *et al.*, 1971).

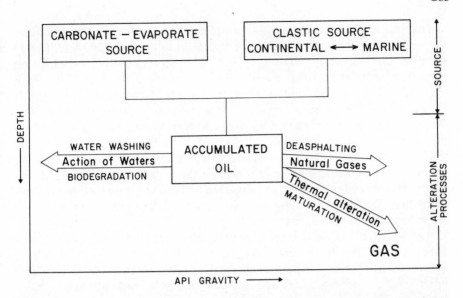

Figure 8.1 The alteration of petroleum (after Evans *et al.*, 1971. Reproduced by permission of Elsevier Scientific Publishing Co.)

However, thermal alteration, if carried to a limited extent is often beneficial as it will at first improve the quality of the oil. Connan *et al.* (1975) have carried out thermal maturation experiments in the laboratory and have shewn that normal, otherwise unaltered oils undergo maturation after exposure to high temperatures. Over a period of 12 months the saturated paraffins increased from 10 to 30% and the pyrobitumen correspondingly increased from a trace to 30%.

Deasphalting

The changes in temperature are often associated with changes in pressure, which may allow alterations in the physical condition of the hydrocarbons. For instance, the heat may encourage the liberation of gas from solution, as may a pressure decrease, while conversely a pressure increase would tend to make any gas redissolve into its associated oil. It has already been noted how some of the higher molecular weight fractions are unstable to heat and will crack to give lighter components. The redissolution of these lighter components in the crude oil due to increase in depth and pressure, or the effect of generated gas being kept in solution by pressure, will eventually lead to the precipitation of asphaltenes in the reservoir and will produce marked changes in the chemical and physical properties of the oil. The API gravity will change and this, accompanied by the precipitation of heavy components, will yield a higher quality, higher API gravity, more produceable, and more valuable oil. This process of natural deasphalting is directly analogous to that used in petroleum refineries in the processing of heavy crudes to remove the asphaltenes.

To cause natural deasphalting, pressure conditions have to be such that no gas cap is formed and that all the gas is kept in solution. Evans *et al.* (1971) discusses the idea of 'gravity segregation' wherein oil pools which, although they have a large vertical extent, have a small temperature difference between top and bottom, and in which the oil gets heavier with depth. This is the reverse of the expected situation. This is supposed to be due to the fact that the oil at the top of the reservoir would tend to have more gas dissolved in it and therefore be less asphaltic than the oil in the deeper horizons. The oil with the dissolved gas and without the asphaltenes will be lighter than the gas-free, asphaltene-rich oil, and therefore the former will sit on top of the latter. This inversion process, as suggested by Evans, would tend to be more likely than the original gravity separation theory which required that the oil settled out via the pore structure of the reservoir (Bailey *et al.*, 1971)

Bailey *et al.* (1974) have pointed out the differences between those oils that have been thermally matured and those which have been subjected to natural deasphalting. Whereas thermal maturation will be a regional phenomenon, deasphalting due to gas injection will be much more localized. Thus in a suite of neighbouring reservoirs some might produce oil with asphaltenes whilst others produce lighter oil and gas without asphaltenes if gas deasphalting had been operating as the main alteration factor in these latter pools. However, if thermal maturation had been the major alteration factor, the type of oil produced by all the reservoirs in the region would be very similar and more related to temperature and depth than to pressure.

Water washing

All the processes discussed so far have tended to make the oil lighter. However, there are other factors which have the reverse effect and they are normally associated with water. The simplest is water washing where moving water acts as a transportation medium for materials in the oil. These water-soluble compounds which can be removed will include the lower molecular weight hydrocarbons, the aromatics, and such polar compounds as fatty acids, with the result that water-washed oil becomes heavier. Obviously, water washing will only be able to remove those compounds which are in the vicinity of the oil–water contact and thus this heavy degraded oil tends to be localized at the oil–water interface, whereas oil higher in the reservoir may well remain unaltered. This is believed to be the cause of many tar layers such as those found in large numbers of Middle East oilfields.

Not many experiments have been carried out on water washing; however, Milner *et al.* (1977) have quoted an experiment by Bayliss in which an oil was degraded by washing with water for over six months. The result was that the oil's light ends were removed by the water.

Bacterial attack

One of the reasons that water washing as such has not been thoroughly investigated is that it tends to occur at the same time as bacteriological degradation. The water brings bacteria, oxygen, and certain nutrients into contact with the oil, the nutrients allowing not only the bacteria to survive but also to feed upon the oil. Microbial action is well known as a means of altering sedimentary organic matter and this alteration may occur in both marine and freshwater environments and under a variety of temperatures and pressures after the death of the organism which is to be consumed. Microbial activity tends to create and intensify reducing conditions and to remove oxygen from the organic matter, thus bringing it to a condition more like that of petroleum. Methane is the only hydrocarbon produced directly from organic matter by microbial action and because of this the role of bacteria in the production of petroleum is uncertain. However, it is known that bacteria under certain conditions can consume organic matter, and thus the bacteria will be able to attack generated oil.

Grant *et al.* (1943) found marine organisms that will oxidize hydrocarbons and they found that the sea water contained between 10 and 1000 such petroleum-oxidizing bacteria per litre, and that were 100–100,000 such bacteria per gram of recent sediment. The oxidation occurred in the presence of hydrogen acceptors, such as free oxygen, nitrate, or sulphate ions, although in the latter case iron is required to be present to react with the hydrogen sulphide which is formed and which would otherwise poison the bacteria. They shewed that the long-chain hydrocarbons were oxidized faster than shorter ones; aliphatic hydrocarbons oxidized faster than cyclic and aromatic compounds; and that unsaturated hydrocarbons oxidized faster than saturated equivalents.

Bailey *et al.* (1973) state that petroleum can be altered by fresh water, either by inorganic oxidation, which only occurs near the surface, or by water washing (see above), or by bacterial degradation. The essential requirements for such biodegradation are water (in which the bacteria live), nutrients, dissolved oxygen, and food. The water is an absolute essential because not only does it produce a carrier for the nutrients and oxygen, but the bacteria themselves cannot live in the oil, only in the water. Williams & Winters (1969) have studied oils that have been partially degraded by meteoric water. The water contained 8 ppm of oxygen and the *n*-paraffins were completely eliminated. In other areas in the same field the *n*-paraffins were less depleted. The altered oils contained more nitrogen than the unaltered, and their increase in optical activity is greater than is expected purely from the removal of alkanes, which indicates that the bacteria add optically active compounds. In general, the more altered oils were from shallower reservoirs. Whilst the oil–water contact is an obvious place for biodegradation to occur, there is normally a fine film of water around the rock matrix through which the bacteria can travel and they can thus attack the oil elsewhere than the oil–water contact. The nutrients are inorganic radicals such as nitrite and phosphate, which are water soluble. Dissolved oxygen is essential because most of the petroleum-consuming bacteria are aerobic and need a constant supply of oxygen which is brought in by the flowing waters which are normally in hydraulic connection with the surface. If the flow of oxygen-containing water stops, the bacteria will die. Food is the last important essential and when crude oil is present it will be the source of food for the bacteria (Zobell, 1964).

The temperature at which the bacteria operate is normally fairly restricted. Often biodegradation will be at its optimum between 60 and 70 °C (Claypool & Rice, 1981) which, after allowing for surface temperature, is equivalent to a 48 °C increase or a depth of 1½–2 km (5,000–6,700 ft). However, the bacteria can remain dormant at temperatures up to 85–90 °C. If there is anything toxic to the bacteria in the water, e.g. hydrogen sulphide or aromatic compounds, the bacteria die and the degradation ceases.

It has been shewn that the surface, when all the conditions are satisfied, bacterial degradation of crude oil is not only rapid but goes to completion. It can be assumed that the same is true in the subsurface.

Bacteria that live on the hydrocarbons have no mouth but only an outer membrane which the food has to penetrate. The apertures through which the hydrocarbons are consumed will, by their size, be the major factor which determines which compounds are eaten. Apertures in the outer membrane of the bacteria are of similar diameter to a methane molecule and hence the normal alkanes are readily devoured; next come the iso-compounds; with the triterpanes being the size at which consumption by the microbe ceases. Aromatic compounds, although some are just small enough to be eaten, are very toxic to the bacteria and therefore normally left well alone. Figure 8.2 shews the effect of bacterial degradation on normal alkanes in crude oils, which are preferentially consumed. Jobson et al. (1972) treated oil samples with bacterial cultures and observed considerable degradation in 3 weeks. One oil dropped in specific gravity from 0.827 to 1.046 (40 to 5 °API) and lost 30% of its n-alkanes. Deroo et al. (1974) studied oils in Alberta where systematic changes were observed. As the oilpools became shallower the degree of alteration increased from normal to very severely altered oils in which even pristane and phytane were destroyed. Cycloalkanes were not affected.

Button (1976) experimented on the influence of clay and bacteria upon the concentration of hydrocarbons dissolved in saline solutions. The clay alone did not reduce the hydrocarbon concentration but in the presence of bacteria, dodecane ($C_{12}H_{26}$) was reduced in concentration from 1.82 µg/l to 0.14 µg/l in 30 days, and thus it would seem that the bacteria are more effective than clays in removing alkanes from solution. However, in the case of aromatic compounds the reverse is true. The bacteria do not eat the aromatic compounds because they are poisonous, but these aromatic compounds are polar and therefore are more readily absorbed on to the

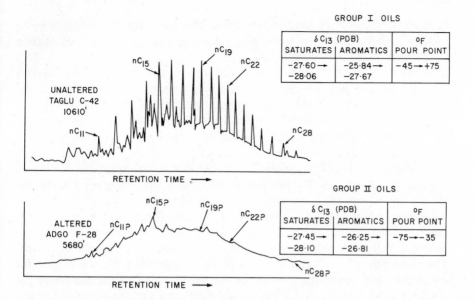

Figure 8.2 Whole oil chromatograms of typical altered and unaltered oils (Burns *et al.*, 1975. Reproduced by permission of the Canadian Society of Petroleum Geologists)

clays than the alkanes. Thus in a mixture of alkanes and aromatics the former will be consumed by the bacteria and the latter will be absorbed upon the clay, but it is worth noting that while the clays will have a maximum capacity for the aromatics, the bacteria, assuming that all the other conditions for their survival are met, will have no such limit.

Hedges (1977) investigated the uptake of dissolved organic molecules by kaolinite and montmorillonite in sea and fresh water. The organic compounds used were glucose, valine, and stearic acid. The former compounds, which are not so polar, were not readily absorbed but the stearic acid was removed from solution in the range 10–1000 ppb. Hedges considered that while some organic compounds can be absorbed by clays, such partitioning would be unlikely to produce the high concentrations of organic matter that occur in recent fine-grained sediments.

Sand, silt, diatomaceous earth, and other absorbents can accelerate bacterial oxidation of hydrocarbons. Oppenheimer (1960) calculated that surface sediments contain about 10^8 bacteria per gram, or 1 mg of bacteria per gram of sediment, and he showed that in sands at 4 °C for 40 days, bacteria could destroy 95% of the organic matter, but in clays only 75% was destroyed. The sands contain a wider selection of bacteria and as the grain size decreases the number of types of bacteria declines but the absolute numbers increase. The smaller particle size means a large surface area and hence more organic matter can be absorbed for the bacteria to feed upon. Clays have smaller pores than the sands and therefore only the smaller bacteria will live in the clay, which reduces the variety of organic compounds that can be consumed.

Bailey et al. (1973) degraded a crude oil with bacteria for a period of 21 days and the normal paraffins up to at least C_{34} were severely depleted; the lower-ring naphthenes and aromatics were attacked at the same time as the intermediate paraffins; but the more condensed cyclic hydrocarbons were not affected. Additional non-hydrocarbon compounds, rich in nitrogen, sulphur, and oxygen, were formed at the same time.

Gould & Smith (1978) reported isotopic evidence for a microbiological role in genesis of crude oil (Barrow Island, W. Australia). The early Cretaceous crude oil is aromatic–naphthenic in character and almost devoid of n-alkanes, whilst the late Jurassic crudes from almost 1,000 m lower are paraffinic–naphthenic with a high wax content. These deeper oils are believed to have been originated from a mixed marine–terrigeneous source. The shallower oil has had a marine source suggested but it is too mature for its depth and thus the possibility of biological degradation must be considered, especially as neither the salinities nor the temperature are too high (90–103 °C for the Jurassic and 63 °C for the Cretaceous).If biological factors were at work in this case, the hydrocarbons would have been converted to carbon dioxide which could, in turn, have reacted with metal ions in solution to give ^{12}C-enriched secondary biogenic carbonate rocks. The $^{12}C/^{13}C$ ratio of such rocks will be different to that of normal primary carbonate rocks. In this case extensive sedimentary secondary biogenic carbonate rocks are found with the early Cretaceous oil, but the carbonate rocks from the Jurassic are normal. This high concentration of secondary carbonates cannot be accounted for by normal diagenetic or geothermal processes as they would produce much more secondary carbonate at the lower level. This accounts for the belief that the Cretaceous oil is biodegraded.

Rubenstein et al. (1977) have investigated the origin of the oil sands of Alberta using computerized GC–MS (a gas chromatograph with a linked mass spectrometer to analyse the column effluent) techniques and the results support the theory that the oil sands and some conventional crudes were formed in the Lower Cretaceous formation and have a common origin but differ in the degree of biodegradation to which they have been subjected. The Alberta tar sands have undergone extensive biodegradation after their formation as conventional crude oil and the bitumens represent the effective end product of this process.

Demaison (1977) considers that tar sands are produced by water washing and bacterial degradation of the fluid, medium-heavy oils. The water washing removes the more water-soluble compounds, especially the aromatics, and the biodegradation removes the paraffins. This action generally occurs when the oil is in contact with oxygen and bacteria-bearing water at a temperature lower than 200 °F. Degradation begins by the formation of an oil mat at the oil–water surface and then spreads.

The effects of bacterial degradation are readily seen, and although most types of organic matter can be affected, it has already been shown that some types are more susceptible to bacterial attack than others. There are some very specific changes in the composition of crude oils, with the paraffins being the most readily attacked and the aromatics and sulphur compounds the least affected. The result is that degraded oils tend to have a much decreased n-paraffin concentration (see Figure 8.2) and they also have an increased sulphur content because not only does the removal of the paraffins result in the original sulphur compounds becoming a greater percentage of the oil, but also the bacteria produce sulphur compounds as part of their metabolic process which are added to the crude oil. For exactly similar reasons, the nitrogen content of crude oils increases upon biodegradation.

Biodegration also affects the optical activity of the crude oil, with the degraded oil rotating the plane of polarized light to a greater extent than for unbiodegraded. The reason is again twofold. Firstly, the prime target of the bacteria are normal alkanes, most of which are not optically active, so the removal of these will enhance any optical activity. Secondly, most biologically produced compounds, including those produced by bacteria, are optically active and thus the bacteria add optically active compounds and remove non-active ones. In addition, bacteria can be selective and only consume one optical isomer in a racemic mixture so that an optical inactive mixture becomes optically active after bacterial attack. Rullkötter & Wendisch report the microbial alteration of 17 α(H) hopanes by the removal of the C-10 methyl group and ring opening. Such analyses may be another method of identifying bacterial attack on oils.

The API gravity decreases with biodegradation, producing heavy, less valuable oils, and the change can be as much as 20° API.

Gas diffusion

Leythaeuser *et al.* (1982) consider that gas diffusion can have adverse effects on hydrocarbon accumulations. They quote the example of the Harlingen gas field in Holland where by diffusive loss through 400 m of shale cap rock the initial amount of methane, in place of 1.93×10^9 std m^3 (cubic metres of gas at standard temperature and pressure) (6.8×10^{10} scf (standard cubic feet)), is reduced by half over a period of 4½ million years.

Thus petroleum in the reservoir can be affected by at least five agencies, gas diffusion, heat, pressure changes, water washing, and biodegradation, and the combined results of the latter four of these are shown in Figure 8.1.

MODIFIED GENERATION THEORY

Price (1980) suggests that changes in oils due to cracking are functions of increasing depth whereas changes in oils due to biodegradation are a function of decreasing depth. Price notes that most reservoirs are shallower, cooler, and contain fewer potential catalysts than do source beds, and he thus suggests that an immature oil in a reservoir is not likely to become more mature within that reservoir by the action of heat and catalysts. The major factor by which generated oils can be altered in reservoirs shallower than the source beds is bacterial degradation. Price suggested that thermal alteration outside the source bed and in the reservoir to convert heavy oil to light paraffinic oil does not occur. He accepts the reverse alteration by the action of bacteria. Hence he suggests that light paraffinic crudes were such when they were trapped and not the result of thermal alteration of heavier oils. Likewise the naphthenic and aromatic crudes are the result of bacterial attack on paraffinic crudes and not untouched primary oils. This is similar to the theory of Philippi (1977), who has proposed a radical new theory of oil generation and maturation. He considers that paraffinic oil is the 'primary' oil and is the oil originally generated from the source material. Other oils are secondary and are generated from this primary oil, i.e. naphthenic oil is an altered primary or paraffinic oil. His evidence for this theory is as follows. Extracts of the organic matter from shales are very similar to primary oil in terms of wax content, wax composition, and heavy iso-paraffin composition, and they are very different to naphthenic oils. He considers the importance of vertical migration and notes that source beds are, or have been, relatively deeply buried for the generation of primary oil, while most producing basins are younger than their sources, which indicates that vertical migration has occurred. If this happens through faults and fractures it is therefore probably an intermittent process. That a common source exists for primary and medium to heavy oils is shown by the odd–even preference, $2C_{29}/(C_{28} + C_{30})$. For heavy-to-medium oils and primary oils this value is the same, but it is very different for immature shales. Thus he argues that heavy-to-medium oils containing normal paraffins without an odd–even preference, have been expelled from mature source beds as primary oils. He also advances similar arguments using naphthene indices instead of carbon preference indices as additional evidence.

Philippi therefore considers that both primary and medium-to-heavy oils are formed in the one source but that the medium-to-heavy naphthenic oils lose their paraffins and that this loss of paraffins is due to microbial action. Microbes need low temperatures and nutrients, i.e. nitrogen-, phosphorus-, and oxygen-containing compounds, and n-paraffins followed by iso-paraffins are consumed the fastest. The light oil is not transformed to heavy oil above certain temperatures because the microbes are killed by the heat. It is Philippi's contention that the oil is formed at depth as primary oil with many paraffins present from the time of generation. This oil either stays at depth and remains paraffinic, or migrates upwards to zones where the bacteria may alter it and by loss of its paraffins becomes heavy-to-medium naphthenic oil. This of course is in complete contradiction to conventional theories which hold that the oil is generated and the effect of depth (as temperature) cracks this heavy oil to light oil and then to gas.

Philippi cites optical activity as evidence for his theory. Microbial action generally tends to select one optical isomer in preference to the other and thus moderate attack will increase the optical activity as the mutual cancellation of opposite rotation of each isomer will disappear. This is in direct contrast to the ideas of Williams & Winters (1969) who believe that changes in optical activity are due to the addition of more optically active compounds produced as by-products of microbial destruction on alkanes.

MIGRATION

Introduction

So far the origins and alterations of petroleum have been discussed. It has been accepted for at least the last 70 years that oil is not generated within the rocks from which it is produced and that it must have migrated to those reservoir rocks from its source. A lot of information supports this concept of migration and such observational evidence includes the following:

1. Gas, oil, and water are found in reservoirs in order of increasing density.
2. The hydrocarbons are normally trapped at the highest point of that reservoir.
3. The properties which enable a rock to serve as a reservoir often develop later than the onset of hydrocarbon generation. The reservoir rock normally has different physical properties to the source rock. In Chapter 6 the conditions of deposition of source and reservoir rocks were discussed and it was shown that under the more energetic conditions of deposition of the large sand-sized particles, which form many reservoir rocks, most organic matter is destroyed. In the low-energy environments of the deposition of source rocks, organic matter has a greater likelihood of surviving.

All this evidence suggests that trapped hydrocarbons were not formed in reservoir rocks, but migrate into such rocks some time after generation. In addition to this migration from source to reservoir, there will also be movement of hydrocarbons within the reservoir. Primary migration involves the transfer of the generated hydrocarbons from source to reservoir, whereas secondary migration concerns the movement of the oil and gas within the reservoir strata.

The mechanism for primary migration has for a long time been a matter of great debate and argument but it is generally accepted that secondary migration occurs by buoyancy of the various fluids involved. Various mechanisms by which primary migration can occur have been suggested, and these include: (a) movement of the oil as a discrete phase, either as globules or as long stringers; (b) in solution in water, in molecular solution, or as a colloidal suspension by the use of solubilizers; (c) as a gas phase; and (d) along a continuous network of organic material which acts like a wick.

Two things common to all methods of primary migration are the presence of water and the need to inhibit the transportation process at the reservoir. This latter phenomenon is known as accumulation. The pore spaces of all these sediments will be filled with a brine, and it is through this solution that the oil has to migrate. Once the reservoir is reached, an efficient method is required so that the oil stops lateral movement and rises under buoyancy to the top of the trap.

Primary migration

Migration of oil in the oil phase

Dickey (1975) considers that the majority of petroleum is formed between 60 and 160 °C which corresponds to depths of between 1,500 and 4,500 m. At such depths most of the shale source rocks would have lost much of their water and porosity. If a good source rock were to contain 500 ppm of generated hydrocarbons, it is likely that the bulk of the smaller pores would be filled by hydrocarbons. These could be expelled by capillary forces assisted by the fluid potential gradient, and by moving to successively larger openings these could move to the reservoir rock. The forces causing such movement would be buoyancy and the force of the moving water. Buoyancy will only be able to assist movement in a predominantly upwards direction. More horizontal movements must be caused by the pressure of the moving water. Dickey maintains that the quantities of oil generated are too great to dissolve in water even with the aid of solubilizers, and thus movement of the oil as an oil phase must occur. This movement can be as small globules or as long thin stringers of oil. In fact, long thin stringers of oil are more likely to move through small pores than are spherical droplets. The theoretical reason for this is dealt with in the section of this chapter relating to secondary migration.

By the time the depths at which oil generation occurs have been reached, the majority of the water which was incorporated into the original sediment will have already been expelled by the forces of compaction (Table 8.1). Thus compaction reduces the amount of water available to flush the oil out of the source rock. For illitic or kaolinitic clays this expulsion of water has occurred within the first 2 km of burial; however, when clays rich in montmorillonite are compacted there is a second expulsion of water as the montmorillonite changes to illite. This conversion occurs during the principal phase of oil

Table 8.1 Changes in porosity during compaction (Welte, 1972)

Depth (ft.)	Porosity (%)
2,000	27
6,560	17
9,840	9
12,120	6
16,400	4

generation and thus water would be available for assisting the migration of generated hydrocarbons. Emery & Rittenberg (1952), Hobson (1961), and Bonham (1980) have shewn that the net flow of water driven out of clays by early compaction in a subsiding basin is downward. Thus if water was to force the oil through the pores, reservoirs would be deeper and not shallower than the source beds. The other effect of compaction is to reduce porosity and make the pore openings much smaller, which will cause restrictions to the flow of hydrocarbons.

Dickey considers that the water that is left after compaction could mostly be structured water. Structured water is bonded to the clay by hydrogen bonds between the hydrogen of the water and the oxygen of the silica (SiO_4) in the clay. These layers can be several layers of water deep but the structuring decreases with distance from the clay surface (Drost-Hanson, 1969). In such situations the remainder of the pore space could be vacant for oil-phase migration to occur within this pore centre network. Dickey suggests that if the shale is water-wet, the saturation of oil at which oil could flow as a continuous phase could be lower than 10%. This assumes that the water acts as a solid part of the rock matrix and the relative permeability of that matrix to oil is greater than to water and thus subsequent compaction will expel oil and not water. In addition, some fine-grained source rocks which are rich in organic matter could leave their surfaces coated with organic matter. Pores which are coated with organic matter will be oil-wet and will allow oil to flow as an oil phase. The driving mechanisms for fluid movement under such conditions will be pressure gradient, buoyancy, and expansion of the oil and water due to temperature increase.

Price (1980) considers that bulk oil-phase movement is possible at high temperatures. The increased temperature will reduce the viscosity of the oil, and increase the pressure, which causes free gas to redissolve in its associated oil and further reduce the oil viscosity. He also suggests that in deep basins pressure will open fault planes and keep them open long enough until the abnormal pressure is relieved by the vertical fluid movement. As the fluids move up the faults, their temperature will fall and any hydrocarbons which were in water solution will come out of solution, and eventually a continuous hydrocarbon phase will exist and fluid pressure will overcome the surface tension forces so that the oil or gas moves as a continuous phase up the fault.

An alternative for oil-phase migration is where the hydrocarbons move in discrete globules, and so long as these globules are smaller than the pore openings there will be no restriction to movement and the globules can move under the effects of buoyancy or be flushed through the rock by the action of flowing water. Welte (1972) quotes the sizes of typical molecules found in crude oil and compares them with the size of pore openings at various shale porosities. At high porosities (15% and above) even the largest asphaltene molecules could pass through the pore openings. Momper (1978) believes that in a compact, low-porosity shale, more than 70% of the pore openings are less than 3 nm in diameter. The figures given by Welte (1972) would restrict the migration through these pores of molecules smaller than the complex multiring compounds. The larger asphaltene molecules could only migrate if the remaining 30% of larger pore entries were interconnected. While many of the asphaltene molecules could be formed in the reservoir by the combination of simpler molecules, the same is not true of the multiring natural product molecules at the upper end of the size range which can pass through average sized pores.

Buoyancy has been suggested as one mechanism for the migration of these oil droplets. This relies upon the fact that oil is less dense than water, especially strong brines, and will normally float on top of such water. However, in the subsurface the oil droplets have also to negotiate the pore network as well as to move through the water. The diameter of these globules will almost certainly be larger than the pore openings, and Hobson (1973) has shewn that very high pressures are required to force a spherical globule through a small pore. If the oil globule is not spherical but elongated, then the required pressure is slightly lowered. If the oil is in the form of a long vertical stringer it is possible that buoyancy of a long column (several metres) of oil could overcome the pressures needed to pass the globule through the pore openings. Such a process involving oil stringers would need fresh oil added to the base of the stringer all the time and thus the expulsion rate would determine the migration rate (see the section on secondary migration for details of the physics of such movements). At great depths and pressures, sufficiently high pressure could be available to force oil droplets through pores. At such depths light hydrocarbon would be generated and the great increase in molar volume would assist the pressure increase. However, at such high pressures the host rocks will be liable to suffer mechanical failure and microfractures will occur which would allow the release of pressure by the migration of the oil.

Barker (1974) has suggested that if oil droplets block the pore openings, these droplets could be forced through the pores when the water expands. This will be limited to small pores where all the exits are blocked by oil droplets and the volume of the water is much greater than of the oil, conditions which may not always exist. Barker suggests that a depth increase of 10 feet will cause sufficient temperature and pressure rise (1 atmosphere) to expel the oil. A separate repressuring will be required at every constriction.

Migration of oil by microfracturing

Reference to microfracturing has already been made while discussing the migration of the oil in the oil phase. Momper (1978) believes that oil-phase migration cannot begin until the source rock has generated at least 850 ppm of oil and that at the peak of generation there will be a 25% increase in molar volume of the organic matter. This volume increase will create pressure build-ups until the mechanical strength of the rock is exceeded and failure takes place. The result of this pressure-induced mechanical failure will be either microfracturing or the reopening of existing fractures which will allow the expulsion of the oil. Once the oil is expelled the fractures will close, and no oil will flow until the pressure again builds up and these fractures reopen or new ones are created. Thus oil is expelled in a pulsed fashion.

Snarsky (1962), Tissot & Pelet (1971), and Lewan *et al.* (1979) pointed out that dense, deeply buried shales do contain such fractures and that they are caused by the overpressuring of either existing pore fluids or of small pockets of oil and gas generated from the kerogen. It has been observed that inclusions of resins within coals can fracture the coals if the resin is liquefied. When this occurs the resin flows in to the cracks it has produced and solidifies. Such cracks are not seen in immature rocks.

This method does not require the presence of water for migration although thermal expansion of water may be instrumental in causing the microfracturing. Fracturing of the rock is believed to be caused by the effects of increased temperature and pressure, resulting from burial, upon pore fluids. These pore fluids can be the interstitial water or the oil and gas generated from the kerogen. Increased burial will raise the temperature and convert the kerogen to fluid hydrocarbons which will have a greater molar volume than the kerogen whence they came. Because the volume these hydrocarbons can occupy is limited, the increase in molar volume will cause an increase in pressure in addition to that caused by the increasing temperature (see the ideal gas equation in Chapter 5). Thus there are two agencies tending to increase the pressure of the generated hydrocarbons and when that fluid pressure exceeds the mechanical strength of the rock, fracturing will occur and oil will flow until the pressure drops sufficiently to allow the fracture to close.

Both these two types of migration, which are both oil phase migration, do not require the presence of water for migration and accumulation of the oil, i.e. the cessation of primary migration, will occur when the oil is unable to travel any further either because of decreases in porosity and permeability or because it has reached the highest part of the reservoir sand, e.g. in an anticline.

The effect of increasing burial will be to raise the temperature of the generated oil, to lower its viscosity, and raise its pressure, assuming the volume remains constant. The first makes oil movement easier and the last can be a driving force for migration. The thermal conductivity of rocks varies: coal is a good insulator but salt, anhydrite, and quartzite are good thermal conductors. The presence of these good conductors of heat may be a benefit to migration.

Microfractures caused in these processes will be considerably larger than the pore openings and thus microfracturing will allow the movement of the very larges of molecules to be found in crude oil. This is the only proposed migration mechanism which will allow molecules as large as the asphaltenes to migrate.

The fluids in the pores immediately before fracturing will be at high pressure in a relatively small volume. The opening of the microfractures will immediately increase the volume and reduce the pressure of the hydrocarbons. There can therefore be changes in the composition of the phases and the system may change from a single- to a two-phase system (both liquid and vapour). The composition of the hydrocarbons in the two phases will depend upon the thermodynamic conditions before and after opening the microfracture (Chapter 5). The microfractures could connect the pockets where the oil is generated with the conventional fault systems where oil-phase migration into a reservoir could occur.

Gas-phase migration

As described above, the opening of microfractures can cause changes in the physcial state of the hydrocarbons. Sokolov *et al.* (1963) and Hunt (1979) quote Sokolov & Mironov (1962) as having shewn that subsurface gases will dissolve large quantities of liquid hydrocarbons under the pressure and temperature conditions which correspond to depths of 6,000 to 10,000 ft. Neglia (1969) has shewn that the Malossa field in Italy is condensate with heavy hydrocarbons dissolved in the gas phase. Rzasa & Katz (1950) indicate that hydrocarbons as large as $C_{18}H_{38}$ will dissolve in the gas phase under the existing conditions (1050 atmospheres and 153 °C). It is quite likely that these gases, as they migrate through the fractures, probably in a generally vertical direction, will dissolve generated oil from adjacent pores, thus increasing the amount of oil which the gas is transporting. Accumulation is probably due to the gas reaching a depth where the reduced temperature and pressure cause retrograde condensation with the formation of an oil phase. Leythaeuser *et al.* (1982) consider that diffusion of gases from source to reservoir can occur. They suggest that 10^9 kg or 5.3×10^{10} scf could have diffused from 1000 km^2 of 200 m thick source rock in the Mesozoic of Western Canada over 540,000 years. Because of the exponential increase in diffusion coefficients with increasing molecular weight (methane = 2.12×10^{-6} cm^2/sec; *n*-decane = 6.08 $\times 10^{-9}$ cm^2/sec), this process will be restricted to the light hydrocarbons and will exclude the petroleum-range compounds.

Kvenvolden & Claypool (1980) suggest that gas and gasolene-range hydrocarbons in a seep in the Norton Sound, Alaska, are extracted from the sediment and transported in the gas phase by upward-migrating carbon dioxide. The mixture is 98% carbon dioxide and less than 0.1% hydrocarbons. The hydrocarbons, having migrated in the gas phase, probably undergo retrograde condensation in the lower pressure and temperature of the shallower environments. The fact that only gas and gasolene-range hydrocarbons are found in this case agrees with McAuliffe (1979), who considers that high molecular weight hydrocarbons are unlikely to be able to travel in the gas phase. The possibility of these larger hydrocarbons being transported in carbon dioxide–saturated water (Momper, 1978) will be discussed later. Kvenvolden & Claypool state that the major component of the seep is the carbon dioxide with a $\partial^{13}C$ value of $-2.7°/oo$, and that this carbon dioxide may have a geothermal origin and come from the decarbonation, by heat or fluids, of carbonate basement rocks.

Migration of oil along a network of solid organic matter (kerogen wick)

McAuliffe (1979) suggested the possibility of migration of oil along continuous networks of insoluble organic matter (kerogen) which is distributed throughout the source rock. As with the methods of migration already discussed, such a migration process would be independent of the quantity, nature, and movement of any water in the shale. McAuliffe has shewn by the use of scanning electron micrographs that such interconnected networks of organic matter can exist in source rocks which contain only 1–6% organic matter. Momper (1978) has suggested that the network is most complete in two dimensions along the bedding planes.

McAuliffe's micrographs only shew a very small section of a source rock and there is no evidence that these networks do extend throughout source rocks. Certainly a minimum amount of kerogen must be available. Hobson (1980) suggests that with a high organic content this network could be in three dimensions, but at lower organic content the network would be two dimensional along the bedding planes. Hobson suggests that even if the network were not continuous there could be a suitable bulge of migrating hydrocarbon on the end of one strand of kerogen which could link across to another fibre of the organic matter, and thus oil flow would continue even with small breaks in the network.

The cause of flow of oil in such a system would be the increased pressures due to the effects of compaction, hydrocarbon generation, and thermal expansion of generated products. The latter two causes will be related to the subsurface temperature. McAuliffe estimated that the minimum level of oil saturation within the rock which would cause hydrocarbon flow is between 2½ and 10%.

Hobson (1980) wonders if the causes of flow could be continued oil generation pushing the oil along the kerogen wick, or whether compaction would squeeze oil along the organic matter network. He notes that the action of a lamp wick is by the oil moving up the wick to replace that which is burnt, while drip-feed lubrication is by syphon action where the oil leaves the wick below the level of the oil in the reservoir. Hobson then shewed that paraffin could move along a wick which was immersed in salt water and that the paraffin was released when it reached the end of the wick. The paraffin would flow if the pressure of the paraffin feeding the wick exceeded that of the water where the paraffin is released. Thus such a system would require a continued pressure applied to the hydrocarbons either from increasing compaction, or the generation of fresh oil, or the thermal expansion of already generated oil.

The constraints on the migration of oil along a network of organic matter include the need to have sufficient organic matter present within the source rock to act as a wick, and the need to have a source of pressure on the hydrocarbons to cause flow. Both these requirements can be met in some source rocks and the only unknown is the rate of flow of oil along such wicks. The need to have sufficient organic matter to assist with migration, either as the wick or as a source of hydrocarbons and thus as a driving force, is one of the reasons that a source rock has to have more than a minimum amount of organic carbon.

Migration of oil in water solution

The molecular solution of hydrocarbons in water has been suggested as a means of transporting generated oil from the source rock to the reservoir. McAuliffe (1966) investigated the solubilities of various hydrocarbons and shewed that small molecules dissolve more readily than large ones. Likewise, he shewed that paraffins are the least soluble group of compounds, followed by naphthenes, then the aromatics, with the heteroatomic ring compounds the most soluble. Price (1976) suggested that the solubility of petroleum hydrocarbons in water at 100 °C is many times greater than in water at 25 °C, and the heavy hydrocarbons shewed the greatest solubility increase with increased temperature. Above 100 °C the solubilities increased dramatically and above 180 °C molecular solution could account for the formation of petroleum accumulations.

Price (1976) suggests that both temperature decrease and increased salt content of the water could cause release of the hydrocarbons. He quotes salinities of 150,000 ppm NaCl starting to cause the release of dissolved hydrocarbons, and 95% liberation of dissolved oil and gas would occur with a salinity of 350,000 ppm. However, many interstitial waters are already concentrated brines and thus these might not dissolve the hydrocarbons in the source, although Price does state that with increasing depth the expelled water (which would carry the oil) becomes fresher (less saline).

Many geologists would consider that the solubility of petroleum hydrocarbons is too low for molecular solution to be an important primary migration mechanism. Price (1976) overcomes objection this in two ways. Firstly, he suggests the high temperature origin of oil (180 °C) where the hydrocarbons are sufficiently soluble in the water, and secondly, by the dewatering of montmorillonite when it changes to illite. This conversion gives water equal to 15.5–17.6% by volume of the sediment. If potassium or aluminium are lacking from the clay this conversion will be delayed until depths are reached at which the subsurface temperature is sufficiently high to dissolve the oil.

Bonham (1980) is also a supporter of the migration theory of molecular solution of hydrocarbons. He suggests that as more sediment units are added to a subsiding basin, the water stays in the same place relative to the depositional surface but rises relative to stratigraphic markers, i.e. the sediments slowly pass through a fixed body of water. In the hotter, deeper basins especially, this could provide an excellent extraction process for removing the generated hydrocarbons. He suggests that any cooling of the water will release the hydrocarbons.

In spite of all these supporters of the molecular solution theory of migration, there are many unanswered questions which cast doubt on this mechanism. The relative solubility of hydrocarbons in oilfield waters is almost in the reverse order to the relative concentration of hydrocarbons in reservoired crude oil. The light aromatic hydrocarbons such as benzene and toluene are more soluble in water than the saturated hydrocarbons and yet it is the saturated hydrocarbons which predominate in most undegraded oils, with the aromatic hydrocarbons being minority components. If oil accumulations represent the proportions in which hydrocarbons can dissolve in water, then all source rocks should be completely depleted in aromatic material. However, source rock analysis shews that aromatic hydrocarbons are still present in source rocks from which migration of oil has occurred. The only way for crude oil composition and molecular solution migration to be compatible is for the nature of the accumulation to depend upon the means by which hydrocarbons are released from the water. The least soluble materials would need to be almost totally liberated from the water while the more soluble compounds remain in solution. Even so, the vast quantities of water required for this mechanism mean that it is unlikely to be a major factor in oil migration. The majority of clays lose their water during compaction well before hydrocarbons are generated and there is therefore very little water available in the sediment at the time of generation to dissolve hydrocarbons.

Bray & Foster (1980) suggest that water saturated with carbon dioxide, possibly at pressures and temperatures of up to 6,000 psi and 100 °C, will dissolve 3% by volume of medium-gravity oil. This is over 100 times more than quoted by Price (1976) for molecular solution of hydrocarbons in water at the same temperature. They consider that saturation of the pore water of the source rock may occur near, and prior to, the onset of strong oil generation. Many geochemists would argue that the principal zone of carbon dioxide generation occurs well before the principal zone of oil generation and thus the carbon dioxide would not be available to help dissolve hydrocarbons.

Migration in water by use of solubilizers

Bray & Foster (1980) suggest that migration in compacting basins is by the expulsion of the carbon dioxide–saturated water carrying the generated oil. In non-compacting basins the migration would occur by the diffusion of the hydrocarbons through the carbon dioxide–saturated water. Accumulation is by the release of the hydrocarbons following the removal of the carbon dioxide from the water. If the water migrated upward to regions of lower pressure, the carbon dioxide would come out of water solution and a liquid hydrocarbon phase would separate. This would set an upper depth limit for such a migration and would result in oils rich in carbon dioxide. Because this latter situation is not common, Bray & Foster suggest that the carbon dioxide is removed from the water by interaction with the rocks. They quote equations for the reactions of smectites and feldspars with water and carbon dioxide to give kaolinite, silica, and various metal bicarbonates. They also give details of experiments to support this theory.

Many polar organic molecules when in water form themselves into aggregates called micelles (Figure 8.3). The action and chemistry of soap in water is typical of these molecules. Soaps are the alkali metal salts of long-chain fatty acids. Fatty

Micelles

$CH_3(CH_2)_N CH_2 COOH$ Fatty acid

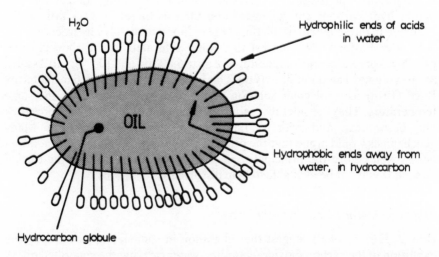

Figure 8.3 Diagrammatic representation of a micelle and its mode of solubilizing hydrocarbons

acids are long alkane chains with a carboxylic acid group at one end. The acidic group is hydrophilic whereas the alkyl chain is hydrophobic, and the formation of the micelle means that the hydrophilic ends will be in water and the hydrophobic alkane chains will stay together and away from the water. Soap acts by picking up dirt, which is often greasy and hydrocarbon-like, and containing it inside the micelles where it is readily miscible with the alkyl ends of the soap molecules. The whole micelle, including its entrapped grease, is soluble in water because of the hydrophilic acidic groups, and so may be washed away with moving water.

It has been suggested that generated hydrocarbons are solubilized by such soap micelles within the formation waters. Baker (1959) suggests that as little as 0.05% soap in the water will allow the hydrocarbons to dissolve and be transported away from the source rock. Cordell (1972, 1973) and Baker

(1962) give more evidence for the transportation of generated hydrocarbons in soap micelles as a major mechanism of primary migration. Many young immature oils often contain fatty acids and other micelle-forming heteroatomic compounds which could have been the surfactants which were involved in solubilizing the oil during its migration. These compounds are not found in the more mature oils and are probably degraded as the oil matures. These heteroatomic solubilizers would have the dual role of hydrocarbon solubilizers during migration and petroleum precursors. There is no doubt that such compounds will allow much greater quantities of hydrocarbons to be dissolved in water. Baker (1962) suggests that the type and size of the micelles will decide the quality of oil in reservoirs by preferential solubilizing of different hydrocarbons. Ionic micelles would have diameters of 6 nm whereas the neutral micelles could be as large as 500 nm.

Tissot & Welte (1978) suggest that because of the size of the pores, any migration process using such micelles would be limited to the first 2000 m. After that depth the pores would be too small to allow the micelles to pass through. The main problems are that the peak of oil generation often does not occur until after 2000 m and the smaller ionic micelles could be absorbed on to the charged surfaces of clay minerals and thus not be able to migrate. Additional factors which make the micellar solution theory difficult to accept are the lack of evidence for surfactants, such as fatty acids, being present in the source rocks in sufficiently large enough quantities and the lack of a suitable mechanism to release the oil from the micelles in the reservoir. Although some work has shown that oil-in-water emulsions, when passed through sands, are broken down and the oil released as the sand grains get finer, this is not entirely satisfactory as it would imply that oil only accumulated in fine-grained sands.

Peake & Hodgson (1967) note that n-alkanes can be accommodated in water in colloidal solutions in amounts much greater than their true solubility. The n-alkanes are accommodated in direct proportion to their abundance and thus any odd–even preference of the alkanes in the source will be preserved during migration and will be observed in an oil accumulation resulting from the dispersal of those colloidal solutions.

Timing of migration and organic richness

The timing of primary migration will depend upon the mechanism by which hydrocarbons are moved from source to reservoir and upon the quantity of hydrocarbons. Tissot & Welte (1978) split primary migration into early and late phases. The late migration phase occurs during the main phase of hydrocarbon generation, at depths between 1,500 and 3,500 m. Late migration occurs after the great quantities of pore water have been expelled during the initial phase of compaction, during which time early migration can be occurring.

The quantity of hydrocarbons generated can determine the delay between the generation of oil and gas and the onset of migration. A very rich source, whose generated products had a much larger molar volume, could initiate flow in a continuous phase either through the pore water or along the kerogen wick, or could cause microfracturing early in the generation process. In a leaner source rock there would be a much longer delay before sufficient quantities of hydrocarbons had been generated which would cause migration to begin. Momper (1978) states that 850 ppm of hydrocarbons must be generated before migration can occur. The time that elapses before this quantity of hydrocarbons is generated will depend upon the amount of organic material present in the rock. A source rock rich in kerogen will generate the 850 ppm of hydrocarbon much more quickly than a lean source rock. If a source rock does not contain sufficient organic matter, migration will never occur. This is one of the reasons that a lower limit of 0.5% of organic carbon for clastic rocks and 0.3% for evaporites is considered necessary before a rock can be considered a source rock.

Secondary migration

Secondary migration is the movement of the hydrocarbon fluids within the reservoir. The factors which cause the accumulation of hydrocarbons, e.g. changes in temperature, pressure, grain size, are normally gradual changes and thus there is no drastic dissolution of hydrocarbons. Hence oil and gas can form as separate phases in the reservoir rock over quite long distances and will eventually migrate to form a pool where their upward movement is limited by strata of permeability too low to allow them to proceed further. It is very unlikely that hydrocarbons are released from their mode of transportation exactly where they are reservoired. The lower limits of permeability which decide the point at which hydrocarbons stop migrating will vary depending upon the nature of the hydrocarbon mixture.

The nature of the seal is of interest because the seal or cap rock need not be a separate and distinct rock unit. A decrease in permeability within one sand unit can be a sufficent seal for one oil although it might not act as a seal for a different hydrocarbon mixture. Shales, which generally have low permeabilities, are good cap rocks and thus a thick shale may not release all its oil to the neighbouring sand. Only the oil generated in the parts of the shale close to the sand will be able to move out of the shale into the sand. Any generated oil deep in the heart of a thick shale will stay there for much longer times, possibly for ever. Thus a source rock may provide a good seal to a reservoir rock immediately below it. The overall ratio and distribution of sand and shale is thus seen to have great importance. Although there must be sufficient shale to source the oil to fill the reservoir there must not be too much shale so that the oil is trapped in the shale by virtue of the shale's low permeability. Thus, alternating beds of sand and shale and moderate thickness can make ideal source and reservoir rock pairs.

Buoyancy, the tendency of the lighter hydrocarbons to rise above the denser salt water, is a primary agency which accounts for vertical secondary migration. The oil and gas globules have to move through small pore openings and the

(a)

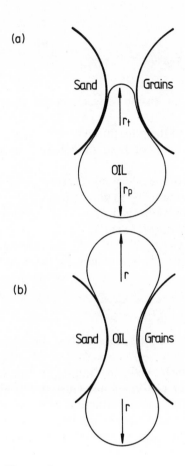

Figure 8.4 Distortion of an oil globule passing through a pore neck

interfacial tension between the hydrocarbon and the water, or capillary pressure, will hinder the movement of the hydrocarbon droplets. The smaller the pore opening, the greater is the capillary pressure. Most globules are of larger diameter than the pore throats through which they have to move and thus the globule has to be distorted during its passage through the neck of the pore. Buoyancy is the upwards driving force and capillary pressure the restraining force.

Once the globule is distorted, i.e. when one end has a smaller radius than the other, the narrow end can enter the neck of the pore. When part of the globule is through the pore, the radii at each can become the same and buoyancy forces will push the globule through the pore (Fig. 8.4). Tissot & Welte (1978) quote Gussow (1954), Hobson (1961), and Hobson & Tiratsoo (1975) in describing the theoretical mathematics involved. The equilibriumn state between buoyancy and capillary pressure is given by:

$$2\gamma\left(\frac{1}{r_\mathrm{t}} - \frac{1}{r_\mathrm{p}}\right) = Z_\mathrm{o}g(\rho_\mathrm{w} - \rho_\mathrm{o})$$

where γ is the interfacial tension between oil and water (dynes/cm)

$\quad r_\mathrm{t}$ is the radius of the pore throats

$\quad r_\mathrm{p}$ is the reservoir rock pore radium (i.e. maximum diameter of oil droplet)

$\quad Z_\mathrm{o}$ is the height of the oil column (cm)

$\quad g$ is the acceleration due to gravity (cm/s^2)

$\quad \rho_\mathrm{w}$ is the density of water (g/cm^3)

$\quad \rho_\mathrm{o}$ is the density of the oil (g/cm^3)

Hobson (1954) and Berg (1975) give the critical height (Z_c) or the minimum height of oil droplet required before movement through the pore begins, as:

$$Z_\mathrm{c} = 2\gamma\left(\frac{1}{r_\mathrm{t}} - \frac{1}{r_\mathrm{p}}\right)\frac{1}{g(\rho_\mathrm{w} - \rho_\mathrm{o})}$$

Thus a long thin stringer of oil is more likely to move through a fine pore than is a spherical globule.

The other form which secondary migration can take often occurs after regional tilting. Barker (1979) quotes Gussow (1954) who used the phrase 'differential entrapment' for the situation where a trap spills hydrocarbons from its bottom into the next higher trap. If this process is continued it will lead to a series of adjacent traps within the same reservoir (Figure 8.5). The

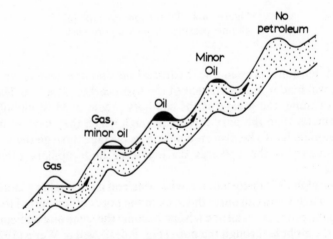

Figure 8.5 Differential entrapment (Barker, 1979. Reproduced by permission of The American Association of Petroleum Geologists)

gas-filled reservoirs will be down dip from the oil-filled reservoirs. The uppermost trap which contains any hydrocarbons will not be completely full and only contains a small amount of oil, well above the spill point. The next trap down will contain just oil to the spill point. The next trap will contain gas on oil (to the spill point) while the lowest trap will be full of gas.

REFERENCES

Bailey, N. J. L., Evans, E. R., & Rogers, M. A. (1971) Evolution and alteration of petroleum in Western Canada. *Chem. Geol.*, **8**, 147.

Bailey, N. J. L., Evans, C. R., & Milner, C. W. D. (1974) Applying petroleum geochemistry to search for oil: Examples from Western Canada Basin. *BAAPG*, **58**(11), 2284.

Bailey, N. J. L., Jobson, A. M., & Rogers, M. A. (1973a) Bacterial degradation of crude oil. *Chem. Geol.*, **11**, 203.

Bailey, N. J. L., Jobson, A. M., & Rogers, M. A. (1973b) Alteration of crude oil by water and bacteria. *BAAPG*, **57**, 1276.

Baker, E. G. (1959) Origin and migration of oil. *Science*, **129**, 871.

Baker, E. G. (1962) Distribution of hydrocarbons in petroleum. *BAAPG*, **46**, 76.

Barker, C. (1974) Some thoughts on primary migration of hydrocarbons. *AAPG--sepm ann. Mtg.*, **1** (April), 4.

Barker, C. (1979) *Organic Geochemistry in Petroleum Exploration*. AAPG Continuing Education Course Note Series No. 10.

Berg, R. R. (1975) Capillary pressures in stratigraphic traps. *BAAPG*, **59**, 939.

Bonham, L. C. (1980) Migration of hydrocarbons in compacting basins. *BAAPG*, **64**, 549.

Bray, E. E., & Foster, W. R. (1980) A process for primary migration of petroleum. *BAAPG*, **64**(1), 107.

Button, D. K. (1976) Influence of clay and bacteria on concentration of dissolved hydrocarbons in saline solutions. *Geochim. Cosmochim. Acta*, **40**, 435.

Claypool, G. E., & Rice, D. D. (1981) Generation, accumulation and resource potential of biogenic gas. *BAAPG*, **65**(1), 5.

Connan, J. K., Letran, K., & Van der Weide, B. (1975) Alteration of petroleum in reservoirs. In *9th World Petroleum Congress, Proc.*, Vol. 2, London. Applied Science Publishers, pp. 171–178.

Cordell, R. J. (1972) Depth of oil origin and primary migration: A review and critique. *BAAPG*, **56**, 2029.

Cordell, R. J. (1973) Colloidal soap as a proposed primary migration medium for hydrocarbons. *BAAPG*, **57**, 1618.

Demaison, G. T. (1977) Tar sands and supergiant oilfields. *BAAPG*, **61**, 1950.

Deroo, G., Tissot, B., McCrossan, R. G., & Der, F. (1974) Geochemistry of the heavy oils of Alberta. In: Oil sands: Fuel of the future. *Memoir 3, Can. Pet. Geol.*, 148–167, 184–189.

Dickey, P. A. (1975) Possible migration of oil from source rock in oil phase. *BAAPG*, **59**, 337.

Drost-Hanson, W. (1969) Structure of water near solid surfaces. *Ind. Eng. Chem.*, **61**, 10.

Emery, K. O., & Rittenberg, S. C. (1952) Early diagenesis of California Basin Sediments in relation to the origin of oil. *BAAPG*, **36**, 735.

Evans, C. R., Rogers, M. A., & Bailey, N. J. L. (1971) Evolution and alteration of petroleum in Western Canada. *Chem Geol.*, **8**, 147.

Gould, K. W., & Smith, J. W. (1978) Isotopic evidence for a microbiological role in the genesis of crude oil from Barrow Island, Australia. *BAAPG*, **62**, 455.

Grant, C. W., Haus, H. F., & Zobell, C. E. (1943) Marine organisms which oxidize petroleum hydrocarbons. *BAAPG*, **27**, 1175.

Gussow, W. C. (1954) Differential entrapment of oil and gas. A fundamental principle. *BAAPG*, **38**, 816.

Hedges, J. I. (1977) Association of organic molecules with clay minerals in solution. *Geochim. Cosmochim. Acta*, **41**, 1119.

Hobson, G. D. (1954) *Some Fundamentals of Petroleum Geology*. Oxford University Press, London-New York-Toronto.

Hobson, G. D. (1961) Problems associated with the migration of oil 'in solution'. *J. Inst. Pet.*, **47**, 170.

Hobson, G. D. (1973) The occurrence and origin of oil and gas. In G. D. Hobson (Ed.), *Modern Petroleum Technology* (4th edn) London, Applied Science Publ. on behalf of Inst. Pet. Great Britain.

Hobson, G. D. (1980) Musing on migration. *J. Pet. Geol.*, **3**, 237.

Hobson, G. D., & Tiratsoo, E. N. (1975) *Introduction to Petroleum Geology*. Scientific Press, Beaconsfield, England.

Hunt, J. M. (1979) *Petroleum Geochemistry and Geology*. W. H. Freeman & Co., San Francisco.

Jobson, A., Cook, F. D., & Westlake, D. W. S. (1972) Microbial utilization of crude oil. *Appl. Microbiol.*, **23**(6), 1082.

Kvenvolden, K. A., & Claypool, G. E., (1980) Origin of gasolene range hydrocarbons and their migration by solution in carbon dioxide in Norton Basin, Alaska. *BAAPG*, **64**(7), 1078.

Lewan, M. D., Winters, J. C., & McDonald, J. H., (1979) Generation of oil like pyrolyzates from organic rich shales. *Science*, **203**, 897.

Leythaeuser, D., Schaefer, R. G., & Yukler, A. (1982) Role of diffusion in primary migration of hydrocarbons. *BAAPG*, **66**(4), 408.

McAuliffe, C. D. (1966) Solubility in water of paraffin, cycloparaffin, olefin, acetylene, cyclo-olefin and aromatic hydrocarbons. *J. Phys. Chem.*, **70**(4), 1267.

McAuliffe, C. D. (1979) Oil and gas migration—chemical and physical restraints. *BAAPG*, **63**, 761.

Milner, C. W. D., Rogers, M. A., & Evans, C. R. (1977) Petroleum transformations in reservoirs. *J. Geochem. Expl.*, **7**, 101.

Momper, J. A. (1978) Oil migration limitations suggested by geological and geochemical considerations. In: Physical and chemical constraints on petroleum migration. Vol. I. Notes for AAPG short course, April 19, 1978. AAPG National Meeting, Oklahoma City.

Neglia, S. (1979) Migration of fluids in sedimentary basins. *BAAPG*, **63**(4), 973.

Oppenheimer, C. H. (1960) Bacterial activity in sediments of shallow marine bays. *Geochim. Cosmochim. Acta*, **19**, 244.

Peake, E., & Hodgson, G. S. (1967) Alkanes in aqueous systems. I. The accommodation of C_{12}–C_{36} n-alkanes in distilled water. *J. Amer. Oil Chemists Soc.*, **44**, 696.

Philippi, G. T. (1977) Depth, time and mechanism origin of heavy to naphthenic crude oils. *Geochim. Cosmochim. Acta*, **41**, 33.

Price, L. C. (1976) Aqueous solubility of petroleum as applied to origin and migration. *BAAPG*, **60**, 213.

Price, L. C. (1980) Crude oil degradation as an explanation of the depth rule. *Chem. Geol.*, **28**, 1.

Price, L. C. (1980) Utilization and documentation of vertical oil migration in deep basins. *J. Pet. Geol.*, **2**, 353.

Rubenstein, I., Strauss, P. O., Spyckerell, C., Crawford, R. J., & Westlake, D. W. S. (1977) The origin of the oil sands of Alberta. *Geochim. Cosmochim. Acta*, **41**, 841.

309

Rullkötter, J., & Wendisch, D., (1982) Microbial alteration of 17α(H)-hopanes in Madagascar asphalts: removal of C-10 methyl group and ring opening. *Geochim. Cosmochim. Acta*, **46**, 1545.

Rzasa, M. J., & Katz, D. L., (1950) The coexistence of liquid and vapour phases at pressures above 10,000 psi. *Trans, AIME*, **189**, 119.

Snarsky, A. N. (1962) *Primary Migration of Oil* (in German): Freiberger Forschungsch. C123, p.63.

Sokolov, V. A., & Mironov, S. I. (1962) On the primary migration of hydrocarbons and other oil components under the action of compressed gases. In: *The Chemistry of Oil and Oil Deposits*. Acad. Sci. USSR Inst. Geol. and Exploit. Min. Fuels, pp. 38–91 (in Russian). English translation by Israel Programme for Scientific Translation, Jerusalem, 1964.

Sokolov, V. A., Zhuse, T. P., Vassoevich, N. B., Antonov, P. L., Grigoryev, G. G., & Kozlow, V. P. (1963) Migration processes of oil and gas, their intensity and directionality. Paper presented at 6th World Petroleum Congress, 19–26 June, Frankfurt.

Tissot, B., & Pelet, R., (1971) Nouvelles données sur les mecanismes de genèse et de migration du pétrole, simulation mathématique et application à la prospection. *Proc. 8th World Pet. Cong. Moscow*, PDI (4).

Tissot, B. P., & Welte, D. H. (1978) *Petroleum Formation and Occurrence*. Springer-Verlag, Berlin, Heidelberg, New York.

Welte, D. H. (1972) Petroleum geochemistry and organic geochemistry. *J. Geochem. Explor.*, **1**, 117.

Williams, J. A., & Winters, J. C. (1969) Microbiological alteration of crude oil in the reservoir. Paper presented at Symposium on Petroleum Transformation in Geologic Environments. American Chemical Society. Division of Petroleum Chemistry. 7–12 Sept., New York City. Paper PETR. 86:E22-G31.

Zobell, C. E. (1964) GAdvances in Organic Geochemistry. New York, Macmillan, pp. 339–356.

Chapter 9

Applied Organic Geochemistry

INTRODUCTION

In preceding chapters it has been shewn how sedimentary organic matter can be converted into petroleum-like hydrocarbons, and that this production of crude oils from the original sedimentary organic matter can occur in several ways. Firstly, the geochemical fossils produce compounds with no or very little alteration and with no major change to the basic skeleton so that they are readily recognizable. Secondly, there are the secondary products which come from the simple breakdown of compounds in the original sedimentary organic matter, e.g. the production of alkanes from fatty acids by decarboxylation. Thirdly, there are the tertiary products which come from the production and subsequent breakdown of kerogen, whose nature and products will depend upon its origin.

The examination of the organic matter in the source rock will allow one to investigate the compounds which have been produced by these three types of mechanism and it will also allow the inspection the remnants of the kerogen and pieces of the original sedimentary organic matter. The kerogen breakdown is a disproportionation reaction giving hydrogen-rich fluid and hydrogen-lean insoluble compounds, and thus both the chemical and physical properties of kerogen change with hydrocarbon production. We can look at the organic matter in the potential source rock and ask four fundamental questions, and the answers to all four need to be in the affirmative for our potential source to be a useful generating source.

The four questions are: 'Is there sufficient organic matter?'; 'Is it the correct type of organic matter?'; 'Is it at the correct level of maturation?'; and 'Has migration of generated hydrocarbons from the source into the reservoir occurred?' A 'Yes' to each of these will give hope that reservoired oil and/or gas might be found in the vicinity of that source bed. However, these questions can be answered much more thoroughly than a simple 'Yes' or 'No' and they can be answered at different stages of the exploration procedure, from the regional review, before any drilling has been done, through to the basin evaluation once production is underway.

In addition to this, a comprehensive reconstruction of palaeoenvironments, which will give source type, together with thermal history, which will give maturation level, will yield a picture of hydrocarbon production of the region and the possible products in different areas. This, if combined with reservoir sands and migration fairways, should increase the success rate when drilling.

The object of the next two chapters is to give an insight into the way that this may be done, together with the problems and pitfalls.

The first problem when collecting samples for geochemical analysis is to avoid contamination (Gibbs, 1970; Douglas and Grantham, 1973; Debyser, 1975). Contact with hydrocarbons or sources of hydrocarbons such as plastics, waxes, etc., must be avoided. It is not always possible to keep contaminants out of samples, due to such things as additives to drilling muds or careless unknowing workers, and a reference collection of all possible pollutants must be kept. This will be especially useful when using gas–liquid chromatography.

Similarly, all solvents used in geochemical analyses should be doubly or triply distilled, analysed for impurities and a log kept of which solvent sample is used for each operation.

If surface or shallow core material is being used great care must be exercised that the material is unweathered (Leythaeuser, 1973).

SURFACE EXPLORATION

The use of hydrocarbon products goes back a very long time, especially the use of material obtained from seeps. 'And God said to Noah ... make thee an Ark of Gopher wood, rooms shalt thou make in the Ark and thou shalt pitch it within and without with pitch' (Genesis, ch.VI, v.13–14).

An oil seep is evidence on the surface of the earth, or in the waters thereon, that at some point below there are hydrocarbons. The size of the seep may vary greatly from minute traces of liquid or gas in water or in soils, to tar springs and the large visible oil seeps. Oil from seeps has been used for centuries, and originally oil-contaminated water was considered of no value as it could not be used for irrigation of plants or for the watering of livestock as the hydrocarbons it contained would poison both flora and fauna. Therefore in the early days of the settler's exploration of the American West, any land that had contaminated water was not wanted by the white settlers and often given to the Indians, a policy which caused problems once the relationship between oil seeps and underground oil accummulations had been realized.

The original oilwell, the Drake well (1859) in Pennsylvania, and the first successful well in Texas (1865) were drilled near oil seeps. Before this the hand-dug wells from which the Burmese obtained oil were dug near seeps, and the first onshore oilwells in Great Britain were drilled on the nearest structure to an observed coastal oil seep. As late as 1949 every oilfield in Iran was associated with surface oil or gas seeps (Link, 1952). Thus the evidence of subsurface oil is contained in these oil seeps; however, the methods by which the oil reaches the surface may be somewhat varied. The microseeps, almost invisible quantities of oil and gas dissolved in water or free in the soils, probably move by primary migration mechanisms, whereas the larger seeps are due to tertiary migration caused by alteration in the strata in which reservoired oil is contained. Typical alterations would be fault systems which produce a pathway from the reservoir to the surface, tilting of the strata, or changes in depth of burial.

In a previous chapter it has been shewn how oil can be readily altered in the subsurface, and once a seep reaches the surface it becomes even more vulnerable, so that the seep may, at first sight, be somewhat different from the oil from which it originated. If the seep has not been in connection with its origin for a long time, it will mature and become very asphaltic or even possibly pyrobitumen. The alteration of the seep will be due to a combination of the following factors. Evaporation of hydrocarbons, even compounds up to the mid-C_{20}'s, will have substantial vapour pressures that will allow them, over varying periods of time, to evaporate. In a few months most of the lighter hydrocarbons will have been removed. Water washing will remove the water-soluble compounds and biological attack will readily consume many components of oil seeps. All the factors that were discussed in connection with subsurface microbial degradation will apply at the surface and the oil seep will be very susceptible to bacterial attack. In addition, certain chemical processes, often due to the effects of sunlight and atmospheric oxygen, allow the formation of polymeric compounds. These often contain oxygen bridges and the bitumen becomes immobile and insoluble.

As far as analysis is concerned, the easiest seeps are the freshest and most fluid ones, where liquid or gas is coming through cracks especially if it is bubbling through water. Besides being the easiest to analyse, these seeps are obviously still in communication with the underground accumulation from which they originated and bear a close relationship to the oil in the reservoir whence they came. However, seeps can be sticky lumps of dark asphaltic material and detailed analysis of paraffins, elemental composition, etc., will indicate whether or not these seeps are still being fed from their source but are being degraded, or whether they are dead seeps; the remains of long-since active seeps. These latter seeps obviously have no value as hydrocarbon source indicators. In more advanced stages of exploration such techniques as computerized GC–MS may allow correlations between the seep and oils to be obtained from the same basin. Care must be taken that one is analysing a genuine crude oil seep and not some human artefact, e.g. dumped oil, or even traces of hydrocarbons in soils from tractors, insecticide solvents, etc.

For use to be made of oil seeps, knowledge of local geology is important. A passage from an oil- or gas-bearing strata will mean the possibility of a seepage and these passageways may be due to a variety of causes. The hydrocarbon-bearing bed may outcrop and a little hydrocarbon, possibly waterborne by any water flowing past the accumulation, can make its way to the surface. Alternatively, hydrocarbon-containing beds which have been bared by erosion or where faulting has occurred to allow communication with the surface or intrusions such as salt, igneous bodies, mud volcanoes, etc., along the edges of which oil can migrate, will allow seeps to occur. In this context it is interesting to note that although Saudi Arabia contains some of the largest oil fields in the Middle East, the absence of earthquake activity has meant that there are very few seeps in that area. Yet to the north in Iraq and Iran where there is earthquake activity, caused by the collision of the Arabian and Eurasian plates, fracturing has occurred and many oil seeps have been reported. This example typifies a worldwide phenomenon and thus, especially during exploration near plate boundaries, the investigation of seeps is especially important.

Geochemical prospecting

The investigation of microscopic seeps is often called geochemical prospecting and will involve the analysis of surface soils, surface waters, the atmosphere, and the sea above submarine basins for traces of hydrocarbons coming from reservoired oil and gas. Also included are the analyses of the inorganic elements which change when associated with hydrocarbons.

The methods include:

1. The analysis of waters which have migrated upwards from the oil-bearing formation (this can also be done with samples of subsurface waters). If the analysed waters do contain hydrocarbons they should be a clue as to whether the sediment contained the correct organic matter for an accumulation to have occurred (Johnson, 1970; for other references see Chapter 2.) Zarrella (1963) noted an increase in the benzene content of brines in the proximity of oilfields and Figure 2.7 gives a summary of the significance of the various dissolved species in relation to oil accumulations.
2. The analysis of hydrocarbon gases in the soil, either free in the soil pore spaces or absorbed on to the soils, will give an indication of anomalous concentrations of such compounds which could be derived from reservoirs underlying the anomaly.
3. The fluorescence of soil samples which would be caused by concentrations of the large molecular weight aromatic compounds (often derivatives of anthracene and phenanthrene) derived in the same way as hydrocarbons in 2 above.
4. The analysis of those kinds of bacteria that live on hydrocarbons will again shew where there are anomalous concentrations of hydrocarbons.

Geochemical prospecting analyses fall into three groups, on the land surface, in the sea and in the air. Geochemical prospecting on the surface of land masses is a combination of the analysis of soils for hydrocarbons with the analysis of waters within the soils for both organic and inorganic constituents. There is no doubt that anomalies of hydrocarbons, bacteria, certain metals, e.g. nickel or vanadium, certain inorganic species, e.g. iodine, do occur in the vicinity of oilfields, but to concentrate too highly on these is unwise because the mechanism which caused them, e.g. the upward diffusion of hydrocarbons, is unknown. There is no guarantee that the observed anomalies will be directly above oil accumulations. For instance, the migration paths may not be direct, especially when time periods of hundreds of millions of years for methane and ethane to diffuse up from 6000 ft are taken into consideration. Aquifers between the reservoir and the surface would tend to transport the hydrocarbons away from the site of the reservoir so that when they left the aquifer and continued to diffuse to the surface they could be a long way from the reservoir, thus confusing the geochemical prospector. Richers *et al.*, (1982) analysed the absorbed, free and headspace gases from soil samples collected from the Patrick Draw field, Wyoming, USA. Landsat imagery had indicated several linear features crossing the field which were interpreted as near vertical extensional fractures. Anomalous concentrations of C_1 to C_7 paraffins appeared near those linear features.

Because of the very small amounts of hydrocarbons that are involved, any light hydrocarbons formed by the first stage of organic matter alteration, i.e. diagenesis, will quite often be larger in quantity than the material diffusing up from below, and great care has to be taken in analysing any hydrocarbons from the soil to discover their age and origin. The lack of higher alkanes, other than methane means that carbon isotope studies are particularly useful.

'Sniffing' techniques

The analysis of hydrocarbons in water and the atmosphere has been carried on for some long while and in both cases some sort of drogue is towed behind and below the investigation vessel and continuous samples of either sea water or the atmosphere are pumped up the connecting line, the hydrocarbons removed and analysed in a manner rather akin to mudlogging. Alternatively sediment samples may be collected and analysed. Very low levels of detection, down to 0.5ppb, can be achieved. One of the major problems is to distinguish between petrogenic and biogenic gas, and in this case the presence of the higher homologues of methane, e.g. ethane, propane, butane, and stable isotope studies are especially valuable. The samples are also often tested for fluorescence to indicate the presence of multiring aromatic compounds. Sackett (1977) suggests that the best results are obtained by 'sniffing' at sea when the sample collection apparatus is near the sea bed. The concentration of the gases is greatest at the sea bed as the gases will not have been able to dissolve in the sea water. In addition, the gases will be picked up near where they left the sediment and before they are swept away by water currents. In summer months, water columns can become vertically stratified and the gases will not reach the near-surface waters. Sackett notes that biogenic gas will have a $C_1/(C_2 + C_3)$ ratio of greater than 1000 and carbon isotope compositions of less than $-55°/oo$. This compares with thermogenic gas where the figures are 0 to 50 and heavier than $-50°/oo$. Thus he considered that the gases with different origins could be detected in underwater seeps and then differentiated (see Figure 10.5) (Bernard *et al.*, 1976).

Stahl *et al.* (1981) have noted that the methane in shallow sediments can be a combination of biogenic gas and thermogenic gas. They have shewn that the thermogenic gas absorbed in near-surface sediments has the same carbon isotope composition as the methane in deep reservoirs. Thus the use of carbon isotopes to identify the origin of methane absorbed in near-surface sediments will considerably improve the results of geochemical surface exploration.

Kvenvolden & Field (1981) have reported the occurrence of thermally produced hydrocarbons from an unconsolidated sediment within a bathymetric depression in the Offshore Eel River Basin of northern California. The evidence that these hydrocarbons were thermogenic included the following. Very high concentrations of hydrocarbon gases from ethane to butane were found with the methane, which had a carbon isotope composition of -43 to $-44°/oo$ (relative to PDB). There were gasolene-range hydrocarbons and complex heavy compounds with a petroleum-like distribution.

These hydrocarbons were believed to have originated deep in the basin and to have migrated to the surface through fractures and faults which developed during the emplacement of a diaper. There were also biogenic gases present as shown by the wetness (Bernard *et al.*, 1976). If this thermogenic material has seeped up from deeper within the basin, the indications are that the conditions for petroleum genesis have existed within that offshore basin. However, exploration and production have not been too rewarding, with only a little gas and no oil production.

By using several detectors, taking samples on a grid pattern and by combining the results using computerized techniques, the source of these seeps on the sea bed can be calculated, and there is no doubt that these sorts of gas sniffers can obtain positive correlations between subsurface faults and surface seeps. Likewise, the source of the hydrocarbons which are being released by the seeps can be correlated with seismic anomalies. Samples can be obtained from the sea bed and analysed in exactly the same manner as the land samples. The geochemical prospecting can be carried out at various stages during the initial basin appraisal. Such investigations can precede seismic surveys or they may follow the seismic search and try to see if hydrocarbons are leaking from structures identified by the geophysical methods. Some contractors will run both geochemical sniffing and geophysical surveys together with an obvious cost saving.

In all of these methods one comes back to the same problem, and that is the means by which the fluids diffused from the reservoir to the surface and either were absorbed near the surface or were released into the overlying water or atmosphere. There is no doubt that these anomalous concentrations and 'plumes' of released gases can be detected and that they do bear some relationship to subsurface structures and oilfields. However, they need to be distinguished from hydrocarbons generated by recent decaying organic matter, i.e. biogenic gas, unless one considers that there are large quantities of biogenic gas trapped in that region. These surface geochemical methods are therefore not always 'pinpointing' methods of locating oil and gas accumulations at depth. They are at worst regional prospecting tools, especially useful where there are intrusive faults or fracture systems, as mentioned earlier, or even permeable beds allowing a pathway for migration to the surface. In such cases petroliferous regions can be differentiated from geologically similar regions which are barren in respect to oil and gas. At best these methods can indicate whether there are anomalous concentrations of hydrocarbons in sediments above previously identified structures. No cap rock is a perfect seal, especially to light hydrocarbons, and a full reservoir might be expected to lose some of its contents to the surface sediments. Thus a positive anomaly above a structure is a good sign although it is doubtful if a structure would be ignored just because there was no obvious geochemical anomaly in the sediments above it.

Organic geochemical analyses

The last type of surface geochemical investigation involves the full range of organic geochemical analyses of potential source rocks which outcrop. These can be analysed with regard to the three basic criteria, amount, type, and maturation of the organic matter. Certainly if the first two criteria, amount and type, are satisfied, for surface samples the last criterion, maturity, is not so important. If a potential source horizon can be shewn to dip so that at some point deeper down it will become more mature, then there is a potential that some of that source interval will have generated oil.

With a knowledge of local geothermal gradients and the local geological structures it is possible to predict the depth at which oil generation, gas generation, and cracking will occur. The types of analyses are exactly the same as those carried out on samples obtained by subsurface coring, the details of which will be given in the following sections. It is the geological interpretation of the results that varies slightly. In a similar manner oil–source correlations can be carried out using unweathered surface samples.

SUBSURFACE EXPLORATION

For source rock identification and assessment four basic questions need to be answered. These are: How much organic matter does the potential source rock contain? What type of organic matter is in that source rock? What is the stage of maturation of that organic matter? Has migration of generated hydrocarbons out of that source rock occurred? The first question can be answered relatively simply; the second and third questions are normally answered by combinations of a series of relatively complex procedures. The last question is probably the most difficult to answer, for once oil has migrated it is difficult to tell whether it has ever been there, but by a combination of techniques some indications can be given as to whether or not migration has actually occurred.

Amount of organic matter

The normal methods to determine the amount of organic matter in a sediment are elemental analysis of the whole rock and the extraction of the solvent-soluble organic matter. In the elemental analysis of the whole rock, the inorganic carbonate is first destroyed by dilute acid and then the organic carbon is burnt to carbon dioxide in an oxygen atmosphere. The amount of carbon dioxide can then be measured, often by a thermal conductivity device which senses changes in the inert carrier gas's thermal conductivity which will be in proportion to the amount of carbon dioxide generated. The total organic carbon (TOC, or C_T) is normally expressed as a percentage of the whole rock.

Some laboratories carry out this analysis on the whole rock from which the extractable, solvent-soluble organic matter has been removed. The problem with this second approach is that the solvent-soluble organic matter, or bitumen as it is sometimes called, is extracted with an organic solvent. Before carbon analysis can be attempted, the rock sample has to be completely free of the solvent and the ease of removal of the solvent will depend on the solvent used. However, it is claimed by some workers that organic carbon analyses carried out on samples from which the bitumen has been removed shew less scatter when plotted against other variables, even though the organic carbon content equivalent to the soluble organic matter is probably not more than 0.1–0.2% of the total. Total organic carbon values can vary probably to within ±0.3%, which is not surprising as the material from which the sample for analysis is taken is not very homogeneous and the size of sample used for analysis will only be 2–3 mg.

The bitumen content within a source rock is estimated by solvent extraction of the solvent-soluble organic matter. This is either done in a soxhlet extraction apparatus or in a high-speed stirring device with an organic solvent; quite normally dichloromethane is used. The bulk of the solvent is then removed by fractional distillation or by a rotary film evaporator and the last of the solvent is removed by evaporation in a warm oven with a stream of nitrogen blowing over the sample. The weight of solvent-soluble extract is normally related to the weight of whole rock or to the total organic carbon. The amount of solvent-soluble extract will depend upon the solvent used and such figures can only be compared when the extraction has been carried out in a similar way using the same solvent. These extracts, when related to total organic carbon, will be misleading if there is significant amounts of dead carbon. Type IV kerogen does not have a hydrocarbon-generating potential and thus will not contribute to the extract. Significant amounts of dead carbon will lower the extract-to-TOC ratio and could possibly falsely indicate that a source rock has no potential for hydrocarbon generation. The organic carbon equivalent to the extract can be measured in the same way as for the whole rock but in this case the hydrogen and nitrogen percentages can also be obtained.

So far it has not been possible to calculate the amount of kerogen present. The total organic carbon of the whole rock can be obtained and from it can be subtracted the carbon equivalent to the bitumen to give the total organic carbon of the insoluble organic material, or kerogen, as a function of the whole rock. To take account of the other elements that are present in the kerogen, e.g. N, O, H, S, the value of the total organic carbon of the kerogen must be multiplied by a conversion factor which will vary with the type of kerogen and its maturity state. Table 9.1 (Tissot & Welte, 1978) gives the conversion factors to estimate total organic matter from the organic carbon content.

Table 9.1 Conversion factors to estimate organic matter from organic carbon

State of maturity	Kerogen type			
	I	II	III	Coal
diagenesis	1.25	1.34	1.48	1.57
end of catagenesis	1.20	1.19	1.18	1.12

Type of organic matter

The methods for determining the type of organic matter fall into two main categories, optical and physico-chemical methods.

In the optical methods the organic matter is separated from the host inorganic rock material, a process which will be described later, and then examined microscopically in transmitted and reflected visible and ultraviolet light. This examination, allows one to recognize the type of material from its structure, e.g. from the remains of plant and animal debris, and its source may be identified.

The physico-chemical methods involve a range of techniques including: (a) elemental analysis of the kerogen after it has been separated from the rock, with the results being plotted on a Van Krevelen diagram; (b) infrared spectroscopy to determine the functional groups present in the kerogen; (c) electron spin resonance; (d) carbon isotope studies; (e) pyrolysis studies on the kerogen or on the whole rock; and (f) detailed examination of the solvent-soluble material.

Degree of maturation of organic matter

The majority of the methods used to determine the degree of maturation of the organic matter are extensions of the methods used to evaluate the organic matter type. The methods include: (a) optical means, involving carbonization of the spores, measurement of the light reflected from polished vitrinite grains, and fluorescence studies; (b) physico-chemical analyses involving the elemental analysis of the kerogen (Van Krevelen diagram); (c) pyrolysis experiments; and (d) detailed examination of the solvent-soluble extract including the composition of the n-alkanes, the CPI, and the amounts of isoprenoid compounds.

Evidence of migration

It is possible to recognize an immature source rock in which there will be almost no hydrocarbons that will have been able to migrate and it is possible to distinguish a mature source rock by the various ways outlined above. As has been shewn, the various kerogens each have a maximum convertibility and an assessment of the degree of maturity will allow an estimate to be made of the amount of conversion that that particular kerogen has undergone. Thus a judgement can be made on the amount of hydrocarbon which should have been generated and this can be compared with the amount left in the source rock, i.e. the amount of bitumen or solvent-soluble organic matter which was extracted from that source rock. Any major discrepancy will imply that migration has occurred.

For a sample from which no migration has occurred there will be a very high solvent extract for a mature source rock, as compared with a very low figure for an immature source rock. Care has to be taken to distinguish between an immature source rock with a low carbon content and low extract, and a mature source rock from which oil has migrated and which also has low carbon and extract. However, by the use of a variety of techniques this problem can be overcome, to the extent that at least an indication of whether oil has migrated can be obtained by a study of all the measurable parameters.

The accomplishment of these analyses involves quite a few stages and procedures and, as will be seen, many of these parameters will give the information which is required. So why do them all? It is dangerous to rely on a particular method because any geochemical system is open to anomalous results. The commonest case is where recycled organic matter, which was previously incorporated in a sediment, has been taken to the metagenesis stage and then erosion of that sediment has occurred and the organic matter has been re-incorporated in a new sedimentary rock. This 'new' rock will probably have a very high carbon content, but detailed examination will shew that it is inert, so a high carbon value is meaningless by itself: the assessment of the type of organic matter is vitally important. However, the opposite mistake can be made. Consider the same sediment quickly examined; too high a value for the maturation could be given, implying no value as a source, but the contemporaneous organic matter could be shewn by detailed investigation to have the correct amount, type, and maturity to produce oil or gas.

Thus for each of the methods to be described, the methodology, the results that can be obtained, the reliability, and the interpretations that can be made from those results will be discussed.

SOLVENT EXTRACTION AND EXAMINATION OF BITUMEN

Samples are treated with an organic solvent to remove the soluble organic matter (bitumen). This is normally carried out using one or another of a series of standardized procedures which involve very high standards of quality control in solvent and sample preparation to ensure the elimination of impurities. Solvents are normally doubly or triply distilled and samples of the prepared solvent are evaporated to near dryness and analysed for high-boiling impurities which would have been concentrated by the evaporation. Care has to be taken to make sure that waxes from fingers, sample bags, etc., or hydrocarbons from drilling fluids or oils used on rigs are eliminated wherever possible, and analysis of all such possible contaminants will allow one to recognize them if they appear in samples (Gibbs, 1970, Douglas & Grantham, 1973). These solvent extractions can be carried out on both cored and unweathered surface material, but with the latter care has to be exercised to ensure that weathering is not present. The effects of weathering can, under the correct conditions, extend to depths of 10 ft with a loss of up to half the mobile hydrocarbons (Leythaeuser, 1973; Clayton & Swetland, 1978).

Samples have their outer surfaces removed to try and eliminate some of these sources of contamination and are then milled to pass a 200-mesh sieve, care being taken that the milling does not generate any heat which might cause the generation of any hydrocarbons, or the loss by the evaporation of any already generated. The milled sample is then placed in a soxhlet extractor and the soluble organic matter extracted with organic solvent. The solvents used vary from laboratory to laboratory, but the common ones include chlorinated hydrocarbons such as chloroform (Trask & Patnode, 1942) and dichloromethane, or benzene and the benzene–methanol azeotrope. For a variety of reasons, including safety and standardization of procedures, the solvents being used are nowadays the chlorinated compounds, especially dichloromethane.

The azeotropic solvents do have the advantage that they will dissolve a much greater range of organic matter than a single solvent and their boiling point is below that of single component solvents, thus easing the problem of removing the solvent without losing too much of the extracted organic matter. However, in the case of the favourite azeotrope, benzene–methanol, there are health problems with the benzene as well as the fact that this solvent dissolves sulphur and sodium chloride which can complicate recovery and interpretation of the extract.

Other methods of extraction include ball-milling of the sample with a solvent, but one can just obtain a paste with little free solvent, so recovery can be difficult and there are also many cleaning problems. Ultrasonic vibrators have been used as have very high-speed mixers to extract the bitumen from the milled rock into the solvent.

After extraction the solvent has to be removed and this is normally done in two stages, firstly to remove the bulk and secondly to remove the final traces of the solvent. For the first stage fractional distillation is often used to remove very carefully only the solvent. The most common alternative is a rotary film evaporator although there have been reports that this method can be less reliable in terms of reproducibility. The second stage is where the last part of the solvent is evaporated in a constant low-temperature oven with a flow of air or nitrogen across the containers. The weight of the extract is calculated and expressed in relation to the rock whence it came, often as a percentage or in ppm (Nooner et al., 1972; Grayson et al., 1970), or the extract can be related to the amount of organic carbon in the rock, in milligrams of hydrocarbons per gram of organic carbon (Figure 9.3). These figures can be interpreted in terms of source potential, as will be demonstrated later in this chapter.

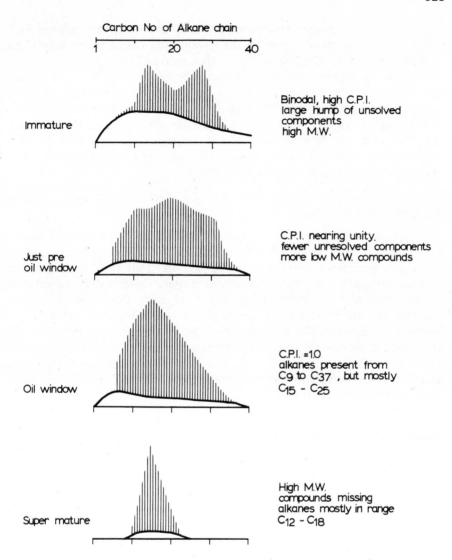

Figure 9.1 Hydrocarbon maturity as shown by gas chromatography of alkanes

Different organic solvents will be able to dissolve different amounts of organic matter. Each solvent will have a total dissolving capacity and at the same time it will be likely to dissolve some compounds more readily than others. A good solvent for such extraction procedures will have a capacity to dissolve many classes of compounds. The soxhlet extraction system has the advantage of washing the rock sample with fresh solvent each cycle and this increases the amount of extract that a given amount of solvent can remove from the rock sample. Benzene–methanol azeotrope is an excellent wide-ranging solvent, but it is flammable and carcinogenic and hence is now little used.

When extract weights are being compared, either directly or combined with other data such as carbon content, the solvent used should be the same. The solvent used should always be specified when quoting extract figures.

Because the lighter ends of the extracted bitumen are of the same boiling point range as the solvent, there will not be complete recovery of those compounds and even some of the higher-boiling components may well, if they have significant vapour pressures, not be completely recovered. It is generally accepted that complete recovery of the bitumen will start about the C_{15} hydrocarbon. Below C_{10} nothing will be obtained and between those values there will be partial recovery of hydrocarbons.

Fractionation of bitumen

The extracted bitumen, once obtained, is fractionated into a series of groups based upon types of compounds that they contain. The most normal groups are: (a) the paraffins and naphthenes (aliphatic material); (b) simpler aromatic compounds; and (c) an ashphaltic-resinous residue. This initial separation used to be achieved by column and/or thin layer chromatography, but more recently the use of high-performance liquid chromatography (HPLC) has become the main method of fractionation. This is similar to gas chromatography, but the mobile phase is a liquid instead of a gas. With this procedure extracts can be quantitatively separated and the alkanes are detected by changes in the refractive index of the column effluent, and the aromatic fractions by ultraviolet fluorescence.

Analysis of the extracted bitumen

The alkane fraction can be divided into normal alkanes, isoalkanes, and naphthenes by urea abduction, the use of molecular sieves or by HPLC used under different conditions, and the analysis of these fractions will give important information about the type of organic matter present and its state of maturity. These analyses are normally executed on a gas chromatograph fitted with capillary columns and nowadays normally linked to mass spectrometer and a computerized library for the identification of the products by comparison of their fragmentation patterns with those of known compounds.

The analysis of the alkanes is of particular significance because, as we have seen, immature source rocks produce alkanes whose spectrum has a pronounced odd–even alkane ratio (high CPI or OEP values) (Scalan & Smith, 1970), which disappears as the source rock becomes more mature. At the same time the distribution maximum of the alkanes shifts to lower carbon numbers (Powell, 1975; Tissot & Welte, 1978; Bray & Evans, 1961). Demaison (1975) noted that this shift to lower molecular weights occurs at the threshold of intense oil generation with the maxima dropping from within the C_{23}–C_{29} range to the C_{13}–C_{19} range, but while this was taking place intermediate binodal distributions were observed. Allan & Douglas (1977) studied the distribution of *n*-alkanes within coal maceral concentrations and have shown that a relationship between maximum vitrinite reflectance and CPI exists and that the CPI decreases as the vitrinite reflectance value increases.

Philippi (1974) discussed the relative influence of marine and terrestrial organic matter and he noted that land-derived matter is rich in the five-ring naphthenes and the higher alkanes (C_{30}, C_{32}, C_{34}), whereas marine material is rich in C_{16} and C_{18} compounds. He therefore considered that for oils of terrestrial origin the ratio (C_{21} + C_{22})/(C_{28} + C_{29}) should lie between 1.2 and 1.6 whereas for oils of marine origin the ratio will be in the range 1.5–3.0.

In immature source rocks the only hydrocarbons that are present are geochemical fossils, mostly isoprenoid compounds, and a few compounds which are the product of low-temperature thermogenic breakdown of the kerogen, but as temperature increases new hydrocarbons are generated. For the immature rock the alkane spectrum is dominated by high molecular weight alkanes in the C_{25}–C_{29} range with a high CPI value, and as the degree of maturation increases the quantity of alkanes increases and these new alkanes are of lower molecular weight with no odd/even preference; and thus the overall alkane distribution moves not only to lower numbers, but also becomes more even, i.e. the CPI or OEP fall. Gas chromatograms of mature source rocks are very similar to those of crude oils, allowing for the loss of the lighter ends of the former during the solvent-removal process, and have a majority of compounds in the C_{15}–C_{19} range with a CPI close to unity. Powell (1978) has shown that the hump in the baselines of such chromatograms is due to large amounts of branched and cyclic alkanes, and is a sign of immaturity. Figure 9.1 shews the changes in the alkane chromatograms as the maturity of their parent source rocks increases, while Figure 9.2 shews the changes in CPI with depth, and as the organic matter matures the alkanes become similar to those in crude oils.

Raschid (1979) has shown that the pristane/phytane ratio increases with increasing organic maturation. He also pointed out that the total amount of phytane decreases during thermal evolution of the organic material, while the pristane/*n*-C_{17} alkane ratio has been reported to decrease with increasing maturity. Both these effects can be ascribed to the lower thermal stability of the branched alkanes as compared with their normal counterparts and to the fact that the isoprenoid compounds are geochemical fossils and no more are generated once the sediment is laid down, although a few can be released from weak association with the kerogen. Meanwhile the normal hydrocarbons are generated in vast quantities during the thermal breakdown of the kerogen and thus the ratio of isoprenoid components to the normal alkanes will decrease with increasing maturity.

The ratio of aromatic hydrocarbons to saturated hydrocarbons tends to decrease with increasing maturation due to thermal cracking, but where bacteriological alteration of the oil is occurring this decrease may not be detectable. The decrease in this ratio is due to the greater generation of

324

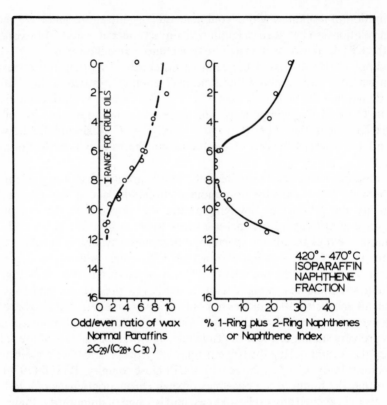

Figure 9.2 Change of odd–even preference and naphthene ratio with depth (Philippi, 1965. Reproduced by permission of Pergamon Press Ltd)

aliphatic hydrocarbons as compared with aromatic hydrocarbons during thermal maturation. The preference of bacteria for *n*-alkanes can mask this effect. The ratios of aromatic and saturated hydrocarbons to extractable organic matter and the ratio of saturated hydrocarbons to extractable organic matter both increase with increasing maturation and this is especially noticeable at the oil window or the principal zone of oil generation. The variation of solvent-soluble organic material has been already discussed. Likewise, the ratio of extractable material to total organic carbon will tend to increase with depth up to the principal zone of hydrocarbon generation (or the oil window). However, increased maturation or the occurence of migration will cause this ratio to fall, making its diagnostic value rather limited. Migration does not have such a great effect upon the ratios of saturated hydrocarbon and saturated plus aromatic hydrocarbons to extractable organic material.

So far these alkane chromatograms have only been telling us about the maturity of the source rock; however, information about the type of the organic matter can also be obtained from the analysis of the alkanes. These gas chromatograms can also indicate which isoprenoid compounds are present and in what quantities, and this information can give clues as to the source type and maturity of the organic matter. Powell & McKirdy (1973a) realized that oils derived from terrestrial organic matter have low pristane-to-phytane ratios.

The alkane spectra can also indicate the type of organic matter from which the alkanes were derived. Harwood (1977) states that oil and wet gas kerogen produce high proportions of n-alkanes in the C_{16}–C_{20} range whilst the dry gas kerogen has a higher percentage of n-alkanes in the C_{20}–C_{24} range. Tissot & Welte (1978) note that immature continental organic matter has a high alkane content which is dominated by high molecular weight n-alkanes with an odd-number preference; while marine organic matter produces smaller amounts of the low to medium molecular weight alkanes and abundant isoprenoid compounds. The n-C_{17} alkane is dominant in brown benthonic algae and the presence of substantial amounts of n-C_{17} is evidence for the involvement of algal material in the formation of the kerogen. The isoprenoids and other geochemical fossils can also be used for oil–oil and oil–source correlations as these fossil compounds are generated, almost entirely, from the original organic matter which also eventually produced the kerogen. Almost no more isoprenoids will be formed from the maturation of the sedimentary organic matter and thus their ratios are approximately fixed. These compounds can also be used in reconstructions of the depositional environment, as certain compounds are more related to terrestrial life than to marine.

The analysis of the naphthenes can be used to give indications of thermal evolution. Philippi (1965) defined a naphthene index (NI) which is the percentage of the one- and two-ring naphthenes in the 420–470 °C isoparaffin–naphthene fractions. The aim of the NI is to quantify the dilution of original biogenic four- and five-ring naphthenes by simpler, lower ring-numbered naphthenes which are generated from the thermal breakdown of the kerogen. The larger the value of the NI the more thermally mature will be the source rock. However, there are problems in comparing source rocks of different types.

Aromatic compounds may be separated from the other components of the solvent soluble organic matter by liquid chromatography (HPLC), and by changing the conditions under which the machine is operated the aromatics can be re-run and separated into individual compounds in the same way as the alkanes may be on the gas chromatograph. As the compounds elute from the column the intensity of the ultraviolet fluorescence is measured and the ratio of the intensities of the three most intense peaks is calculated for each compound. Because these ratios are constant for any particular compound, they can be used as a method of identification by comparison with the data for known compounds. This sort of comparison may be computerized so that the instrument is fully automatic and an analysis of the aromatic compounds obtained. Again, as maturation increases the aromatic compounds become smaller, contain fewer rings, and are simpler. This is another way of assessing thermal maturation.

Radke *et al.* (1982) have studied the variation of aromatic compounds with maturation. The compounds especially investigated were phenanthrene, various isomeric methyl phenanthrenes and some dibenzothiophenes. They established a methylphenanthrene index (MPI) which is the sum of the ratio of four isomeric methyl phenanthrenes to phenanthrene. The compounds used were the 1-methyl-, 2-methyl-, 3-methyl- and the 9-methyl phenanthrenes. Similar ratios were obtained for the dimethylphenanthrenes & phenanthrene (DPI) and methyldibenzothiophene and dibenzothiophene. A linear relationship between MPI and vitrinite reflectance was observed within the oil window (0.6–1.3% R_o). Both MPI and DPI increased with depth up to a maximum which corresponds to the maximum on the oil generation curve (Figures 9.3, 9.4 & 9.9). MPI is little influenced by facies changes.

Analysis of geochemical fossils

More advanced analyses of the alkane and aromatic constituents of the extracted bitumen are afforded by the use of gas chromatographs coupled to mass spectrometers with computers for library storage of reference material and comparison of the data of fragmentation patterns. The mixture is injected into the gas chromatograph and separated as normal on a capillary column, but the column effluent is stripped of the carrier gas and the separated hydrocarbons are fed into a mass spectrometer, where they are fragmented and the fragmentation patterns recorded. Organic compounds within a mass spectrometer and under the same conditions always fragment in the same way and produce charged species which are separated on basis of their mass/charge ratio. Thus a picture is built up of the amount of each species of each mass/charge ratio, and this should be the same for the same compound analysed under the same conditions. The particle with the highest mass/charge ratio is the parent ion, i.e. the original molecule with one electron knocked off, and a conventional chromatogram may be drawn using the intensities of these parent ions. At the same time the compound is identified by comparing its fragmentation pattern with those in the computer library.

Seifert & Moldowan (1978, 1980) used the GLC–MS analysis system of steranes, terpanes, and monoaromatics to investigate the migration, maturation, and the source of crude oils. Among the criteria quoted are the concentrations of the C_{29}–C_{32}, β-β-hopanes, high values of which indicate immaturity as these compounds are thermally sensitive and will break down if heated too much. Likewise, the ratio of the 17β H to 17α H trisnorhopanes is a maturation indicator as the C_{27} 17β hydrogen compound is known to be unstable at the temperatures needed to generate petroleum. (The α and β refer to the stereochemistry of the hydrogen atom at the 17 position on the steroid nucleus. See Figure 1.6). Thus if it is present in the bitumen the source rock has not been heated to sufficiently high temperatures to have generated oil. The ratio of moretane to hopane is also a maturation indicator because it has been shewn that the moretane concentration falls and the hopane concentration rises with increasing maturity, and samples must contain less than 6% moretane for the source rock to have been in the petroleum-generating zone for long enough to allow oil to have been generated. Seifert & Moldowan also stated that the yield of saturated products in the yields of pyrolysis experiments must be greater than 50 ppm or there would not be sufficient organic matter present to generate oil. Other indicators such as the ratios of certain terpanes have been shewn to be source-input specific and thus these sorts of analyses can give indications of the type of organic material present and the level of organic maturation.

The use of natural product geochemical fossils is now becoming widespread in the determination of source, temperature, and burial histories. Mackenzie *et al.* (1980) use the changes in porphyrins to determine the degree of maturity of sedimentary organic matter. They noted that the nickel porphyrins either disappear or are converted to lower molecular weight forms at shallow depths and they have completely disappeared by the depth at which similar changes occur to the vanadium porphyrins. They have related these changes to other parameters used to determine the organic maturity and have found them a useful additional method.

The analysis of geochemical fossils will also give information on the type of source material and thus on possible depositional environments. Marine organic matter, which is mostly phytoplankton with a little zooplankton and some benthic algae, contributes to the sedimentary organic matter n-alkanes and n-fatty acids in the range C_{12}–C_{20}, often with the most abundant compounds being C_{15} and C_{17} (these are often synthesized by algae but can also originate from the C_{16} and C_{18} fatty acids). Other compounds derived from this same source include C_{15}–C_{20} isoprenoids, abundant steroids, and triterpenoids such as the hopanes, as well as some carotenoids. In contrast, continental organic matter is mostly composed of the remains of higher plants incorporated into the sediment in deltaic or similar environments and this contributes to the sedimentary organic matter high molecular weight alkanes (C_{25}–C_{33}), often with a high odd carbon number preference, together with some cyclic material, cycloalkanes, and aromatics. The isoprenoids and the hopanes also occur in this type of sediment. Lastly, in lacustrine or paralic environments there is often plant material which has been seriously altered by bacterial degradation. The organic matter in sediments which results from this type of environment is enriched in lipids, especially the long-chain normal, iso-, and anteisoalkanes in the range C_{40}–C_{50}, and this results in waxy oils (Hedberg, 1968). There is little in the way of isoprenoid and cyclic hydrocarbons with the exception of the hopanes.

Thus the analysis of the bitumen is able to give one information as to both the type of organic matter and the level of maturation of that organic matter, and Seifert & Moldowan also indicate that some information as to migration and to 'in reservoir' maturation can be obtained by very detailed examination of the solvent-soluble extract.

Nishimura (1982) has investigated the 5β isomers of stanols and stanones which can give information as to the quality of the sedimentary organic matter and the oxicity state of the depositional environment. Mackenzie *et al.* (1981) have shewn that there is an increase in aromatization and side chain cracking in monoaromatic steroid hydrocarbons with increasing maturity. Rullkötter & Wendisch (1982) report the biodegradation of $17\alpha(H)$-hopanes with removal of the C-10 methyl group and ring opening. If this occurs on a significant scale some care will need to be exercised in interpreting geochemical fossil (biological marker) analyses.

Analysis of low molecular weight hydrocarbons

So far all these analyses have been carried out upon the bitumen extracted from the rock;; however, the light hydrocarbons, i.e. C_1–C_8, have also been found to be thermal maturity indicators for source rocks. It is just that the techniques that are required to obtain the compounds without loss of all or some of them are more difficult due to the volatility of these low molecular weight compounds. They can be analysed, either as head space or cutting gases (C_1–C_4) or by the gasolene-range analysis (C_4–C_8) of cores and cuttings. It is essential that the techniques involved in the collection of these samples ensure that all components are contained within the sample.

Bailey *et al.* (1974) have analysed gas cuttings and the gasolene-range hydrocarbons from Western Canada. They have shown that in the diagenetic stage the gas is predominantly methane with very little gasolene-range compounds at all, but by the time the catagenetic stage is reached the gas contains abundant hydrocarbons in the C_2–C_4 range and the gasolene fraction is also very rich. In the metamorphic zone, however, the gas is mostly methane with almost no gasolene-range hydrocarbons. These analyses can be plotted both vertically and areally so that one may see not only the sedimentary sections that are likely to contain hydrocarbons, but also one can identify in the subsurface the ranges in which oil is generated and at which oil is cracked and wet gas and then dry gas generated.

INSTRUMENTAL ANALYSES

In the preceding section while discussing the analysis of the organic material extracted from the sediments, the use of major instrumental analysis techniques was assumed. These are the use of the gas chromatograph, especially combined with the mass spectrometer with a computer link-up and the use of high-performance liquid chromatography systems. Other instrumental analyses are now described.

Elemental analysis

Organic carbon

The original and probably most important elemental analysis that is carried out is that of the total non-carbonate, or organic, carbon content of potential source rocks. This is often designated the abbreviations C_T or TOC. The analysis involves determining how much carbon there is present within a sediment and in what form that carbon occurs. The actual analysis is carried out on 2–3 mg of finely milled sample in any one of a number of commercially available instruments which can often analyse other elements at the same time or after minor modification. These other elements include oxygen, sulphur, hydrogen and nitrogen.

It is generally accepted that an amount of 0.5% organic carbon for clastic rocks or 0.3% in a carbonate is the minimum required for the production of hydrocarbon accumulations. This amount of carbon is required not just to generate sufficient oil to fill a potential reservoir, but to assist in migration. When the insoluble organic matter produces fluid hydrocarbons there is a volume increase which will often cause a pressure increase and will aid migration. Alternatively, if migration of generated oil occurs by some sort of wick mechanism, there needs to be enough organic matter to form this continuous or near-continuous wick. Hence the importance of having a sufficient quantity of organic carbon with regard to both generation and migration of reasonable quantities of oil. However, these figures should be considered as minimum values and obviously the higher the carbon content of a potential source rock, the greater is its source potential. All other things being equal, a rock with 10% of organic carbon will generate more oil than a rock with 1% carbon, and may start to expel the generated oil earlier. However, it is not only how much carbon that is present but also the form in which that carbon occurs which is important. This can be assessed in a number of ways, one of which involves the total organic carbon and solvent-soluble extract.

One of the commonest ways to express the relationship between these two parameters is in terms of the ratio of extract to carbon, often milligrams of extract per gram of organic carbon or just the ratio of the extract and carbon percentages. This is shewn in Figure 9.3, the tabulation of organic geochemical parameters. As can be seen, there is often a change of the amount of soluble organic material present in relation to total organic carbon, with depth. Albrecht *et al.* (1976) have shown that the ratio of hydrocarbons (mg) per gram of total organic carbon, when plotted against depth, increases to a maximum and then decreases. These diagrams demonstrate that the saturated hydrocarbons increase much more than the aromatic components and they can also be used to indicate the zones of intense oil generation and the zone of the destruction of liquids and the formation of gas (Figure 9.4). As will be seen later, diagrams like these can be related to other organic geochemical parameters.

One of the more common ways of expressing this data is to plot the weight percentage of the extract against the total organic carbon content (Figure 9.5). On these diagrams several major trends can be identified and these are mostly related to the degree of maturation of the organic matter within that rock.

The amount of material which can be removed from a rock specimen will

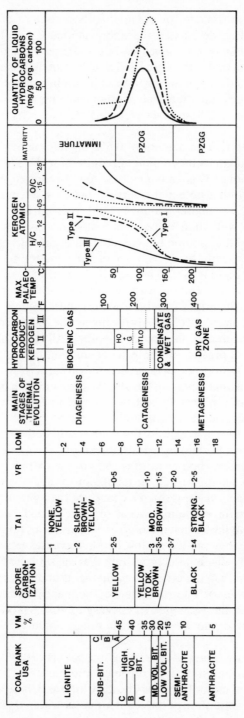

Figure 9.3 Maturation indices (after Tissot & Welte, 1978; and Hood *et al.*, 1975. Reproduced by permission of Springer-Verlag, Heidelberg, and The American Association of Petroleum Geologists)

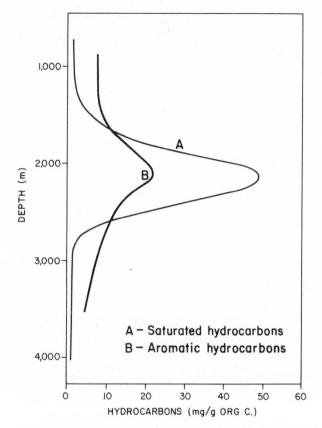

Figure 9.4 Evolution of saturated and aromatic hydrocarbons with depth (Albrecht *et al.*, 1976. Reproduced
by permission of Pergamon Press Ltd)

depend upon the solvent used in the extraction process. Hence for these
correlations to be meaningful they should be carried out on samples extracted
with a solvent which dissolves the maximum amount of as many classes of
compounds as possible. The graphs in Figure 9.5 were drawn using data
obtained using benzene–methanol azeotrope as the solvent.

The presence of inert carbon within the source rock will increase the total
organic carbon, but not the extract value. Such samples may well plot in
anomalous positions. Consider the fate of a source rock with a given amount of
original organic carbon: the actual value does not matter, but for convenience
let us say 3%. It is known that the majority of this carbon is converted into
kerogen which eventually breaks down to give the mobile compounds, which
are rich in hydrogen and are able to migrate, and a graphite-like pyrobitumen
skeleton which is lean in hydrogen. A small amount of the carbon is accounted
for by the geochemical fossils. Let us now follow this organic matter through
the various stages of its alteration using Figure 9.5.

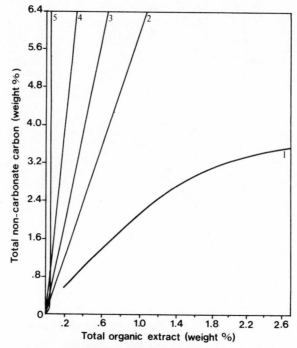

1   Unexpelled source rock (micro-reservoir, Kuh-e-Bangestan - Gurpi sequence.)

2   Expelled mature source rock (Kazdhumi Fm.)

3   Immature oil-type source rock (Kimmeridgian Shale)

4   Immature gas-type source rock (Shemshak Fm.)

5   Senile source rock (Kashafrud Fm.)

Figure 9.5   Total organic carbon versus total organic solvent soluble extract

In its immature state there will be very little extractable hydrocarbon, and what there is will consist of geochemical fossils together with any early generated material coming from biogenic or low-temperature thermogenic reactions. The amount of extractable material will depend upon the kerogen type. In Figure 9.5, lines 3 and 4 correspond to this immature situation. Line 3 represents an immature source rock which can contain large amounts of organic carbon but gives very little extractable material. The immature gas-type kerogen (line 4) generates smaller amounts of liquid hydrocarbon per unit amount of organic carbon at this stage.

As maturation proceeds the kerogen will break down and produce more and more generated hydrocarbons and there will then exist a mixture of generated hydrocarbons, and some partially converted kerogen, becoming more aromatic and pyrobituminous in nature. If migration does not take place, due for example to lack of permeability or insufficient overburden pressure, there is a microreservoir situation which is illustrated by line 1 in Figure 9.5. When the rock is crushed for extraction the mobile hydrocarbons are freed and can be removed. For a fixed amount of carbon there is a high percentage of extract, but this type of material is not a good source rock because the oil has not migrated to the reservoir and the oil can not be removed from the source rock for commercial use. The amount of extractable hydrocarbon will decrease with increasing exposure to temperature. Should a reservoir rock be analysed and plotted on this diagram, it will be located below line 1. However will be easily distinguishable from the microreservoir case because in the latter case there will still be some of the original kerogen and the pyrobitumen. This means that the organic carbon content of the extract will account for only a proportion of the total organic carbon of the whole sample. However, in a reservoir rock, where there is only the oil and neither of the other components are present, the organic carbon content of the extract will be approximately the same as the organic carbon content of the whole rock.

In a mature sample where migration has occurred, some of the hydrogen-rich compounds will have migrated away from the source rock and thus the organic carbon content of the source will fall from its original value, and its extract quantity will be reduced. This situation is illustrated by line 2. When organic maturity reaches the metamorphosed stage the original organic matter is completely converted into secondary hydrocarbons, which by this stage are almost completely gas plus a small amount of residual liquid, and the graphitic pyrobituminous residue. The quantity of the extract for this stage seems to be almost constant and independent of the amount of organic carbon, as shown in line 5. The carbon is mostly contained in the pyrobituminous residue and the volume of extract is probably only a function of the surface area of the pores and is totally unrelated to the carbon content.

Thus for a fixed amount of organic carbon the extract amount will vary with the maturity of the sample and for an immature kerogen the extract volume will be low, but the quantity of extract increases with maturation to a maximum during the catagenesis stage. For two similar samples there will be less extract per unit amount of organic carbon where migration has occurred. Further maturation will destroy the generated hydrocarbons and exhaust the kerogen, leaving a graphitic pyrobitumen and a very little extractable material on the surface of the pores. Lines 3 and 4 represent the diagenesis stage, lines 1 and 2 represent the trend of samples in the catagenesis stage, while line 5 represents the metagenesis stage. Although this diagram gives indications of maturity and thus to some extent the products of kerogen maturation, it tells one very little about the type of organic matter present in the source rock. Thus the amount of extractable material in a source rock will increase with maturity until migration begins and then it will fall. On Figure 9.5 this is represented by the sequence: lines 3/4, line 1, line 2, line 3/4, line 5. Thus a sample in either the diagenesis stage or in late catagenesis will have the same ratio of extract to organic carbon (Figure 9.3). In this situation another parameter is required to indicate the maturity level of such a specimen.

334

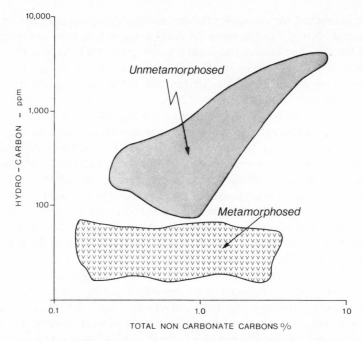

Figure 9.6 Hydrocarbons versus total organic carbon (Claypool *et al.*, 1978. Reproduced by permission of The American Association of Petroleum Geologists)

Any maturation parameter which involves the bitumen is likely to reach a maximum value and then decline. For example the ratio of hydrocarbons per gram of organic carbon (Figure 9.4) where in the zone of intense oil generation that parameter may have the same value on two occasions, one before and one after the peak of oil generation.

In general, the higher is the extract value, especially of hydrocarbons, in relation to the organic carbon content, i.e. the higher the E/C_T ratio, the better is the source potential of that source rock. The only exception is the microreservoir case (Erdman, 1961; Philippi, 1965).

Claypool *et al.* (1978) studied the organic geochemistry, incipient metamorphism, and oil generation of the shale members of the Phosphoria formation of the western interior USA. They have used an extraction and fractionation programme similar to that described above and have produced the following results. They plotted the hydrocarbon content against total organic carbon (Figure 9.6) as well as the saturated/aromatic ratio against the ratio of hydrocarbons to total organic carbon (Figure 9.7) and these diagrams show how immature and mature sedimentary organic matter can be differentiated.

Kerogen analysis

So far all the elemental analyses that have been described are on the whole rock

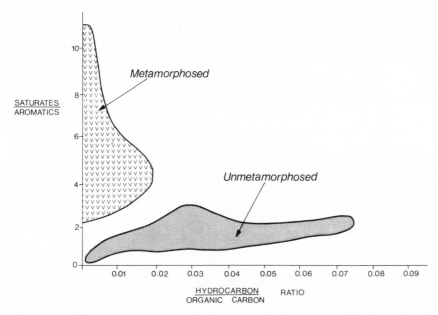

Figure 9.7 Maturity of source rocks by aromatics/saturates and hydrocarbons/ total organic carbon (Claypool *et al.*, 1978. Reproduced by permission of The American Association of Petroleum Geologists)

sample after milling and before solvent extraction, but it is also possible to analyse the isolated organic material, the mineral-free kerogen concentrate.

Down (1939) studied oil shales by removing the mineral matter and then analysing the residual organic material for the elements carbon, hydrogen, nitrogen, sulphur, and oxygen. Dancy & Giedroye (1950) oxidized kerogen to various volatile components and by analysis of these inferred the structure of the original kerogen. Laplante (1974) isolated kerogen from sedimentary rocks by treatment of the rock with hydrochloric and hydrofluoric acids. (More details of this procedure will be given in the section on optical methods.) Laplante found that in general the carbon content of the kerogen rose with depth, the oxygen content fell, while the hydrogen percentage rose to a maximum at about 10,000 ft and then fell. This is in line with the maturation and alteration of the kerogen as described in the last chapter (Figure 9.8). The maximum hydrogen content of the kerogen will determine whether oil or gas is produced, with the value of 6% hydrogen being, according to Laplante, the critical figure above which oil is produced and below which gas is the major product. The horizons at which hydrocarbons are generated are indicated by a reversal in the rise of hydrogen percentage with depth and the threshold of intense oil generation can be plotted out and will lie below the oil accumulation zone.

Claypool *et al.* (1978) have studied the colour and the atomic H/C ratios of kerogens and noted that the black kerogens have an atomic H/C ratio less than unity whereas the ratios for yellow to brown kerogens are greater than 1.0. At the

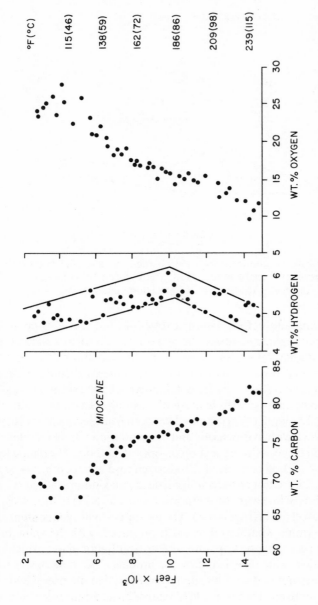

Carbonisation of kerogen with depth, South Pecan Lake field, Cameron Parish, Louisiana. (Laplante, 1974. Reproduced by permission of The American Association of Petroleum Geologists)

Figure 9.8 Carbonization of kerogen with depth (Laplante, 1974. Reproduced by permission of The American Association of Petroleum Geologists)

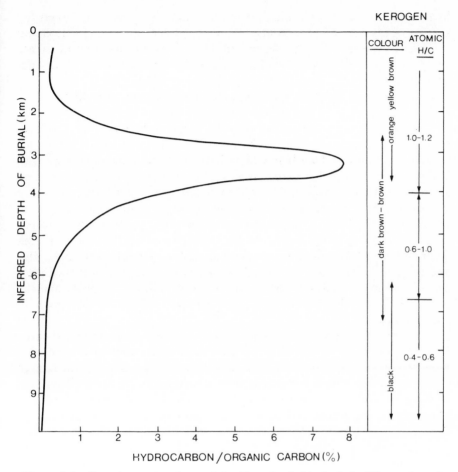

Figure 9.9 Organic geochemistry of the Phosphoria black shale (Claypool *et al.*, 1978. Reproduced by permission of The American Association of Petroleum Geologists)

same time they investigated the effects of depth on the production of hydrocarbons, as shewn by the ratio of hydrocarbons to total organic carbon, and related these to the colour and atomic H/C ratio of the kerogen (Figure 9.9). As can be seen, the peak of hydrocarbon generation occurs when the organic matter is a dark orange to brown and the H/C ratio is approaching unity, thus shewing that these latter parameters can give information on maturity levels which is related to the amounts of hydrocarbon produced. Hence the value of elemental analyses of separated kerogen.

Lijmbach (1975) stated that the H/C ratio of source material was an indication of its oil- or gas-generative potential. The humic material with a low H/C ratio would appear only able to source gas, while material with higher initial H/C ratios are indicated as potential oil sources. Ishiwatari *et al.* (1978) state that terrestrially derived material constitutes a poor source for hydrocarbons because of its highly oxidized nature. Harwood (1977) reported that kerogens with original atomic H/C ratios greater than unity are oil and gas sources whereas those kerogens with H/C ratios between 0.8 and 1.0 are wet-gas sources, while those with an H/C ratio of less than 0.8 are dry-gas sources.

So far the analyses of the kerogen, particularly in terms of its hydrogen to carbon ratio, have been interpreted as source indicative or maturity indicative by different workers and there may be some ambiguity. For instance, Claypool's comments that brown kerogens (i.e. immature) have H/C ratios greater than unity and are below their generative maximum, compares with Harwood's comment that kerogens with H/C greater than unity are oil and gas sources. Tissot & Welte (1978) have separated out the maturity and source factors as indicated by kerogen elemental analysis, by including the analysis of oxygen to the analysis for carbon and hydrogen. These are elements expressed as atomic H/C and O/C ratios and they are plotted on the Van Krevelen diagram, named after the first person to use it during the study of coals. This approach is based upon the detailed studies of kerogen, described in a previous chapter, and upon the known differences in the chemistry and the structure of each different kerogen type together with how these vary with maturation.

Thus, by the analysis of a mineral-free kerogen concentrate for three elements and with these expressed as atomic ratios plotted on the Van Krevelen diagram (Figure 6.25), it is possible to indicate not only the type of kerogen that is involved, but also its state of maturation. Hence the types and possible quantities of hydrocarbons generated can be assessed using a knowledge of the products of the thermal alteration of the various kerogens at each level of maturation. These were described fully in Chapters 6 and 7. The only place on the diagram where type of kerogen becomes difficult to determine is in the metamorphic stage where the elemental composition ratios of the three kerogen types are very similar. This is, of course, to be expected as they are all approaching a similar form of graphitic pyrobitumen. However, this is of little importance since all kerogens produce gas in this zone and the type can still be estimated by visual examination of the kerogen. It is worth noting that just because source rock analyses place a kerogen in the metagenesis zone, one need not write off that kerogen as a source of reservoired oil and gas in the locality. For that kerogen to have reached the metagenesis zone it must have passed through the principal zones of oil and gas production and it is possible that the fluid products generated in those zones migrated away from the source and are reservoired elsewhere. A reconstruction of the history of the sediments could indicate when the now overcooked or super mature source bed was producing oil and it could be possible to infer likely sands which were then available to accommodate the migrating hydrocarbons. The timing of fault movements could also be important in assessing the chance of generated oil migrating before the source bed is buried too deeply for hydrocarbons to survive. An investigation of the subsequent history of any potentially suitable sands could be profitable. Thus an overcooked source rock could have generated hydrocarbons and it is the subsequent fate of those hydrocarbons which will have determined, whether or not they have survived. Thus from this one diagram it is possible to estimate the type of hydrocarbons that will be produced, as this will depend upon both kerogen type and the degree of maturation of the kerogen, with both these factors being evaluated at the same time.

Table 9.2 gives an indication of the atomic compositions of kerogens at different stages in their maturation (Tissot & Welte, 1978).

Table 9.2

Zone	Type	I		II		III	
	Ratio	H/C	O/C	H/C	O/C	H/C	O/C
oil		1.45	0.04	1.25	0.08	0.8	0.18
wet gas		0.7	0.05	0.7	0.05	0.6	0.08
dry gas		0.5	0.05	0.5	0.05	0.5	0.06

The main problem with this type of analysis is that the dried, mineral-free kerogens have a tendency to absorb water, carbon dioxide, and oxygen from the atmosphere. Great care has to be taken when analysing kerogen concentrates to try to ensure that the kerogens are kept in such a way that they cannot absorb these compounds which can alter and invalidate the results of kerogen analysis. The presence of inert or recycled carbon within a kerogen concentrate can give misleading results and lower the assessment of the generative potential. This can be overcome by the removal of this 'dead' carbon by such methods as gravity separation (Kinghorn & Rahman, 1980).

Harwood (1977) has plotted the generation of hydrocarbons, as expressed by the extract in terms of the original kerogen weight, against the kerogen atomic H/C ratio (Figure 9.10). By determining the extract and the H/C ratios of samples, it is possible to indicate their generating potential in terms of both quality (i.e. oil, wet or dry gas), and of quantity, as the extract scale can be converted to a scale of barrels per acre-foot of rock containing a given percentage of organic carbon.

Rogers et al. (1974) have used the elemental analysis of the reservoir bitumens as a guide to thermal maturation. They considered that for mature oil zones the atomic H/C ratio of these reservoir bitumens was greater than 0.58 for the wet gas; for condensate zones the ratio lay between 0.58 and 0.53; while for the dry gas metamorphosed zone the ratio was less than 0.53. They also correlated this with the solubility of the bitumens in carbon disulphide.

Stable isotope studies

These studies are used to determine both type and degree of maturation of

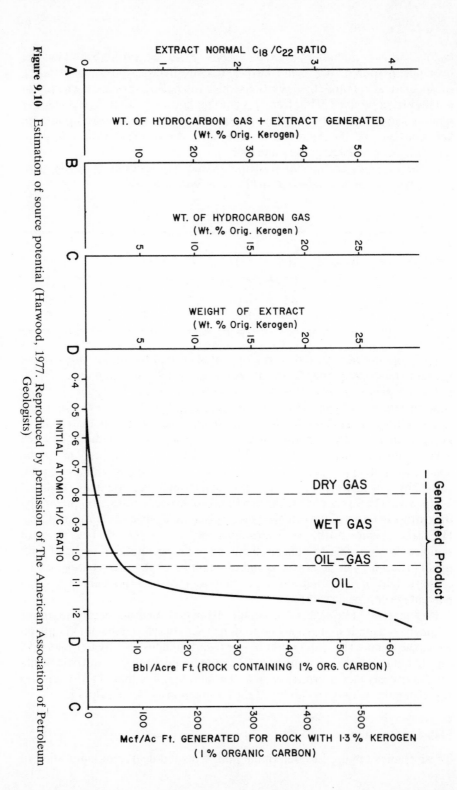

Figure 9.10 Estimation of source potential (Harwood, 1977. Reproduced by permission of The American Association of Petroleum Geologists)

sedimentary organic matter. The use of stable isotopes to determine the source, and hence type, of kerogen is possible because terrestrial plants use atmospheric carbon dioxide in their photosynthesis, and atmospheric carbon dioxide is isotopically lighter than the carbon in the bicarbonates of sea water which becomes incorporated into marine organic life. Thus the carbon in terrestrial plants is enriched in ^{12}C relative to marine plants, that is, it has a more negative $\partial^{13}C$ value. Hence the $\partial^{13}C$ values of kerogens and crude oils are probably a reflection of the relative contributions made by terrestrial and marine organic matter to the source of petroleum, i.e. the kerogen. For similar reasons plants which utilize atmospheric nitrogen absorb isotopically lighter nitrogen than do marine plants. Thus nitrogen isotope studies can also give indications of kerogen and hydrocarbon source (Stuermer et al., 1978).

Examination of the kerogen will thus indicate the probable origin of the organic matter and its eventual maturation products. A wide variety of uses can be made of $\partial^{13}C$ studies on both sedimentary organic matter and petroleum, including assessments of the environment of the source of the organic matter as well as comparisons between various oils and between source rocks and oils. Studies on the carbon isotope distribution in kerogens shew that terrestrial organic matter is enriched in the lighter isotope ^{12}C compared with marine organic matter in the order of 5–10 °/oo, and in ancient sediments the $\partial^{13}C$ of marine derived kerogen is normally 3–5°/oo higher than that of land-derived kerogen.

Stuermer et al. (1976) used the cross-plot of the $\partial^{13}C$ against the atomic H/C ratio as a simple method of indicating the source of sedimentary organic matter (Figure 9.11). With crude oils the isotopic composition of the different components varies: the normal alkanes are most deficient in ^{13}C but the aromatics and heterocompounds are most enriched, and the whole oil reflects the relative distributions of its component fractions. Work on the C_{15}-plus fractions of non-biodegraded crudes has given some source information. On a total crude the $\partial^{13}C$ value is related to the relative amounts of its compound-type groups which gives little information as regards origin or maturity. However, some successful applications have been made to gases. Biogenic gases can be differentiated from thermogenic gas by ^{13}C studies. For biogenic gas the $\partial^{13}C$ values range from −55 to −75°/oo whereas gas from liptinitic kerogen has $\partial^{13}C$ values between −40 and −58°/oo. This should be compared with coal and deep-dug thermogenic gas, where the range is from −25 to −40°/oo. It should be noted that these values can vary to some extent depending upon the original type of organic matter.

Schoell & Milner (1979) have used cross-plots of the $\partial^{13}C$ and the ∂D ($\partial D = [(R_{sample}/R_{SMOW})^{-1}] \times 10^3$ where R_{sample} is the isotopic ratio of the sample and R_{SMOW} is the isotopic ratio of a standard Mid Ocean Seawater such that R_{SMOW} ($= D/H_{SMOW}$) = 1.050 D/H NSS-1 which is the U.S. National Standards Bureau reference sample No. 1. Craig 1961) values of the C_{15+} and the C_{15-} fractions for crude oil correlations and they have shewn that correlations based on these values are more discriminative than the usual

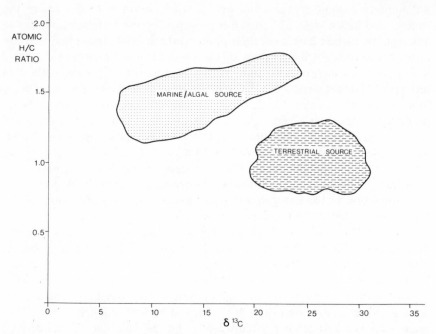

Figure 9.11   Organic type estimation using carbon isotopes (Stuermer *et al.*, 1976.
Reproduced by permission of Pergamon Press Ltd)

$\partial^{13}C/\partial^{13}C$ correlations. Because the alkanes are most affected by heat, it is
the cross-plots of these fractions which are used in maturation studies, whereas
the aromatic fractions are the best for oil–oil correlations. Yeh & Epstein
(1981) consider that both the δD and $\delta^{13}C$ values of crude oils appear to be
determined by the isotopic composition of their precursors. They could not
relate δD values to the age of the oil although $\delta^{13}C$ values did alter
systematically with oil age. No correlation between δD or $\delta^{13}C$ values and API
gravities was discerned and Yeh & Epstein thus consider that maturity and
chemical composition have little influence on the isotopic compositions of
normal crude oils. Rigby *et al.* (1981), having studied crude oils and brown
coals from the Bass Strait region of Australia, conclude that the changes in the
isotopic composition of the coals with increasing maturity shew that the
original δD values which were characteristic of the contributing plant materials
were more readily erased than the corresponding $\delta^{13}C$ values. Ross (1980)
used a cross-plot for carbon isotope values (ordinate) against the optical
rotation (logarithmic absicca) as diagnostic of the genetic character of oils in
the San Juan Basin. This plot (Figure 9.12) is also useful in interpreting the
effects of bacterial alteration and thermal maturation. Thermally mature oils
follow a trend towards the lower left on this plot, whereas biodegraded move
towards the bottom right. Claypool *et al.* (1980) used isotopic data to
differentiate biogenic non-associated dry gas from thermogenic gas in the Cook
Inlet Basin of Alaska using the values given above.

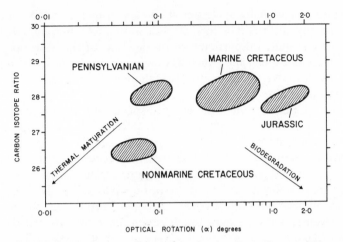

Figure 9.12 Optical rotations as maturatio indicator (Ross, 1980. Reproduced by permission of Penn Well Publishing Co.)

Redding *et al.* (1979) have noted that during coalification, hydrogen-bearing gaseous constituents, which are depleted in deuterium, are released from the organic matter and thus the residue should be enriched in deuterium at a rate which is proportional to the increase in maturity. However, for North American and European coals there was no correlation between D/H ratios and maturity, but these coals could be differentiated from Gondwanaland coals from South Africa and Australia because these latter coals were enriched in deuterium, thus shewing that deuterium enrichment can give clues to the source of organic matter.

Coleman *et al.* (1981) have shewn that methane oxydizing bacteria can alter the hydrogen and carbon isotopic compositions of residual methane with the result that the origin of such gas could be falsely attributed. Changes in the isotopic carbon and hydrogen compositions were temperature dependant but at every temperature the enrichment in the heavier isotopes is 8–14 times as great for the ∂D value than the $\partial^{13}C$. This could allow for identification of gas which has been subjected to microbial oxydation.

The use of carbon isotope data in relation to maturation studies is more limited; however, gases associated with oil in a reservoir tend to have 10 to 22°/oo less ^{13}C than their associated oil and if the gases are formed by the thermal maturation of reservoired crude oil, then the $\partial^{13}C$ value will increase with time because the earliest formed gases have the lowest $\partial^{13}C$ values (Silverman & Epstein, 1958).

Infrared spectroscopy

This can be used to examine concentrates of sedimentary organic matter for clues to both the type of the kerogen and its degree of maturation. Infrared spectroscopy is based upon the absorption by organic compounds of radiation in the infrared region of the spectrum. As the spectrum is scanned, each type of functional group absorbs at its characteristic wavelength, the amount absorbed being, up to a limit, proportional to the amount of the absorbing species present. Thus the spectrum gives information as to the types of functional groups present in the organic matter, and these will vary partly with type and partly with maturation.

As far as type is concerned the infrared spectrum will give information in relation to the relative amounts of aliphatic ($-CH_2-$, $-CH_3$) groups, aromatic structures, especially the polycyclic compounds, and the amount and type of heteroatomic groups present. When used for maturation studies infrared spectroscopy can indicate how much of a particular functional group there is present. This is important with functional groups whose abundancies alter with increasing maturation; for instance, carbonyl groups are present in the diagenetic stage, but are almost totally absent from the onset of the intense generation of hydrocarbons, whereas alkyl groups will decrease all through the catagenetic stage but aromaticity will increase at the same time. A particular level of absorption for a given functional group (e.g. aliphatic) may be due to an immature kerogen of one type with little aliphatic groupings, or to a mature sample of another kerogen type which originally had a lot of aliphatic material, but has lost them during maturation. Thus care needs to be exercised using the infrared spectra of kerogens but it can be a useful extra diagnostic instrument (Dypvik *et al.*, 1979).

Electron spin resonance

This is a technique which has been used to measure the state of maturation of kerogens. It measures the intensity of a signal which results from unpaired electrons. These unpaired electrons are present in free radicals within the kerogen and are produced when bond-splitting reactions occur. Normally free radicals are very reactive and they tend to react together to form new molecules in which these unpaired electrons are shared to give a stable electronic configuration. However, in a large, insoluble, immobile organic matrix, such as kerogen within a sediment, these free radicals can remain with their unpaired electrons because they are not free to move and meet each other and then react. The cracking of alkyl chains from the nucleus of kerogen will tend to produce such free radicals.

If the matter containing the free radicals is placed in an intense magnetic field and subjected to microwave radiation, the free radicals will resonate and alter the frequency of the applied radiation, and this change in frequency can be measured so that the number of free electrons, the location of the resonance frequency and width can all be determined. It has been asserted that these parameters can be used to determine degree of maturity of kerogen (Marchand *et al.*, 1968, 1969). This is because the paramagnetic susceptibility is proportional to the number of free radicals and this will increase with thermal evolution of the kerogen and as more groups, such as alkyl chains, are being broken off the kerogen. In general there does seem to be an increase in the signal with depth and temperature up to the beginning of the metagenesis stage (Durand *et al.*, 1977). This occurs at a level equivalent to about 2% vitrinite reflectance but then the value decreases, which obviously causes interpretation problems as a given signal can represent two quite different levels of maturation, one in the pre-metagenesis stage and the other in metagenesis.

Another problem is that the signal will depend upon the amount of material which is capable of being split off from the kerogen to give free radicals, which will be type dependent. The more aliphatic kerogens will be able to produce more free radicals. However, when one considers the stability of such produced free radicals the reverse is true, i.e. they are more stable on type III kerogens, because these are inherently more aromatic, a property which allows the free radicals greater chance to be stabilized.

The method is therefore potentially suitable for maturation studies before the metagenesis zone (i.e. in the diagenetic and catagenetic regions) and because the number of these free radicals seems to depend, not only upon the maturity of the kerogen but also upon its type, there is a factor which is due to regional and depositional conditions that needs to be taken into account. To overcome this, the method requires sophisticated setting up and standardization together with correlation procedures to account for these regional and depositional factors. The great capital cost of this method together with all these other factors restrict its suitability, and its use is very limited.

Pyrolysis methods

The artificial maturation of kerogen, whether with or without its accompanying inorganic rock matrix, in the absence of oxygen is called pyrolysis. The name covers a number of related methods which are quantitative extensions of the pyrolysis of oil shales to produce commercially usable quantities of hydrocarbons, but on much smaller scales to assess both organic matter type and level of maturation. The majority of pyrolysis methods used to examine kerogen involve the heating of the kerogen whilst it is still in the rock, which is normally in a powdered form.

Variable-temperature pyrolysis

Barker (1974) pyrolysed rocks in a stream of nitrogen which was used to sweep the volatile pyrolysis products away from the sample and into a flame ionization detector which analysed the amount of carbon in the effluent with respect to time, and thus to temperature, since the oven was heated to warm up at a uniform rate. The use of oxygen-free nitrogen prevents oxidation of the organic material.

As has already been noted, the organic matter in a sedimentary rock is composed of the already generated mobile material which can be removed with solvents, and the partially converted kerogen. These will be present in varying amounts depending upon the maturation level and, to a lesser degree, the occurrence of migration. Thus when a sample is heated at a steadily increasing rate the first thing that occurs is that the volatile hydrocarbons are boiled off at about 130 °C. The kerogen breakdown is a thermally induced reaction and thus artificially putting energy, in the form of heat, into the kerogen will 'mature' the kerogen and produce a second group of volatile hydrocarbons. This second group, being maturation products, do not occur until a higher temperature, normally around 400 °C.

With regard to organic matter type, Barker considered that the presence of large quantities of methane in the second peak was indicative of a kerogen with gas potential whereas higher hydrocarbons in the second peak indicated an oil-prone source. This type of equipment probably has more value in terms of maturation studies, because Barker noticed that as the depths from which the samples originated became deeper, the relative size of the first peak increased owing to the fact that more hydrocarbons had already been formed in the rock because of extra energy caused by the extra heat related to the deeper depth of burial. At the same time, the second peak became smaller and moved to higher temperatures because the more easily generated material had been released from the kerogen and appeared in the first peak, and thus much more energy (i.e. higher temperature) is required to convert the remaining kerogen into hydrocarbons.

Generative potential of source rocks is given by the ratios of the peak sizes. For immature rocks the ratio of the first peak to both peaks will be small and this ratio increases with increasing maturity; similarly, the temperature of the second peak increases with increasing maturation. Samples run on cuttings from wells will give indications when the zone of intense hydrocarbon generation is being approached and when the dry-gas zone occurs (Figure 9.13).

Tissot et al. (1974) described an improved version of such a pyrolysis apparatus where carbon dioxide from the breakdown of the kerogen is trapped and, after the first two peaks have been recorded, the amount of carbon dioxide is measured. This carbon dioxide comes from the oxygen-containing parts of the kerogen, especially the carbonyl linkages. As in the Barker system the first peak is composed of the already generated hydrocarbons which are simply volatilized by the heat (peak P_1) and the second peak (P_2) is the hydrocarbons generated during pyrolysis. The new third peak (P_3) represents the CO_2 generated from the organic matter. Again the maximum temperature of the second peak is recorded (T_{max}) and used diagnostically.

Tissot, like Barker, suggests that the ratio $P_1/(P_1 + P_2)$ is the transformation ratio and it gives the relative amount of hydrocarbons generated to the total potential hydrocarbon generation. This ratio, together with the temperature of the second peak, can be correlated with other indices of maturity, such as vitrinite reflectance. The peak temperature is itself a clue as to the maturity of the organic matter, because with increasing depth the value of T rises. However, this relatively simple system is complicated by the

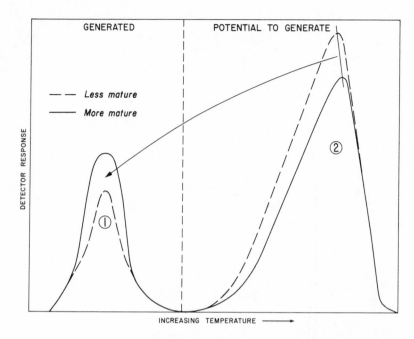

Figure 9.13 Pyrolysis methods of maturation assessment (Baker, 1974. Reproduced by permission of The American Association of Petroleum Geologists)

fact that the value for T will depend upon the type of organic matter involved and this may be the greatest influence on the value of T_{max}. The presence of clay minerals also tends to increase the value of T_{max}, thus giving a misleadingly high indication of the maturity of the sample.

So far these pyrolysis methods have only investigated the volatile organic matter and have ignored the graphitic residue. While this can not be measured by pyrolysis, it can be measured in the total organic carbon measurements and taken into account. The use of the Van Krevelen diagram has previously been described. This is based upon the elemental analysis of kerogen and will give information about the type and maturation level of the kerogen. The main problem is the lengthy procedures required to remove the inorganic rock materials. Espitalié et al. (1977) correlated the atomic H/C ratios and O/C ratios of many samples with the 'hydrogen index' and 'oxygen index' respectively. These indices are derived from the pyrolysis of the whole rock sample and the total organic carbon value. The hydrogen index is P_2/TOC and the oxygen index is P_3/TOC. Because straight-line relationships between the atomic ratios and the appropriate indices were obtained, the atomic ratios in the Van Krevelen diagram were replaced by these new indices. The new diagram (Figure 9.14) can be used in exactly the same way as the Van Krevelen diagram to determine organic matter type and maturity. Because the inorganic material does not need to be removed, this analysis is quicker and cheaper.

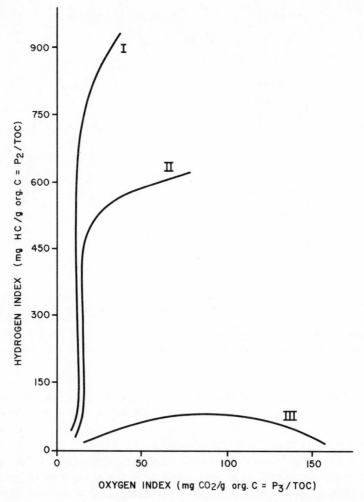

Figure 9.14 Kerogen type and maturatio by pyrolysis (Tissot & Welte, 1978. Reproduced by permission of Springer-Verlag, Heidelberg)

Tissot & Welte (1978) suggest that a semi-quantitative assessment of the genetic potential of source rocks can be achieved using such instruments by taking the first two peak areas expressed in kilograms of hydrocarbons per tonne of rock and using the following scale: less than 2 kg/tonne (2000 ppm), not an oil source but some gas potential; 2–6 kg/tonne (2000–6000 ppm), a moderate source rock; greater than 6 kg/tonne (6000 ppm), a good source rock.

There are, however, certain queries about the results obtained from such instruments under certain circumstances. The third peak, which gives information about the oxygen content of the kerogen, can also contain carbon dioxide from inorganic carbonates, even though these should not decompose at the temperatures involved. Therefore care must be taken to ensure that the collection of organic carbon dioxide takes place at temperatures at which inorganic carbonates are not decomposed. This will involve a knowledge of the mineral constituents present in the sample and their behaviour when heated, since metal carbonates decompose at different temperatures. Siderite yields carbon dioxide at about 400 °C and dolomite above 500 °C; hence the need for care as the instrument will heat samples to above 500 °C.

The material generated in the second peak may contain large amounts of aromatic material which is not observed in undegraded oils. The presence of these aromatic compounds is an indication that the kerogen is gas-prone and will produce few liquid hydrocarbons. However, in such cases the large P_2 peak will give a large 'hydrogen index' which is indicative of an oil-prone kerogen. The aromatics are probably produced from the lignin and tannin components of gas-prone kerogens. Because these pyrolyses take place in a hydrogen-deficient medium, the products will be hydrogen deficient and unlike those produced by the thermal alteration of that material in nature. Another difficulty with this technique is that the first peak should only contain already generated hydrocarbons. However, most extraction techniques only quantitatively recover hydrocarbons above C_{15}, because the lower compounds are lost either in transit, storage, or processing. Thus only the less volatile compounds remain. The generated hydrocarbons can contain compounds up to C_{36} and above, and under the mild pyrolysis conditions equivalent to the first peak, the number volatilizing will be limited and many higher boiling point compounds will remain in the sample. These will eventually distil over to be incorporated in the second peak. The alkane $C_{15}H_{32}$ has boiling point of 268 °C and, although some will vaporize and be detected under the first-stage heating, not all the already generated solvent-soluble compounds will be included in the first peak. Thus a large amount of this material, which can include compounds such as $C_{36}H_{78}$ (b.p. 497 °C), will be included in the second peak which is supposed to be material generated solely during pyrolysis. This problem therefore tends to make samples look more immature than they actually are and the heavier are the extractable compounds, (i.e. the more immature), the worse the problem (Clementz, 1979).

Espitalié et al. (1980) have shewn that clay minerals can trap hydrocarbons and at a higher temperature and after longer times they are converted to lighter hydrocarbons and a graphitic residue. The effect is most noticeable with low organic matter amounts and highly active mineral matrices, e.g. clastic-clayey sediments. This will obviously affect the relative sizes of peaks 1 and 2 with consequences for both kerogen type and maturation interpretation. The pyrolysis of kerogens, with and without various minerals present, allows correction graphs to be obtained so that the values of $P_1 + P_2$ unaffected by the mineral matter are obtained. Horsfield & Douglas (1980) have also observed the effect of minerals on the pyrolysis of coals and kerogens. In this case it was stated that the relative quantities of the products produced during pyrolysis will alter depending upon the minerals present.

Large amounts of dead carbon will also give misleading results as the total organic carbon value will be increased but the first two peaks will have no contribution from this dead carbon. The oxygen and hydrogen indices obtained from samples with large quantities of recycled organic matter will be erroneously low and give a false indication of the generative potential of that sample. If the sample is being macerated for the optical examination of the organic matter (see below), it is preferable to analyse the mineral and dead carbon-free kerogen concentrate for carbon, hydrogen, and oxygen. These will give atomic H/C and O/C ratios which will be much more accurate than pyrolysis in these circumstances.

There is also the possibility of contaminants being present in the sample and these can include hydrocarbons from drilling muds, hydrocarbon spills, plastic bags, coatings, cotton bags, or from the wax on operators' fingers, and all these will appear in the first peak, making the sample look more mature than it actually is. Because of the small quantity which is involved, even some grease from a finger can make a significant difference to the results. These types of problems shew the sorts of errors which can creep into pyrolysis methods and certainly comparisons of the relative magnitude of the first two peaks produced in variable-temperature pyrolysis do not tally with the amounts of organic carbon in the soluble and insoluble fractions of organic matter in rocks.

A more recent pyrolyser, the Oil Shows Analyzer, has been developed to overcome some of the problems discovered with earlier instruments. In order to decrease the quantity of already generated material in the second peak (P_2) the heating range of the first peak has been extended to 320 °C instead of 200 °C. This has the additional advantage of allowing the first peak to be split into hydrocarbon gases evolved below 80 °C (P_o) and free hydrocarbons evolved from 80 to 320 °C (P_i). At the same time the total organic carbon content is ascertained by oxydation of the kerogen residues. This allows the instrument to directly calculate the hydrogen index (P_2/TOC), the total production index ($P_o + P_i$)/TOC as well as the gas and oil production indexes $P_o/(P_o + P_i + P_2)$ and $P_i/(P_o + P_i + P_2)$ respectively.

In spite of the problems, pyrolysis methods can be very useful, either as a substitute for Fischer assay (Espitalié *et al.*, 1977) by plotting hydrogen index against *T* max or by plotting peak areas or the transformation or production ratios against depth to indicate the zones of intense hydrocarbon production. The methods are quick, relatively cheap, and can be used on site (Clementz *et al.*, 1979).

There are other methods of pyrolysis. Claypool & Reed (1976) described a thermal analysis technique for source rock evaluation. In Tissot's system care is taken to try and keep all the hydrocarbons in the sample until the time of analysis. Claypool & Reed remove everything up to C_{15} by sample drying at 105 °C. They increase the temperature of the oven at a different rate but, as with the methods of Barker and of Tissot, two peaks are obtained, for 'volatile' and 'cracked' hydrocarbons. The oven is heated from 30 to 800 °C at a rate of 28 °C/min, with the products swept into a flame ionization detector by the nitrogen carrier gas. The first peak (already generated material) is seen in the 30–400 °C range and a second peak appears between 400 and 800 °C, although the first peak is not always seen. The temperature of the second peak maximum is a function of the maturity of the sample. The different temperature ranges for the two peaks is a function of the different rates of heating as compared with other similar systems.

The area of the first peak was plotted against the C_{15+} solvent-soluble hydrocarbons and the area of the second peak against total organic carbon, and in both cases linear relationships were reported. Thus two peak areas may be plotted against one another and a composite diagram produced (Figure 9.15). From this diagram the source potential of the samples can be determined in terms of bbls (Barrels, 1 barrel = 42 U.S. Gallons) per acre-foot or MCF (million cubic feet) per acre-foot. However, the same limitations and qualifications about the interpretation of these results apply to this equipment as to the similar devices described earlier.

In all these systems of pyrolysis so far discussed, the relative amounts of already generated material to the total genetic potential is an index of source rock maturation, which is supposed to be independent of the type of kerogen involved (Barker, 1974; Tissot & Welte 1978). The ratio of already generated hydrocarbons to total genetic potential (already generated plus hydrocarbons generated by cracking during pyrolysis), i.e. $P_1/(P_1 + P_2)$, increases with depth of burial or temperature. A large positive deviation from this progressive increase with depth (i.e. a sudden increase in $P_1/(P_1 + P_2)$) indicates the presence of strata where hydrocarbons are being generated. A large negative deviation from the progressive increase in the transformation ratio $P_1/(P_1 + P_2)$ with depth indicates a horizon where the generated hydrocarbons have migrated.

Another use of programmed pyrolysis is to produce kerogen facies maps whereby the relative amounts of the different principal kerogen types are obtained by statistical analysis of the hydrogen and oxygen indices. If the kerogen is a mixture, its composition, in terms of Tissot's three kerogen types, can be obtained from the obtained values of the hydrogen and oxygen indices.

The volume of generated oil that has migrated from the source rock can be estimated using programmed pyrolysis. The type of kerogen, or the relative composition of mixed kerogens, can be obtained from such pyrolyses. The total generative potential of such a kerogen or kerogen mixture can be estimated from theoretical values (Chapter 8). Peak P_2 represents that material that has not yet been converted to mobile hydrocarbons and if this is subtracted from the theoretical maximum potential of that kerogen, the amount of hydrocarbon already generated is obtained. This value can be

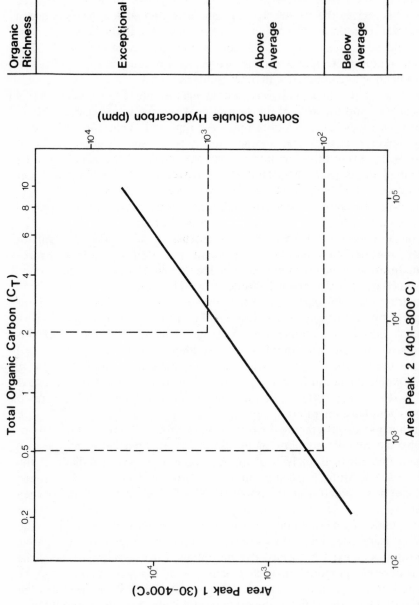

Figure 9.15 Source rock potential assessment (Claypool & Reed, 1976. Reproduced by permission of The American Association of Petroleum Geologists)

compared with the first peak (P_1) and any difference should represent the amount of generated hydrocarbon which has migrated.

Fixed-temperature pyrolysis

There are other forms of pyrolysis, normally at fixed temperatures, such as that of Giraud (1970) where the rock has all the solvent-soluble organic matter removed by exhaustive solvent extraction. The samples are heated at a relatively low temperature (approximately 200 °C) and the organic matter which is generated is trapped in a tube immersed in liquid nitrogen. At the end of the required pyrolysis time the trap is warmed and the contents flushed directly on to the column of a gas chromatograph where they can be analysed. Continental sediments tended to produce methane and benzene derivatives, where as marine sediments gave higher molecular weight, more paraffinic compounds. These chromatograms are the equivalent to the second peak in programmed pyrolysis. The fact that gas-prone and oil-prone organic matter give different but significant quantities of pyrosolate adds to the uncertainties about the diagnostic value of the second peak in programmed pyrolysis. It is the composition of the peak, not its size, which is the important factor.

The range and type of hydrocarbons produced by the cracking of the kerogen are a very good indication of the source potential of the sample. Sources of paraffinic oils produce a chromatogram which primarily consists of doublets of alkenes and alkanes up to C_{30}. There will be very little naphthenic or aromatic material in such chromatograms. The sources of mixed-base oils will have similar chromatograms to that described above but with a substantial envelope of unresolved naphthenes. Condensate sources produce very little material above C_{18} and the doublets are less pronounced as the alkene components are often depleted. Gas source rocks produce large amounts of methane and low molecular weight aromatic compounds.

Work is being carried out to try to combine the two sorts of methods whereby in a temperature-programmed pyrolysis the components of each peak are analysed to give indications of source type, molecular weight range (for correlation and typing purposes), as well as spotting contamination, etc. Huc & Hunt (1980) have described a pyrolysis instrument which produces two peaks, but the pyrolysis oven effluent is split so that half passes through an FID (flame ionization detector) to produce the two peaks while the other half is analysed on a gas chromatograph.

Such instruments will probably be of more value in source rock analysis than the simple two- or three-peak devices as the maturity of the compounds produced can be ascertained, and it is their nature as well as their volume which indicates the source potential of the kerogen whence they came. A gas kerogen could produce a large volume of methane and low molecular weight aromatics and be interpreted as oil prone on a two- or three-peak pyrolyser on the basis of its hydrogen index. These instruments still cannot indicate the presence of recycled carbon, substantial amounts of which will indicate reduced potential if the results of such pyrolyses are combined with total organic carbon. The amount of recycled carbon can only be determined by acid maceration, density separation, and visual examination.

An alternative method of fixed-temperature pyrolysis to the trapping system used by Giraud is for the pyrolysis products to be passed directly on to the column of the gas chromatograph. Larter & Douglas (1979) describe such an apparatus in which kerogens and coal maceral concentrates have been studied. They have shewn that the different kerogen types produce different pyrograms and that the pyrogram of one kerogen is sufficiently different from the pyrogram of other kerogens for unknown samples to be identified. The more aromatic kerogens have pyrograms dominated by aromatic and phenolic compounds, whereas the more aliphatic kerogens have pyrolysates principally composed of aliphatic hydrocarbons (see above). This kerogen-type analysis can be confirmed by the study of the ratio of m-xylene to n-octane produced in such experiments. A ratio below 0.3 is typical of alginites, while sporinites lie between 0.3 and 1.0 and vitrinites give ratios greater than 1.0 and often over 20.

Horsfield & Douglas (1980) have noted that if these pyrolyses are carried out with whole rock samples then the pyrolysates will depend upon the minerals present. The relative amounts of material produced by pyrolysis of kerogen will depend upon the type of accompanying mineral matter as well as the organic matter type and level of maturation, and it is questionable if such pyrolyses are of use to determine organic matter and maturity. This type of analysis is the basis of the combined system of programmed pyrolysis with the analysis of the compounds produced. The ability to obtain a detailed examination of the components of the first two peaks of a programmed pyrolysis widens the scope for source, maturation, and product estimation. (Huc & Hunt, 1980).

Gransch & Eisma (1966) proposed the 'carbon ratio' as a means of evaluating the degree of maturation of a source rock. They defined the carbon ratio as C_R/C_T, where C_T is the total organic carbon content of the rock and C_R is the carbon remaining after pyrolysis at 900 °C for 1½ hours. They stated that ratios of less than 0.6 indicated immature source rocks containing low aromatic kerogens with high genetic potential, whereas ratios of greater than 0.6 indicate either a high potential source rock after maturation or a high aromatic kerogen with a low genetic potential.

The theory involved in this method is relatively simple. It has already been demonstrated that there can be three types of organic carbon within a source rock, viz. the secondary mobile or bitumen; the secondary immobile or pyrobitumen; and the primary kerogen which is yet to be converted into the first two. As the source rock becomes more mature, the percentage of the pyrobitumen increases until eventually it is not only practically the only form of carbon present but it resembles graphite, which is inert at 900 °C in nitrogen. Hence in the metagenesis zone, irrespective of the kerogen type with which one is dealing, there will be only relatively small changes in the organic carbon content due to pyrolysis, and the carbon ratio will be approaching unity. For immature kerogens there will be much larger changes upon pyrolysis as the insoluble kerogen is converted to bitumen and pyrobitumen. (The bitumen is removed by the stream of nitrogen and may be analysed as described above.) Hence is this case the carbon ratio will be small. However, the exact magnitude of carbon ratio for immature kerogens will depend upon the kerogen type. For type I the ratio is 0.2, rising to 0.4 for type II, and 0.6 for type III. Gransch & Eisma (1966) state that the thermal maturation of humic kerogens starts at a carbon ratio of about 0.6, whereas for the non-coaly matter the carbon ratio starts at 0.2 or less. Rad (1982) has plotted the ranges of carbon ratio for the three kerogen types (Figure 9.16). Because all kerogens when approaching maturity have carbon ratios nearing unity, the carbon ratio method can, to some extent, indicate kerogen type.

The main problem with this method is that a given value of carbon ratio can apply to each kerogen type, but at different levels of maturation. However, if another assessment of the level of maturity is available, such as vitrinite reflectance, or space coloration, then the type of kerogen can be obtained. Conversely, if the carbon ratio and the type of kerogen is known, its level of maturity can be assessed. Thus the carbon ratio is a useful link between kerogen type and maturity and can be used to discover the one if only the other is known, or as a check between other assessments of kerogen type and maturity.

Tissot *et al.* (1974) shewed that the pyrolysis products of terrestrial kerogen can contain upwards of 50% carbon dioxide and water and therefore a higher percentage loss of oxygen and hydrogen than of carbon would cause an apparent increase in the residual carbon value. This could result in a carbon ratio greater than unity because there could be an increase in carbon relative to the other residual components after pyrolysis if these other components have decreased more than the carbon. This will often occur with samples from coaliferous horizons. To overcome this problem the amount of carbon lost during pyrolysis can be measured either from the quantity of generated hydrocarbons plus the liberated carbon dioxide, or more easily by taking into account the weight lost during pyrolysis, as described by Rad (1982). By using a corrected figure for C_R, a more accurate carbon ratio may be obtained. Thus the concept of carbon ratio, while not an absolute indicator of maturity or kerogen type, is a useful connecting link between the two, and if carbon ratio and one of the other two is known, the third may be calculated.

There are several ways in which the carbon ratio may be measured and it is important not to confuse them. Rad (1982) used powdered rock samples otherwise untreated for both total organic carbon measurement and pyrolysis. Gransch & Eisma (1966) used powdered rock samples which had been solvent extracted to remove the soluble bitumen and then acid treated to remove inorganic carbonates. Heroux *et al.* (1979) measured carbon ratio using the mineral-free kerogen concentrates. Obviously each method will give a different value for the carbon ratio of a particular sample and it is essential to

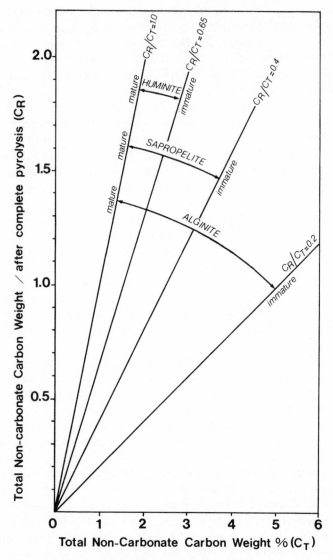

Figure 9.16 Carbon ratio. (Reproduced by permission of Scientific Press Ltd)

ensure that the right reference frame is used to obtain correct interpretations and that similar methods are used in comparative studies.

Many of these pyrolysis methods which have been successfully used with whole rock are now being carried out on mineral-free kerogen concentrates. This sort of work is still new, and few detailed results are available, but initial reports are encouraging.

Pyrolysis methods are numerous and at first sight seem to produce all the information required for the petroleum geochemist. However, there can be difficulties in obtaining consistent results of a high enough standard. Some problems have already been mentioned, such as the rate of temperature increase, the inorganic rock matrix, contamination, and loss of free hydrocarbons, especially with small samples. Great care is required with these methods in order to minimize these problems and to produce consistent, meaningful results. There is no doubt that pyrolysis methods are a valuable aid to petroleum geochemical exploration methods, but as with all geochemical methods they should not be used alone as they may be misleading.

OPTICAL METHODS

Introduction

Examination of sedimentary organic matter using the optical microscope requires pieces of the organic matter at least 1 μm or larger. For quantitative work using photometers, the minimum diameter should be at least 3 μm. The matter can be viewed in transmitted or reflected, visible white or ultraviolet light (fluorescence) and the information yielded by these varying techniques is often complementary. In the past a lot of the reflected-light investigations were on polished surfaces of the whole rock; however, this technique is being replaced by viewing the mineral-free kerogen concentrate either as a strewn slide or mounted in a plastic base which is then polished. In this way a much greater concentration of organic matter can be examined without the complications of the mineral elements (Saxby, 1970).

Dancy & Giedroye (1950a, b) have studied the removal of the minerals and the effect that the required reagents have upon the kerogen, and they believe them to be minimal.

Since the majority of studies are carried out on kerogen concentrates, it is necessary to remove the mineral matter. This is normally done by a process known as acid maceration. The rock is crushed to about ¼-inch sized pieces. A smaller size will destroy the spores and break the larger pieces of organic matter, such as the vitrinite macerals, while larger lumps will require much longer treatment. The pieces are then treated with a 10% hydrochloric acid solution to destroy the carbonates and this is followed, after removal of all the chloride ions, by washing with various strengths of hydrofluoric acid to destroy the silicates. This treatment may involve gentle heating and agitation to assist the reaction. After each acid wash, the acid is decanted and fresh acid added until all the rock is disseminated. The sample is then washed to neutrality with distilled water and any remaining heavy mineral matter can be finally removed by floatation and centrifugation using a heavy, dense liquid such as zinc bromide solutions (Barss & Williams, 1973; Bostick & Alpern, 1977; Kinghorn & Rahman, 1980).

The sedimentary matter is a mixture of different sorts of material, each with its own chemical composition and specific gravity. The specific gravity of each component will depend on its composition, which is related to its origin and level of maturation. This fact allows sedimentary organic matter to be split into groups according to density. Figure 9.17 shews how the three kerogen types and their level of maturation can be identified from the range of specific gravity of their components (Kinghorn & Rahman, 1980). As each kerogen type matures it becomes denser with, at any given maturation, type I kerogen being lighter than type II, which is in turn lighter than type III.

Most kerogens are a mixture of two or more kerogen types and this fractionation method allows the relative proportions of each type to be ascertained. This can be done for all the source rock in a basin which is in the 'oil window' and, because each kerogen type has a different convertibility and different products, a much more accurate picture of the products which the mixed kerogen is likely to produce can be obtained.

Kerogen type assessment

The first use to which the optical examination of separated organic matter concentrates can be put is to assess the kerogen type. This involves recognizing the structure of the organic debris, its shape, and its colour. Transmitted white-light microscopy is especially good for picking out structured components such as spores, pollens, bits of cuticle, or structured algal material as these can be viewed and their shape and structure assessed. These components are transparent and therefore easily identified in transmitted light, but the technique can also be used to view the outline shape and structure of the opaque materials, the vitrinites and inertinites, etc. These latter macerals are land derived as compared with marine-originated algal and plankton species, and thus the relative amounts of each in a kerogen can be assessed and an overall picture of the source potential of the kerogen obtained. A more detailed examination of the opaque material, especially the detailed structure, can be obtained using a reflected-light microscope.

The other advantage of the reflected-light microscope is that it can differentiate vitrinite, which can have some oil and a good gas potential; from inertinite, which is dead carbon and has no hydrocarbon generation potential.

The use of fluorescence is especially useful in identifying the amorphous liptinitic materials which have some oil or gas potential. Because these materials are amorphous and the detailed structure too fine to see, it is difficult to tell whether they are the planktonic algal amorphous matter which has a high hydrocarbon generation potential, or the amorphous remains of organic material which has been attacked by bacteria and left as an amorphous mass having no hydrocarbon potential. Ultraviolet or blue light will cause the liptinitic amorphous material (high oil potential) to fluoresce, unlike the amorphous material which has no oil potential. Powell et al. (1982) consider that the main disadvantage of the optical examination of kerogen is the inability to distinguish between the hydrogen-rich and the hydrogen-lean amorphous kerogens. Only fluorescence can distinguish between these oil-prone and gas-prone amorphous kerogens. Fluorescence can also be used

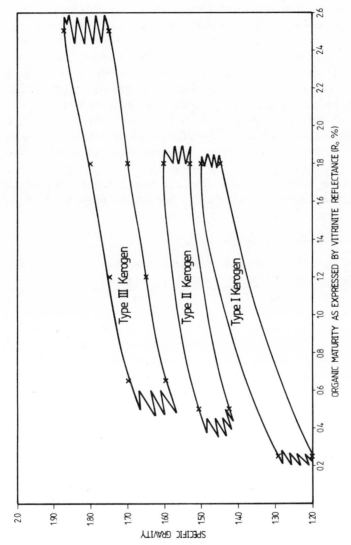

Figure 9.17 Variations of specific gravity of kerogen with maturation

with reflected light microscopes to distinguish between hydrocarbon prone and inert macerals.

Thus by viewing the concentrates of sedimentary organic matter in transmitted and reflected, white and ultraviolet light it is possible to differentiate the various organic matter types. Type I kerogen mainly consists of structured algal material with amorphous matter occasionally the result of microbial reworking. Type I can also contain spores, pollen grains, etc., which have been blown off the land and settled through the water to become incorporated in the sediment. Type II also contains spores, pollens, and cuticle material, but its main constituent is sapropelic amorphous matter plus some structured algal or planktonic remains. There will also normally be a small terrestrial humic-based component, the magnitude of which depends upon the conditions of deposition, such as nearness to shore line, deltas, river speeds, prevailing winds, etc. Type III kerogen is the humic, opaque, coaly material which includes vitrinites and inertinites, but it may also contain sapropelic material, some algal remains, etc. It is rare to obtain a kerogen which is solely composed of one type, and most kerogens are mixtures which vary across and up and down the source bed interval (Figure 7.4).

Kerogen maturation assessment

The other use to which the optical examination of organic matter is put is to determine the level of organic maturation. The processes by which sedimentary organic matter is converted into oil and gas have already been discussed. These processes are often called bituminization, which is a diagenetic process with biochemical and geochemical stages. In the biochemical part of bituminization the deposited organic matter is converted by bacterial and chemical processes during burial into a complex polymer which has a lower nitrogen and oxygen content than its precursors.

Burial causes the sediments to compact and to expel water, and the kerogen is converted into hydrocarbons by the thermal cracking related to the burial. This is the second stage, the geochemical bituminization, or the early metamorphic stage (Teichmuller & Teichmuller, 1968). Coalification is the equivalent process by which organic matter is transformed into coal or peat via the different stages of brown coal (lignites), sub-bituminous and bituminous coals, to the anthracites and the meta-anthracites. This process is characterized by an increase in the carbon content of the matter and decreases in the oxygen and hydrogen contents. Coalification also involves two stages. There is an early or biochemical stage during which the biological processes take place under the normal conditions of temperature and pressure, which ends at the production of soft lignite, and a second or geochemical stage which occurs after the deeper burial of the organic matter and is due to the influence of increased depth and elevated temperatures. This late stage lasts from the start of the hard brown coal (sub-bituminous) up to the meta-anthracite stage (McCartney & Teichmuller, 1972; Castano, 1973; Alpern, 1976).

Thus coalification and bituminization are related processes, and both have biochemical and geochemical stages. In both cases there is the production of fluids (oil and gas) and a solid residue (coal or pyrobitumen) caused by aromatization and condensation. The major difference between the two is caused by the differences in the original organic materials which were included in the sediment. This process of organic metamorphism is accompanied by progressive changes in the physical and chemical properties of the organic matter. These changes include the darkening of the organic matter, a decrease in light transmission, a decrease in electrical resistivity, a loss of structural detail, increases in the indices of reflection, refraction, and lustre, together with changes in their spectral characteristics, and an increase in carbon content. Consequently a measurement of the progressive changes in any one of these properties could be used as an indicator of the organic metamorphism and the related processes of coalification and bituminization. The optical properties, especially reflectance, translucency, and colour, can be used to determine coal rank and the different stages of oil and gas formation.

Carbonization is a process of purely thermal destruction or decomposition of the organic matter in the sediment with a consequent increase in the relative carbon content and an end product of graphite or anthracite. During this process the different organic components respond to the temperature in different ways, and thus some plant microfossils containing the waxy substance sporopollenin offer good measures of palaeothermal history. Among these microfossils, the spores and pollens are the most thermally sensitive (Cooper & Murchison, 1971; Gutjahr, 1966).

Once in the metamorphic environment, the spores are affected by increase in temperature and change their colour from yellow to pale yellowy-brown (diagenesis), through shades of brown (catagenesis), eventually to black (metagenesis). As the carbonization proceeds the translucency of the spore coat decreases, the absorption of light increases and thus, with careful and precise measurements, an accurate temperature correlation can be made. These measurements of translucency or colour can be related to the degree of organic maturation or to the elemental composition or to maximum palaeotemperature (Grayson, 1975; Gutjahr, 1966).

Care has to be exercised in order to obtain meaningful results. Suitable spores which are plentiful over the whole area to be studied are required in order to have comparable results. Local and regional tectonic heat-generating movements such as folding and faulting may play an important part in the thermal evolution of the organic matter and need to be recognized during optical examination to prevent misleading conclusions being reached.

Depth of burial normally affects the diagenesis of the organic matter, as increasing depth means increasing temperature. However, it is difficult to predict a precise depth of burial for the complete destruction of organic matter (from the point of view of petroleum prospecting) as the procedure is associated with other parameters that can influence the normal effect of the thermal gradient. In general, very deep drilling is not favourable for petroleum prospecting, not because of lack of technical ability or the disappearance of porosity, but because of the low tolerance of oil to elevated temperatures (Evans & Staplin, 1971). The total organic matter may not be uniformly altered, and Correia (1971) has shewn that spores and pollens will change colour much more readily prior to the alteration of the co-existing acritarchs, which themselves alter before the co-existing chitinozoa. It has been found that even under similar conditions the various types of spores and pollens carbonize differently owing to the various chemical characteristics of their exine structures, and thus the need to find a spore that occurs in all the samples from the particular region under investigation.

The composition of the host rock is an important factor as the different minerals have different thermal conductivites. The thick sediments with low heat flows are more prospective than those with high heat flows. Correia (1971) has shewn that microfossils are more readily altered in clay environments than in calcareous ones as the clays, although not necessarily homogeneous, contain more catalysts. The carbonates, even though more homogeneous, are less favoured catalytically and hence have less effect upon spores and pollens. A quartz-rich sediment with a high thermal conductivity may more greatly affect the organic matter and give rise to darker-than-expected spores, and similar results may occur due to locally high thermal gradients within the same depositional basin. Recycled organic matter, including spores and pollen grains, may also be darker in colour and exhibit different ultraviolet fluorescence characteristics than the contemporaneous material. These are often best identified in reflected light.

The advantages of the measurement of the carbonization of spores and pollens are that the method is quick and cheap and there is a wide occurrence of spores in sedimentary rocks. The notable limitation of this technique only occurs when studying the whole range of organic metamorphism because the spores turn black at about the medium-volatile bituminous rank of coal. However, since crude oils are metamorphosed at this stage, producing methane, hydrogen sulphide, carbon dioxide, and graphite (Staplin, 1969), this is not too much of a disadvantage to petroleum geochemists. Opposed to this is the fact that carbonization can be used to detect palaeotemperature levels from the very early stages of organic metamorphism, while other techniques such as vitrinite reflectance cannot. Thus the two techniques together are able to cover the whole range of organic metamorphism.

Spores are unicellular reproductive bodies consisting of protoplasm, cellulose, and an outer waxy wall (the exine). The incorporation of spores in sediments takes place during sedimentation and during the alterations that affect the rock and its contents, the incorporated spores split open and lose their protoplasm and cellulose so that only the most resistant part, the exine, surviving all stages of thermal maturation is to be found embedded in sedimentary rocks.

The spores deform and elongate, probably due to the results of compaction and they can break along 'shrinkage cracks' to result in split spores. Besides the colour of the spore, its physical condition is quite often a very good indication of its thermal history (Figure 7.6).

As a technique for estimating the maturity of oil generating kerogen its chief advantage is that sporopollenin is chemically closer to sapropel than is the vitrinite and it may give a truer indication of the level of organic maturation. Spores and pollens are present in more kerogens than vitrinite macerals which are used in reflectance measurements. Even if spores are absent, the method can be applied to the amorphous kerogen so the method is of wider application than vitrinite reflectance. However, certain anomalies to exist and in some cases polynomorphs have been oxidized and bleached and thus they retain a pale colour even at high levels of maturation; conversely, in bituminous and radioactive shales the uptake of heavy minerals by the organic matter may result in the spores becoming more strongly coloured than would be expected from their level of thermal maturation.

The spores, pollens, and other organic matter, having been obtained from the inorganic mineral matter as described earlier, are mounted in glycerin jelly as a strewn slide under a cover slip. All the methods of transmitted light investigation can be applied to all the types of organic matter which are transparent.

Transmitted light examination

There are four main methods of measuring the level of organic maturation using spores and other sedimentary organic matter. The first involves an assessment of the physical condition of the structured elements. As these components, especially the spores and pollens, are exposed to increasing temperature they wrinkle, split, crack, and deform. Even the opaque material, the vitrinite macerals, will lose the sharp crisp outlines and edges which they have when immature (Figure 7.7). The second method is the measurement of the amount of fluorescence caused by excitation with blue or ultraviolet light. Although this has great advantages as a qualitative tool, its practical difficulties render it less useful quantitatively as the amount of fluorescence seems to depend not only on the degree of carbonization, but also upon the methods of preparation and the time that elapses between sample preparation and examination. Thus such fluorescence measurements are extremely difficult to carry out so as to achieve reproducible results. However, it is a procedure that is under constant investigation and subject to improvements and refinements.

The third method, colorimetric measurement on the organic matter, involves the estimation of the colour of the organic matter when viewed in transmitted light. The colours can be related to the levels of maturation as discussed earlier. Based on these colour changes, Correia (1971) and Staplin (1969) proposed two scales which with the structural alteration cover the range of the organic maturation Thermal Alteration Index (TAI) and Level of Organic Maturation (LOM) (See Figure 9.3).

In the fourth method, translucency measurements, progressive organic metamorphism is measured by recording the translucency of the spores as they become more opaque. This involves measuring with a photometer the amount of light passing through a spore. Since the degree of carbonization differs in different taxa or groups of spores, selection must be made of a standard specimen which has good vertical and horizontal distribution and abundance in the stratigraphic unit under study. Normally a minimum of 30 measurements on separate spores are made (Grayson, 1975), all in the same location on the spore coats. The translucency of kerogen may be related to the elemental analysis of the kerogen and the zones of hydrocarbon generation (Table 9.3). There can, however, be problems in correlations between organic matter in different sediments. The average of the measurements is the carbonization value assigned to the sample. Measurements thus obtained can be used to plot isocarbonization maps which can be used to draw a regional picture of thermal history and petroleum prospects.

Table 9.3

Translucency %	Kerogen carbon weight (%)	Generative potential
80	65	
75		
70		pregeneration
65		
60	70	
55		
50	75	– – – – – –
45		early generation
40		– – – – – – –
35	80	
30		
25	82	peak of
20		generation
15	83	
10		
5	84	– – – – – – –
		post generation

Of these four methods, the first two can also be carried out using a reflected-light microscope and useful and sometimes complementary information obtained. So far all the methods have discussed the optical examination of the structured organic matter, but observations on the non-structured matter can also be of value. This normally involves either fluorescence studies, colour estimations, or translucency measurements (Figure 7.5). If blue or ultraviolet light is shone upon organic matter, the kerogen may be excited and emit radiation. This fluorescence is intense in immature samples and decreases until it has completely disappeared at the end of the principal zone of oil generation. At the same time, the colour spectrum of this fluorescence moves towards the red with increasing maturity. There exists a direct relationship between sporinite fluorescence, vitrinite reflectance, coal rank, and thermal maturation which means that fluorescence can be used as a maturation indicator. This method can be used to determine oil-prone kerogens as described earlier. This is because the highly liptinitic hydrogen-rich kerogens will shew fluorescence whereas the hydrogen-lean particles, such as gas-prone, amorphous kerogen or vitrinites, do not fluoresce. Recycled organic matter can also be distinguished because it is hydrogen deficient and does not fluoresce.

The examination of the colour of amorphous material colour is carried out on a strewn slide by choosing the least coloured material as the maturation indicator. This is because any recycled material is darker than contemporaneous kerogen and the colour depends, in part, on the thickness. By choosing the thinnest part, a truer colour is obtained (this problem does not arise with the colorimetric measurement of spore coloration.) The colours may be slightly darker than for spores of similar maturation, but fall in approximately the same ranges. The same problems of bleaching and false darkening by heavy-metal absorption are encountered in the examination of amorphous matter as for spores. Peters *et al.* (1977) have used the colour of kerogen as a guide to its maturation. Kerogens from recent sediments have been heated at different temperatures and for different times and then viewed in transmitted light. The activation energies were calculated for each colour change and these shewed that the principal zone of oil generation occurs when the kerogen is between dark and very dark brown. The activation energies increase with the maturity of the organic matter, and Peters *et al.* shewed that the spore and kerogen colours do differ a little. Claypool *et al.* (1978) have shewn that thermally immature sediments are characterized by yellow to brown organic matter with atomic H/C ratios greater than 1.1, whilst thermally mature sediments have brown to dark brown kerogens whose H/C ratios are between 0.6 and 1.1. Incipiently metamorphosed sediments have very dark brown or black kerogens with H/C ratios in the range 0.4 to 0.6.

Reflected light examination

The optical study of organic matter in both transmitted and reflected light can give indications of both kerogen type and level of maturation. There is one reflected technique yet to be discussed, and that is vitrinite reflectance. This is very popular and some of that popularity can be attributed to the fact that it gives the result as a percentage which is easily storable, usable, and related to other numerical parameters, which is much more difficult with the methods described above.

Macerals develop from the tissues of plants from which coal forms and these macerals can be distinguished into three main groups: vitrinite, inertinite, and exinite. Table 9.4 gives subdivisions of the maceral groups with their origins. With increasing rank the macerals alter in their physical and chemical properties; the reflectance and carbon content rise while the amount of volatile matter falls. These changes, especially that of the reflectance, can be correlated

Table 9.4 Maceral groups of hard coals

Group	Maceral	Origin
vitrinite	telinite collinite vitrodetrinite	wood and bark tissues
exinite	sporinite cutinite resinite alginite liptodetrinite	spore and pollen exines, cuticles, resins, and algae
inertinite	micrinite macrinite semifusinite fusinite sclerotinite inertodetrinite	primarily from wood and bark tissues which have suffered a different biochemical history to those forming the vitrinite macerals; a small contribution from fungal tissues in certain coal seams

with coal rank. The vitrinite macerals are considered to be the most reliable indicators of the degree of organic metamorphism because they have the lowest ash content, are the most abundant and homogeneous, as well as being the most sensitive species for carbonization studies. Reflectance can be measured on very small grains and thus coaly intrusions interspersed in the inorganic sediment can be used in the absence of coal seams to measure organic metamorphism (Castano, 1973; Hood *et al.*, 1975; Foscolos *et al.*, 1976).

The reflectance of a polished planar surface is defined as the fraction of the normally incident light reflected from that surface and depends upon the refractive and absorptive indices of the material and the surrounding media. Organic matter obtained from the maceration process in a finely disseminated form is dried and mixed with a clear-setting plastic powder or an epoxy resin and made into a briquette which is then polished to an optically flat, scratch-free surface.

The microscope together with its photomultiplier are fitted with stabilized power supplies and readings are taken in monochromatic light. The photomultiplier output is calibrated against a set of standards of known reflectivity so that the photomultiplier output can be converted directly into percentage reflected light. The vitrinite macerals display bireflectance and the stage is rotated so that a maximum value is recorded. Because the vitrinite particles in the sediment may differ from one another in optical properties, the mean of at least 100 readings is normally taken (International Committee for Coal Petrology, 1971; American Society for Testing and Materials, 1972). The measurements are taken using oil immersion objectives.

Because coalification and petroleum genesis (bituminization) are mainly dependant upon temperature, each stage of the petroleum maturation can be matched against a particular coal rank. Likewise, the alteration of oil from heavy ashphaltic to light paraffinic and finally its destruction to gas can be correlated with the vitrinite reflectance (Vassoyevich *et al.*, 1970, 1974; Alpern, 1975; Hood *et al.*, 1975). Studies have shewn that oil formation usually starts at the sub-bituminous coal stage, about a maximum reflectance in oil of 0.55% (R_o max) (up to this is equivalent to the diagenetic stage of kerogen conversion). The expulsion of oil comes later in the highly volatile bituminous coal range (i.e. 0.7% R_o max). It is difficult to define the upper limit of the principal zone of oil accumulation since oil occasionally migrates upwards for considerable distances and some petroleum reservoirs are in areas where the vitrinite reflectances are only in the order of 0.3% R_o max. However, the lower boundary of the principal zone of oil formation can be accurately defined and it is generally accepted that it matches approximately with the medium volatile bituminous coal stage (1.3–1.5% R_o max). This cut-off can be somewhat elastic and can depend upon local conditions. This phase (0.7–1.5% R_o max) is the start of the catagenesis stage and corresponds to the 'oil-window' or the principal zone of oil formation. Following this is the second half of the catagenesis zone or the principal zone of gas formation, where gas starts to be formed but where there is still oil present. Above reflectances of 2% (semi-anthracite) or 3% (anthracite) is the dry gas zone, i.e. the metagenesis stage. During this stage, oil is cracked to lighter and lighter products and these are in order, oil, condensate, wet gas, and dry gas.

If a sufficient number of samples shewing lateral and vertical distributions in a sedimentary basin are investigated and the vitrinite reflectance of the organic matter in these samples is determined, isoreflectivity maps can be produced, based on isoreflectivity lines for individual stratigraphic horizons. From these maps the principal zones of oil and gas formation can be defined and consequently the regional occurrences and the probable compositions of the oil and gas can be estimated. The presence of reworked organic matter within the sediment under investigation is shown by the multipeaked vitrinite reflectance histogram which is obtained in such cases.

However, this type of procedure totally ignores the type of organic material and this information will have to be obtained from another source and added to give a more complete picture of the probable reservoired hydrocarbon products, since the type of organic matter can have a significant bearing on the products of maturation. Likewise, not all samples of sedimentary organic matter contain copious amounts of vitrinite macerals. Type III will be rich in them but type II will be much leaner, and there will be very few, if any, in type I kerogen. Thus not only can finding sufficient macerals be difficult, but obtaining them over the whole basin could be nigh impossible. Carbonates with mainly marine amorphous kerogen have almost no vitrinites, nor do anhydrites and evaporites. If a sequence has no vitrinites, it is sometimes possible to interpolate from data above and below. However, Sikander & Pittion (1978) have made reflectance measurements on ashphaltic pyrobitumen which resembles vitrinite in some physical properties, and have indicated ways of overcoming this problem. The main source of oil is amorphous sapropelic organic matter which is quite different from vitrinite and thus using the changes in vitrinite as a guide to the alteration of sapropelic matter may be fraught with all kinds of hazards.

Care has to be taken as the reflectance of vitrinites may vary between adjacent lithologies. The vitrinite of the Carboniferous of north-eastern England has substantially lower reflectivities than coals of similar age and rank in Germany and this has to been ascribed to possible impregnation of the vitrinite macerals by resins.

Claypool & Mereweather (1980), from a study of the organic composition of shales, shew a relationship between time, temperature, and vitrinite reflectance and thus indicate how vitrinite reflectance and burial history can give approximate palaeotemperatures.

Maturation profiles can be obtained using vitrinite reflectance (Figure 7.1). The slope of the profile depends upon the geothermal gradient and the exposure time. However, these simple profiles can be altered by unconformities, intrusives, faults, and recycled organic matter. Figure 9.18(a) shows the effect of unconformities upon the maturation profile. The offset in the graph indicates the depth of section that has been lost during the erosion. The different slopes suggest different thermal histories for the two sediments. The shallower the slope, the greater the probability of a slower sedimentation rate and lower exposure time.

Intrusives cause the effects of contact metamorphism to be superimposed on those of normal burial. The contact metamorphism affects the maturity of the intruded rocks to a thickness of approximately twice the thickness of the intrusive body (Figure 9.18(b).

Faults can be recognized on maturation profiles providing they are geologically young and have a large vertical displacement. Figure 9.18(c) shows reverse faulting, where the younger and less mature rocks have been taken down into a higher-temperature area. Eventually the effect of the extra heat will mature the organic matter and the breaks in the profile will cease to be recognizable.

Recycled organic matter, which has had a previous thermal history, is quite common in sediments. Because of this previous thermal history, recycled organic matter will mature more slowly than organic matter which was fresh when incorporated into the rock. Thus the maturation profile for a recycled vitrinite will be steeper than for primary vitrinites. Care must be taken to exclude secondary organic matter from all maturity assessments (Figure 9.18(d)).

Once such maturation profiles have been established, the maturity at any depth can be closely estimated even if direct assessment of maturity is impossible for a particular horizon. The amount of erosion that has occurred since maximum burial can be obtained by extrapolating the profile back to a vitrinite reflectance value of 0.2% (Dow, 1977).

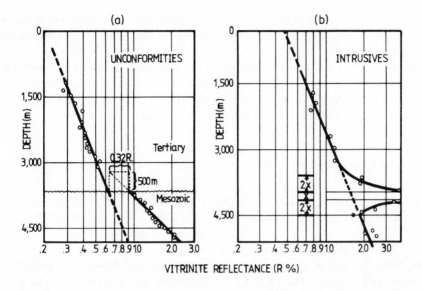

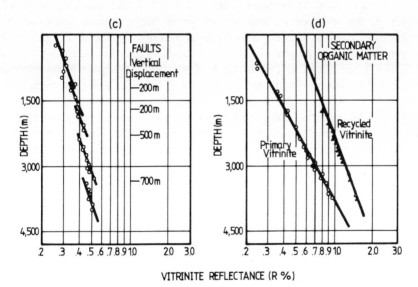

Figure 9.18 Use of vitrinite reflectance (Dow, 1977. Reproduced by permission of Elsevier Scientific Publishing Co.)

THERMAL HISTORY OF SEDIMENTS

The maturation indexes so far described have been related to the maximum palaeotemperature to which the sediment has been subjected. However the important factor in the thermal alteration of kerogen is the quantity of heat available and this involves the time the organic rich sediment has spent in each temperature regime. It is for this reason that the time–temperature index was devised.

Waples (1980) compares Lopatin's (1971) time–temperature index with the thermal alteration index (TAI) of Staplin (1969) and vitrinite reflectance. Lopatin calculates his time-temperature index (TTI) by summing the individual indices of maturation for each 10 °C temperature interval:

$$\text{TTI} = \sum \tau = \sum_{n_{\min}}^{n_{\max}} (\Delta T_n)(r^n)$$

where $n_{\max}$ and $n_{\min}$ are the highest and lowest temperature intervals encountered, n is a temperature factor (90–100 °C = -1, 100–110 °C = 0, 110–120 °C = $+1$, etc.), $r = 2$ (to represent a doubling of the reaction rate every 10 °C), and T_n is the length of time spent in interval n, given by the difference in age at which the sediment enters and leaves the particular temperature interval. The total TTI is the sum of the interval TTIs which are calculated by multiplying r^n (actually 2^n, where n is the temperature factor from -8 at 20–30 °C through 0 for 110–110 °C, and onwards) and ΔT_n (which is the number of millions of years the sediment spent in that temperature range). The Lopatin formula contains the factor 2^n because of the general assumption that the rate of a chemical reaction doubles for every 10 °C rise in temperature. Following the discussion in Chapter 7, this is not an unreasonable assumption to make. The breakdown of kerogen occurs in a series of reactions, each requiring a different activation energy and total energy. The most labile components will require the least energy to separate them from the nucleus of the kerogen, and as the maturation process proceeds more and more energy will be required to remove the remaining groups from the kerogen. The early products may either not be bound by full chemical bonds and will be weakly absorbed onto the kerogen, or they may be joined via bonds such as ether linkages which require less energy to break than carbon–carbon bonds. The last groups to be removed from the kerogen will require most energy, which is supplied as heat due to increased burial. The laws of chemical kinetics are expressed by the Arrhenius equation (reaction rate $= A \exp(-E_a/RT)$). As the reactions to remove compounds from the kerogen require higher energies and activation energies, both E_a and T increase. Because of the different increase in these parameters, the overall reaction rate approximately doubles for every 10 °C rise in temperature.

Estimation of time–temperature history of sediments

It is normal to plot the age of the sediment (in millions of years) against depth and to superimpose the subsurface temperature, to shew the time the sediment spent in each temperature interval.

Lopatin suggested that definite TTI values correspond to the different stages in the oil generation process and these are given in Table 9.5. Waples (1980) has shown that TTI can be successfully correlated against other maturity parameters, such as TAI, vitrinite reflectance, bitumen/TOC ratios, CPI, H/C atomic ratios, API gravity, etc. Table 9.5 shows the values of TTI for the various stages in the generation of petroleum and how they correspond to TAI and vitrinite reflectance (Waples, 1981; Cohen, 1981)

Table 9.5

Generation stage	TTI		R_o	TAI
onset of maturation		15	0.65	2.65
peak oil generation		75	1.00	2.9
end of oil generation		160	1.30	3.2
upper limit for oil with API° 40		2,500	1.75	3.6
upper limit for oil with API° 50		21,000	2.00	3.7
upper limit for wet gas		21,500	2.20	3.75
upper limit for dry gas		65,000	4.8	4.00

The TTI can be used to determine whether the proposed source rock in a basin has been under the correct time–temperature regime to have generated oil or gas. Not only that, but important information with regard to timing can be obtained. It is possible to use TTI data to predict during what time the source rock was in the temperature 'oil-window' and a comparison of this with age of reservoir rocks, migration fairways, etc., will show what chance there is of oil being trapped once it is formed. Likewise, investigation of TTI for the potential reservoir will indicate whether that interval has been so buried that any oil in it at that time will have been thermally cracked, or subjected to bacterial attack.

If there is a poor correlation between the TTI and other maturity parameters, it may be that the information used to calculate the TTI was faulty due to lack of knowledge of geological history, for example a poor assessment of uplift and erosion during and after tectonic events. Thus the method can be used to reconstruct tectonic events.

The grid used to calculate the TTI was shown in Figure 7.12; Table 9.6 give the calculation of the TTI for that interval. The result is 180.274. Thus any organic matter within that rock will be past the upper limit of oil generation, but well before the upper limit of the preservation of any generated oil. A source rock represented by that line could have generated oil which, if a suitable trap was available at the time of generation, could have been preserved. Oil in that interval would have been generated from about 55–2.5 million years ago with a peak of generation 17.5 million years ago. If the line represented a reservoir rock, any oil contained would not have been destroyed.

The effect of igneous intrusions and varying geothermal gradients can be taken into account using this method. Figure 9.19 shews how the time–temperature grid is altered when a sediment of lower thermal conductivity is

Table 9.6 Calculation of TTI for sediment shown in Figure 7.12

Temperature (°C)	n	time T (my)	2^n	$2^n \times T$ interval	total
30– 40	−7	5	0.0078	0.039	0.039
40– 50	−6	5	0.0156	0.078	0.117
50– 60	−5	5	0.0313	0.157	0.274
60– 70	−4	20	0.0625	1.25	1.524
70– 80	−3	10	0.125	1.25	2.774
80– 90	−2	30	0.25	7.50	10.274
90–100	−1	10	0.5	5.00	15.274
100–110	0	15	1.0	15.00	30.274
110–120	1	25	2.0	50.00	80.274
120–130	2	5	4.0	20.00	100.274
130–140	3	10	8.0	80.00	180.274

deposited. These time–temperature grids must not be confused with geological cross-sections. The above discussion has concerned the improvement of maturation assessment techniques from a simple estimation of maximum temperature to a calculation of time–temperature index which takes into account the quantity of heat available to alter the kerogen. The next improvement which needs to be tackled is to take into account the quantity of

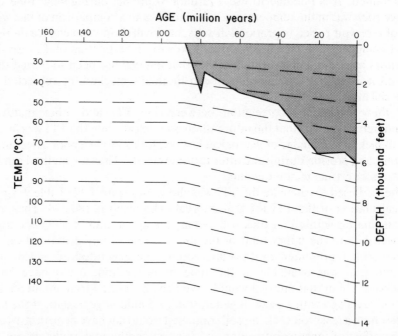

Figure 9.19 Time and temperature of oil generation. (From *Organic Geochemistry for Exploration Geologists*, 1981, Douglas Waples, Burgess Publishing Company, Minneapolis, Minnesota. Reprinted by permission)

kerogen involved. All units of the quantity of heat (e.g. calorie or BTU) involve the amount of heat required to raise a given mass of water by one degree (Celsius or Fahrenheit respectively). The same quantity of heat will raise a larger quantity of water by a smaller amount and vice versa. Thus the amount of kerogen in relation to the quantity of heat available should be relevant in determining the degree of thermal maturation. So far this variable is uninvestigated.

An outline of the range of techniques available for the assessment of the type, amount and maturation of sedimentary organic matter has been given. From the methods a number of parameters can be obtained which can be related to one another, especially when these parameters are used for maturation level assessments. Heroux *et al.* (1979) give a very comprehensive review and correlation of all the methods of assessing maturation. Figure 9.3 gives a comparison of some of the commonest methods.

REFERENCES AND FURTHER READING

Albrecht, P., Vandenbrouche, M., & Mandengue, M. (1976) Geochemical studies on the organic matter from the Doula basin (Cameroon) I. *Geochim. Cosmochim. Acta*, **40**, 791.

Allan, J., & Douglas, A. G. (1977) Variations in content and distribution of *n*-alkanes in a series of carboniferous vitrinites and sporinites of bituminous rank. *Geochim. Cosmochim. Acta*, **41**, 1223.

Alpern, B. (1975) Introduction, in pétrographie de la matière organique des sédiments, relations avec la paléotempérature et le potential pétrolier: Paris, CNRS Colloq. Int. pp. 191–193.

Alpern, B. (1976) Fluorescence et réflectance de la matière organique dispersée et évolution des sédiments: *Cent. Rech. Pau. Bull.*, **10**, 201.

American Society for Testing and Materials (1972) Standard method for microscopical determination of the reflectance of the organic components in a polished specimen of coal, *ASTM D 2798-72*.

Bailey, N. J. L., Evans, C. R., & Milner, C. W. D. (1974) Applied petroleum geochemistry to search for oil. *BAAPG*, **58**, 2284.

Barker, C. (1974a) Pyrolysis techniques for source rocks. *BAAPG*, **58**, 2349.

Barker, C. (1974b) Programmed temperature pyrolysis of vitrinites of various coal ranks. *Fuel*, **53**, 176.

Barss, M. S., & Williams, G. L. (1973) Palynology and nano-fossil processing techniques. *Geol. Survey Paper, Canada*, 73–26.

Bernard, B. B., Brooks, J. M., & Sackett, W. M. (1976) Natural gas seepage in the Gulf of Mexico. *Earth Planet. Sci. Lett.*, **31**, 48.

Bostick, N. H., & Alpern, B. (1977) Principles of sampling, preparation and constituent selection for microphotometry in measurement of maturation of sedimentary organic matter. *Journal of Microscopy*, **109**, 41.

Bray, E. E., & Evans, E. D. (1961) Distribution of *n*-paraffins as a clue to recognition of source beds. *Geochim. Cosmochim. Acta*, **22**, 2.

Castano, J. R. (1973) Application of coal petrographic methods in relating level of organic metamorphism to generation of petroleum. *BAAPG*, **51**, 772.

Claypool, G. E., Love, A. H., & Maughan, E. K. (1978) Organic geochemistry, incipient metamorphism and oil generation in Black shale members of the Phosphoria Formation of Western Interior USA. *BAAPG*, **62**, 98.

Claypool, G. E., & Mereweather, E. A. (1980) Organic composition of some upper cretaceous shales. Powder River Basin, Wyoming. *BAAPG*, **64**(4), 488.

374

Claypool, G. E., Paterson, J. M., & Swetland, P. J. (1978) Petroleum source bed evaluation of Jurassic Twin Creek limestone, Idaho-Wyoming thrust belt. *BAAPG*, **62**, 1075.

Claypool, G. E., & Reed, P. R. (1976) Thermal analysis techniques for source rock evaluation. *BAAPG*, **60**, 608.

Claypool, G. E., Threlkeld, C. N., & Magoon, L. B. (1980) Biogenic and thermogenic origins of natural gas in Cook Inlet Basin, Alaska. *BAAPG*, **64**(8), 1131.

Clayton, J. L., & Swetland, P. J. (1978) Subaerial weathering of sedimentary organic matter. *Geochim. Cosmochim. Acta*, **42**, 305.

Clementz, D. M. (1979) Effect of oil and bitumen saturation on source rock pyrolysis. *BAAPG*, **63**, 2227.

Clementz, D. M., Demaison, G. J., & Daly, A. R. (1979) Well site geochemistry by programmed pyrolysis. *11th Offshore Technology Conference*, April/May 1979.

Cohen, C. R. (1981) Time and temperature in petroleum formation: Application of Lopatin's method to petroleum exploration: Discussion. *BAAPG*, **65**(9), 1647.

Coleman, D. D., Risatti, J. B., & Schoell, M. (1981) Fractionation of carbon and hydrogen isotopes by methane oxidizing bacteria. *Geochim. & Cosmochim. Acta*, **45**, 1033.

Cooper, B. S., & Murchison, D. G. (1971) The petrology and geochemistry of sporonite. In: Brooks *et al.* (eds), *Sporopollenin*, Academic Press, London.

Correia, M. (1971) Diagenesis of sporopollenin and other comparable organic substances : application to hydrocarbon research. In: Brooks *et al.* (eds), *Sporopollenin*, Academic Press, London.

Craig, H. (1961) Standard for reporting concentrations of deuterium and oxygen-18 in natural waters. *Science*, **133**, 1833.

Dancy, T. E., & Giedroye, V., (1950a) Further researches on the determination of the chemical composition of oil shales. *J. Inst. Pet.*, **36**, 593.

Dancy, T. E., & Giedroye, V. (1950b). Further investigation of the chemical constitution of the kerogen of oil shales. *J. Inst. Pet.*, **36**, 307.

Debyser, T. M. (1975) Contamination of recent sediments. *Geochim. Cosmochim. Acta*, **39**, 531.

Demaison, G. J. (1975) Relationship of coal rank to palaeotemperatures in Sedimentary rocks. In: Alpern, B. (ed.), *Petrographie de la matière organique des sédiments, relation avec la paléotemperature*. Paris: Editions du centre National de Recherche Scientifique, p. 217.

Douglas, A. G., & Grantham, P. T. (1973) Docosane in rock extracts. A possible contaminative source. *Chem. Geol.*, **12**, 249.

Dow, W. G. (1977) Kerogen studies and geological interpretations. *J. Geochem. Expl.*, **7**, 79.

Down, A. L. (1939) Analysis of the kerogen of oil shales. *J. Inst. Pet.*, **25**, 230.

Durand, B., Marchand, A., & Combaz, A. (1977) Etude de kérogènes en résonance paramagnétique électronique. In: Campos, R., Goni, J. (eds), *Advances in Organic Geochemistry*, Empresa Nacional Adaro de Investigaciones Mineras.

Dypvik, A., Rueslatter, H. G., & Thronden, T. (1979) Composition of organic matter for N. Atlantic Kimmeridgean shales. *BAAPG*, **63**(12), 2222.

Erdman, J. G. (1961) Some chemical aspects of petroleum genesis as related to source bed recognition. *Geochim. Cosmochim. Acta*, **22**, 10.

Espitalié, J., Laporte, J. L., Madoc, M., Marquis, F., Leplat, P., Paulet, J., & Boutefen, A. (1977) Méthode rapide de caractérisation des roches mères, de leur potential pétrolier et de leur degré evolution. *Rev. Inst. Fr. Pet.*, **32**, 23.

Espitalié, J., Madec, M., & Tissot, B. (1980) Role of mineral matrix in kerogen pyrolysis. Influence on petroleum generation and migration. *BAAPG*, **64**(1), 59.

375

Evans, C. R., & Staplin, F. L. (1971) Regional facies of organic metamorphism in geochemical exploration. *3rd Internat. Geochemical Exploration Symposium, Proc. Canadian Institute of Mining and Metallurgy*, Special Vol. II, pp. 517–520, Montreal.

Foscolos, A. E., Powell, T. G., & Gunther, P. R. (1976) The use of clay minerals and the inorganic and organic geochemical indicators for evaluating the degree of diagenesis and oil generating potential of shales. *Geochim. Cosmochim. Acta*, **40**, 953.

Gibbs, G. W. (1970) Collection and storage of samples in organic geochemistry. *Geochim. Cosmochim. Acta*, **34**, 629.

Giraud, A. (1970) Application of pyrolysis and GLC to geochemical characterization of kerogen in sedimentary rocks. *BAAPG*, **54**, 439.

Gransch, J. A., & Eisma, E. (1966) Characterization of insoluble organic matter of sediments by pyrolysis. In: Hobson, G. D. & Spears, G. C. (eds), *Advances in Organic Geochemistry*, Pergammon Press, Oxford.

Grayson, F. J. (1975) Relationship of palynomorph translucency of carbon and hydrocarbons in clastic sediments. *Pétrographie de la matière organique des sédiments, relations avec la paléotempérature et potentiel pétrolier*. Paris, CNRS Colloq. Int.

Grayson, M. A., Levy, R. L., & Wolf, C. J. (1970) Organic analysis of the Pueblito de Allende meteorite. *Nature*, **227**, 148.

Gutjahr, G. C. (1966) Carbonization measurements of pollen grains and spores and their application. *Leidse Geologische Mededelingen*, **38**, 1.

Harwood, R. J. (1977) Oil and gas generation by the laboratory pyrolysis of kerogen. *BAAPG*, **61**, 2082.

Hedberg, H. D. (1968) Significance of high wax oils with respect to the genesis of petroleum. *BAAPG*, **52**(5), 736.

Heroux, Y., Chaynon, A., & Bertrand, R. (1979) Compilation and correlation of major thermal maturation indicators. *BAAPG*, **63**, 2128.

Hood, A., Gutjahr, C. C. M., & Heacock, R. L. (1975) Organic metamorphism & the generation of petroleum. *BAAPG*, **59**, 986.

Horsfield, B., & Douglas, A. G. (1980) The influence of minerals on the pyrolysis of kerogen. *Geochim. Cosmochim. Acta*, **44**(8), 1119.

Huc, A. Y., & Hunt, J. M. (1980) Generation and migration of hydrocarbons in offshore South Texas Gulf Coast sediments. *Geochim. Cosmochim. Acta*, **44**(8), 1081.

International Committee for Coal Petrology (1971) *International Handbook of Coal Petrology*, 1st supplement to 2nd edition. Paris, CNRS.

Ishiwatari, R., Ishiwatari, M., Kaplan, I. R., & Rohrback, B. G. (1977) Thermal alteration experiments on organic matter from recent marine sediments in relation to petroleum genesis. *Geochim. Cosmochim. Acta*, **41**, 8151.

Ishiwatari, R., Rohrback, B. G., & Kaplin, I. R. (1978) Hydrocarbon generation by thermal alteration of kerogen from different sediments. *BAAPG*, **62**, 687.

Johnson, A. C. (1970) How to hunt oil and gas using inorganic surface geochemical methods. *Oil & Gas J.*, **68**, 110.

Kinghorn, R. R. F., & Rahman, M. (1980) The density separation of different maceral groups of organic matter dispersed in rocks. *J. Pet. Geol.*, **2**(4), 499.

Kvenvolden, K. A., & Field, M. E. (1981) Thermogenic hydrocarbons in unconsolidated sediment of Eel River Basin, offshore Northern California. *BAAPG*, **65**(9), 1642.

Laplante, R. E. (1974) Hydrocarbon generation in Gulf Coast tertiary sediments. *BAAPG*, **58**, 1281.

Larter, A. G., & Douglas, A. G. (1979) Low molecular weight aromatic hydrocarbons in coal maceral pyrolysates as indicators of diagensis and organic matter types. In: Krumbein, W. E. (ed.), *Environmental Biogeochemistry and Geomicrobiology*. Ann Arbor Science Publishers Inc.

Leythaeuser, D. (1973) Effects of weathering on organic matter in shales. *Geochim. Cosmochim. Acta*, **37**, 1137.

376

Lijmbach, G. W. M. (1975) On the origin of petroleum. *9th World Petroleum Congress*, Vol. 2, pp. 357–369.

Link, W. K. (1952) Significance of oil and gas seeps in world oil exploration. *BAAPG*, **36**(8), 1505.

Lopatin, N. V. (1971) Temperature and geological time as factors of carbonization. *Akad. Nauk., SSSR, Ivz. Ser. Geol.* No. 3, p. 95.

Mackenzie, A. S., Hoffman, C. F., & Maxwell, J. R. (1981) Molecular parameters of maturation in Toarcian Shales Paris basin, France—111. Changes in aromatic steroid hydrocarbons. *Geochim. & Cosmochim. Acta*, **45**, 1345.

Mackenzie, A. S., Quirke, J. M. E., & Maxwell, J. R. (1980) Molecular parameters of maturation in the Toarcian shales, Paris Basin, France. II. Evolution of metalloporphyrins. In: A. G., & Maxwell, J. R. (eds), *Advances in Organic Geochemistry*.

Marchand, A., Libert, P. A., & Combaz, A. (1968) Sur quelques critieres physico-chimique de la diagenesis d'un kerogen. *Compt. Rend.*, Ser. D., **266**, 2316.

Marchand, A., Libert, P. A., & Combaz, A. (1969) Essai de caracterization physico-chimique de la diagenesis de quelques roches organiques, biologiquement homogenes. *Rev. Inst. Petro.*, **24**, 3.

McCartney, J. T., & Teichmuller, M. (1972) Classification of coals according to degree of coalification by reflectance of the vitrinite component. *Fuel*, **51**, 64.

Nishimura, M. (1982) 5β-isomers of stanols and stanones as potential markers of sedimentary organic quality and depositional palaeoenvironments. *Geochim. & Cosmochim. Acta*, **46**, 423.

Nooner, D. W., Oro, J., Gilbert, J. M., Ray, J. L., & Mair, J. E. (1972) Ubiquity of hydrocarbons in nature. *Geochim. Cosmochim. Acta*, **36**, 953.

Peters, K. E., Ishiwatari, R., & Kaplin, I. R. (1977) Colour of kerogen as an index of organic maturity. *BAAPG*, **61**, 504.

Philippi, G. T. (1965) Depth, time and mechanism of petroleum generation. *Geochim. Cosmochim. Acta*, **29**, 1021.

Philippi, G. T. (1974) Influence of marine and terrestrial source material on the composition of petroleum. *Geochim. Cosmochim. Acta*, **38**, 947.

Powell, T. G. (1975) An assessment of the hydrocarbon source rock potential of the Canadian arctic islands. *Geol. Survey of Canada*, paper 78–12.

Powell, T. G., Creaney, S., and Snowdon, L. R. (1982) Limitations of the use of organic petrographic techniques for identification of petroleum source rocks. *BAAPG*, **66**(4), 430.

Powell, T. G., & McKirdy, D. M. (1973a) Relationships between pristane and phytane, crude oil composition and geological environment in Australia. *Nature*, **243**, 37.

Powell, T. G., & McKirdy, D. M. (1973b) Effect of source material, rock type & diagenesis on n-alkane content of sediments. *Geochim. Cosmochim. Acta*, **37**, 623.

Powell, T. G., & McKirdy, D. M. (1975) Geologic factors controlling crude oil composition in Australia and Papua–New Guinea. *BAAPG*, **59**, 1176.

Rad, F. K. (1982) Hydrocarbon potential of the Eastern Alborz Region, Iran. *J. Pet. Geol.*, **4**(4), 419.

Radke, M., Welte, D. H., & Willsch, H. (1982) Geochemical study on a well in the Western Canada Basin: relation of the aromatic distribution pattern to maturity of organic matter. *Geochim. & Cosmochim. Acta*, **46**, 1.

Raschid, M. A. (1979) Pristane–phytane ratios in relation to source and diagenesis of ancient sediment from the Labrador shelf. *Chem. Geol.*, **25**, 109.

Redding, G. E., Schoell, M., Monin, J. C., & Durand, B. (1979) Hydrogen and carbon isotopic composition of coals and kerogen. In: Douglas, A. G., Maxwell, J. R. (eds), *Advances in Organic Geochemistry*, p. 711.

Richers, D. M., Reed, R. J., Horstman, K. C., Michels, G. D., Baker, R. N., Lundell, L., & Marrs, R. W. (1982) Landsat and soil-gas geochemical study of Patrick Draw oilfield, Sweetwater County, Wyoming, *BAAPG*, **66**, 903.

Rigby, D., Batts, B. D., & Smith, J. A. (1981) The effect of maturation on the isotopic composition of fossil fuels. *Organic Geochemistry*, **e**, 29.

Rogers, M. A., McAlary, J. D., & Bailey, N. J. L. (1974) Significance of reservoir bitumens to thermal maturation studies. *BAAPG*, **58**(9), 1806.

Ross, L. M. (1980) Geochemical correlation of San Juan basin oils—a study. *Oil & Gas J.*, **78**(44), 102.

Rullkötter, J., & Wendisch, D. (1982) Microbial alteration of 17α(H)-hopanes in Madagascar asphalts: removal of C-10 methyl group and ring opening. *Geochim. & Cosmochim. Acta*, **46**, 545.

Sackett, W. M. (1977) Use of hydrocarbon sniffing in offshore exploration. *J. Geochem. Expl.*, **7**, 243.

Saxby, J. D. (1970) Isolation of kerogen in sediments by chemical methods. *J. Chem. Geol.*, **6**, 173.

Scalan, R. S., & Smith, J. E. (1970) An improved measure of the odd–even preference in the normal alkanes of sediment extracts and petroleum. *Geochim. Cosmochim. Acta*, **34**, 611.

Schoell, M., & Milner, C. W. D. (1979) Isotopic crude oil correlation. *Abstracts of 9th International Meeting on Organic Geochemistry, Newcastle*, Sept. 1979.

Seifert, W. G., & Moldowan, J. M. (1978) Applications of steranes, terpanes and monoaromatics to the migration, maturation and source of crude oils. *Geochim. Cosmochim. Acta*, **42**, 77.

Siefert, W., & Moldowan, J. M., (1980) The effect of thermal stress on source rock quality as measured by hopane stereochemistry. In: Douglas, A. G., & Maxwell, J. R. (eds), *Advances in Organic Geochemistry*, p. 229.

Sikander, A. H., & Pittion, J. L. (1978) Reflectance studies on organic matter in lower paleozoic sediments of Quebec. *B. Can. Pet. Geol.*, **26**, 132.

Silverman, S. R., & Epstein, S. (1958) Carbon isotopic comparisons of petroleums. *BAAPG*, **42**, 998.

Stahl, W., Faber, E., Carey, B. D., Kirksey, D. L. (1981) Near surface evidence of migration of natural gas from deep reservoirs and source rocks. *BAAPG*, **65**(9), 1543.

Staplin, F. L. (1969) Sedimentary organic matter, organic metamorphism and oil and gas occurrence. *B. Can. Geol.*, **17**, 47.

Stuermer, D. H., & Payne, J. R. (1976) Investigation of seawater & terrestrial humic substances by ^{13}C and proton NMR. *Geochim. Cosmochim. Acta*, **40**, 1109.

Stuermer, D. H., Peters, K. E., & Kaplin, I. R. (1978) Source indicators of humic substances and protokerogen. *Geochim. Cosmochim. Acta*, **42**, 989.

Teichmuller, M., & Teichmuller, R. (1968) Geological aspects of coal metamorphism. In: *Coal and Coal-bearing Strata*. Elsevier, New York, p. 233.

Tissot, B., Durand, B., Espitalié, J., & Combaz, A. (1974) Nature and diagenesis of organic matter. *BAAPG*, **58**, 499.

Tissot, B. P., & Welte, D. H. (1978) *Petroleum Formation and Occurrence—A New Approach to Oil and Gas Exploration*. Springer-Verlag, Berlin-Heidelberg-New York.

Trask, P. D., & Patnode, H. W. (1942) *Source Beds of Petroleum*. Tulsa, Okla., AAPG.

Van Krevelen, D. W. (1961) *Coal*. Amsterdam, Elsevier Publ. Co.

Vassoyevich, N. D., *et al.* (1970) Principal phase of oil formation. *Int. Geol. Rev.*, **12**, 1276.

Vassoyevich, N. B., *et al.* (1974) Principal zone of oil formation. In: *Advances in Organic Geochemistry*. Technip, Paris, p. 309.

Waples, D. W. (1980) Time and temperature in petroleum formation. Application of Lopatin's method to petroleum exploration. *BAAPG*, **64**(6), 916.

Waples, D. W. (1981) Time and temperature in petroleum formation. Application of Lopatin's method to petroleum exploration. Reply. *BAAPG*, **65**(9), 1649.

378

Zarella, W. M., Mousseau, R. J., Coggleshall, N. D., Norris, M. S., & Schrayer, G. J. (1967) Analysis and significance of hydrocarbons in subsurface brines. *Geochim. Cosmochim. Acta*, **31**, 1155.

Chapter 10

Applications of Organic Geochemistry

CORRELATION OF OILS AND SOURCE ROCKS

Introduction

The correlation of crude oils and source rocks is of great importance in the continued exploration and assessment of the hydrocarbon potential of a basin. If it is possible to correlate a crude oil and a source rock it can be feasible to infer migration pathways which, when combined with geological knowledge of the region, may allow other potential traps to be identified. However, if no correlation between an oil and a known source exists, or if there are two oils which can be shewn to have different sources, then one is dealing with a multiple source which will be more favourable to the discovery of hydrocarbon pools than a single-source basin. Each source in a basin, providing it is mature, can have provided the hydrocarbons for one or more reservoirs and until each oil pool is correlated with the source and any others from that same source, there will still be potential for oil discoveries in that basin. In this context, negative correlations are as important as positive ones because they again give indications of the mutiple source rock situation. If two oils do not correlate then there must be a source for each. Likewise, if an oil and source rock do not correlate it not only means a second source is required for the known oil, but the known source may well have sourced other reservoirs. The correlations may also include the analysis of oil seeps, from which the same conclusions may be drawn.

Crude oil, once generated, can be significantly altered and changed by water washing, biodegradation, gas deasphalting, and thermal maturation. These factors have to be taken into account when carrying out analyses for correlation purposes, because two oils which had the same origin may superficially look very different, depending upon the subsequent fate of their reservoirs. If one reservoir and its oils are shallowly buried, the effects of water washing and biodegradation will make it heavy, sulphurous, lacking in paraffins, etc.; whilst the oil from the same source in a deeper reservoir, possibly on the opposite side of a fault, will be lighter, more paraffinic, and less asphaltic due to thermal maturation and gas deasphalting.

However, differential migration will also alter crude oils so that they, at first instance, look different. Many of the oil components, especially the larger, more polar compounds, can be absorbed on to the minerals through which they are migrating, and hence the type of rock which provides the migration fairway and the distance over which migration has to occur will affect the quality of the reservoired crude oil, with attendant problems for correlation studies.

The effects of these alteration factors upon crude oil can be summarized as follows. The normal paraffins become greater in quantity and lighter in quality with increasing maturity, but overmaturation will cause the quality to eventually fall. The odd–even preference (CPI) will decrease with increasing maturity and the position of alkanes of maximum abundance decreases from between C_{25} and C_{31} to within the range C_{15} to C_{20} for the optimum maturity of crude oils. Once the most abundant compound in the alkane spectrum is below C_{15}, the zone of thermal cracking has been reached and the quantity of hydrocarbons, especially liquid ones, will decrease. The alkanes are affected in the opposite way, by biodegradation, which selectively removes alkanes and adds heavy, sulphurous asphaltic compounds to the oil. The alkanes can be 100% removed and thus biodegraded oils are easily recognized. However, their correlation will have to involve other components. The alternative components are often the biological markers (geochemical fossils) but Rullkötter & Wendisch (1982) have indicated that some caution is required as bacteria can also alter the structure of such molecules as hopanes by methyl group removal and ring opening.

Heavy components, such as asphalts, can be removed either during gas deasphalting when they are precipitated by the resolution of free gas, or by absorption on clay minerals during migration. The overall result is that the gravity of the oil (°API) will decrease due to water washing, biodegradation, or deasphalting, but increase due to thermal maturation. Migration will have a less obvious effect but it will tend to remove the heavy polar compounds and to make the oil lighter.

The overall result is that care has to be taken when analysing the results of tests carried out to correlate oils. As an example, consider the normal alkanes in two oils, which may have very similar alkane compositions. This may be due solely to thermal maturation and be totally unrelated to their source. Similarly, two oils from the same source may have completely different alkane distributions due to different histories subsequent to generation. For similar reasons, tests involving seeps and bitumen from source rocks will need special care. To overcome these problems a wide variety of investigations are normally carried out, which include alkane spectrum, CPI, analysis of geochemical fossils, nickel/vanadium ratios, stable isotope studies, as well as the investigation of the naphthene and aromatic compounds present.

Summary of previously described methods

In Chapter 3 the physical methods of correlating crude oils were described, and principal among these were methods based upon the US Bureau of Mines Hempel distillation. In this method the crude oil is fractionally distilled at atmospheric pressure and then reduced pressure in to fractions having a 25 °C temperature range. The average boiling point, specific gravity, and viscosity are measured and the oil can be classified into broad groups. Other indicators such as correlation index can be obtained for each fraction, and crude oils which have similar patterns of correlation indices are said to be related. Bass (1963) used correlation indices to differentiate oils from different sediments while studying the composition of crude oils in Colorado and Utah. Other properties which can be used are the cloud and pour points and the asphaltene content, either as measured by precipitation or as assessed by the carbon residue of the residuum. These can be combined with correlation index. Colour and odour have also been used as a correlation method.

Later in Chapter 3, the more recent correlation methods of Sachanen and of Tissot & Welte were described. These use analyses of the types of function groups present within the crude oil to define the oil. The amount of aliphatic, naphthenic, and aromatic material together with the sulphur content and various ring analyses are used in all these methods.

The methods based upon the physical properties, especially those based on the Hempel distillation, have been carried out for many years and a large amount of data has accumulated and is available for comparison purposes. However, the tests based on physical properties and, to a lesser extent those methods based on functional group analysis, have the disadvantage that they study the bulk parameters of the oils as they are found. They do not take into account the fact that the oils may have had different sources and their subsequent history has made them similar. Thus, whilst these correlations are quite good in terms of production evaluation they have less use in oil genesis studies. The other problem, especially with the distillation procedures, is the large quantities of oil that are required. Hence these tests are immediately ruled out as a means of examining source rock bitumen extracts.

Oil–oil correlations

Chapter 9 described a large number of organic geochemical techniques which were used to assess the type and maturity of the organic matter. Because of the nature of some of those methods, they can be used in oil–oil and oil–source correlations. The particular techniques which are of value in oil correlation studies are detailed analysis by gas-liquid chromatography of the bitumen fraction of extracts and of the oils, biological markers, stable isotope studies, and trace metals.

Gas–liquid chromatography of hydrocarbons

Gas–liquid chromatography can give indications as to whether or not a relationship between oils exists but the effects of post-generative alteration can lead to misleading results as described above. The analysis of hydrocarbon distributions, while not absolute evidence, can be useful confirmatory information. However, analyses of the naphthene and aromatic fractions can be more helpful, especially when either ring analyses are used or individual biological marker compounds are investigated. Steranes and other four-ring naphthenes have been successfully used in the correlation of altered oils. The other use to which gas chromatography can be put is in calculating various compound ratios, e.g. pristane/phytane; pristane/n-C_{17}; cyclopentanes/ n-paraffins; C_{27} steranes/C_{28} steranes. Stable isotope and nickel/vanadium ratios have also been shewn to change very little and are often used in correlation studies, not infrequently combined with other data. Erdman & Morris (1974) have used C_6–C_{10} component ratios to correlate crude oils.

Analysis of geochemical fossils

Pym *et al.* (1975) reports significant differences in the distribution patterns of the hopanes in crude oils from different geological areas. Oils from the Middle East were differentiated from Nigerian crudes because each had a different set of hopane compounds. Koons *et al.* (1974) used carbon isotope ratios of the saturates, paraffin content, light hydrocarbons, and the C_{15+} hydrocarbons, especially the steranes, to separate reservoired crude oils into two distinct families. Williams (1974) used carbon isotope ratios, correlation index curves, and hydrocarbon composition to identify three different types of oil in the Williston Basin (Figure 10.1). Powell & McKirdy (1975) used wax content, correlation indices, pristane/phytane ratios, and geological environment to classify Australian and Papuan oils into three categories.

Deroo *et al.* (1977) investigated oils from the Western Canadian basin and found they belonged to three separate families. They used techniques which included percentage composition of alkanes, aromatics, naphthenes, and asphaltenes; the relative distribution of naphthenes with one and two rings compared with naphthenes containing three or four rings (Figure 10.2); the distribution of isoprenoid compounds relative to normal alkanes, the thiopene compound distribution, and sulphur content. Seifert (1977) has used biological markers to correlate crude oils. The parameters that he used included sterane and terpane content as well as carbon isotopes and the ratio of branched to normal alkanes in the C_{14} to C_{30} range. Seifert (1978) pyrolysed shale samples, which he had previously extracted, and shewed correlations based upon steranes and terpenes which were in the extracts, were also present in the pyrolsate.

The analysis of the ring structure and carbon number of compounds by mass spectrometric analysis can aid differentiation between different oils. The aromatic and naphtheno–aromatic compounds, especially the tetracyclic molecules which are probably derived from steroids, are the most commonly used. Besides being suitable for oil–oil correlations, these analyses can also be used for oil–source correlations (Tissot *et al.*, 1974). Rullkötter & Welte (1979) have used low electron voltage gas chromatography–mass spectrometry of the aromatic fractions for oil–oil and oil–condensate correlations.

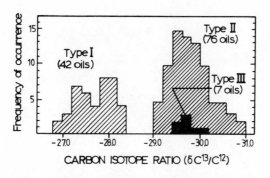

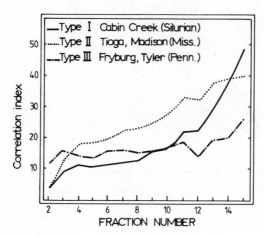

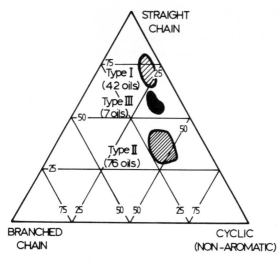

Figure 10.1 Geochemical data for major oil types in the Williston Basin (after Williams, 1974. Reproduced by permission of The American Association of Petroleum Geologists)

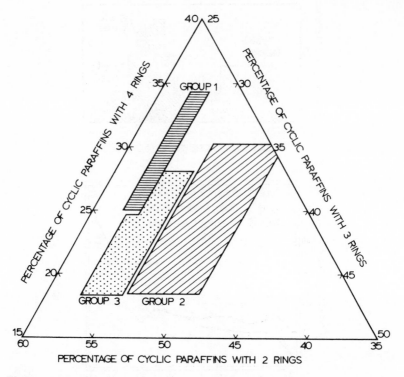

Figure 10.2 Oil correlation by cycloparaffins (Deroo *et al.*, 1977. Reproduced by permission of the Canadian Society of Petroleum Engineers)

Isotope studies

Barker (1979) quotes Sutton (1977) who has used carbon isotope studies on the saturated and aromatic fractions to identify three different oil types (Figure 10.3). Ross (1980) uses a cross-plot of the carbon isotope values (ordinate) against optical rotation (logarithmic abscissa) as diagnostic of the genetic potential of oils in the San Juan Basin (Figure 9.12). This type of approach to correlation will give the relation between oils based on one of the most successful methods, namely carbon isotope studies. The other parameter used is the optical properties of the oil, which are related, in part, to the biological marker-type compounds, and thus the combination seems an excellent choice as a correlation parameter.

Schoell & Milner (1979) used the cross-plots of $\partial^{13}C$ and ∂D values of the C_{15+} and C_{15-} fractions for crude oil correlations. Because the alkanes are most affected by heat the cross-plot involving the alkanes is the most useful for maturation studies, but the diagram based on data using the aromatic compounds is very useful for oil–oil correlations. Redding *et al.* (1979) have used the deuterium content of coals to differentiate coals from different sources. Thode (1981) uses sulphur isotope studies to identify three different major oil types in the Williston Basin, North Dakota. Migration has little effect

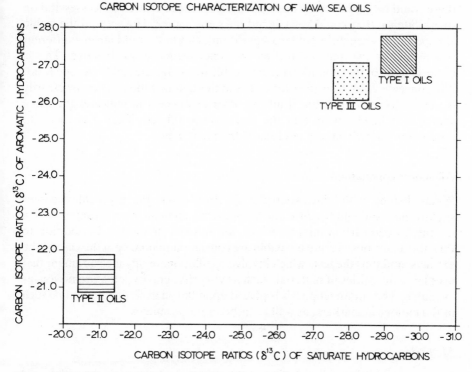

CARBON ISOTOPE CHARACTERIZATION OF JAVA SEA OILS

TYPE I OILS

TYPE III OILS

TYPE II OILS

CARBON ISOTOPE RATIOS (δ^{13}C) OF AROMATIC HYDROCARBONS

CARBON ISOTOPE RATIOS (δ^{13}C) OF SATURATE HYDROCARBONS

Figure 10.3 Oil correlation by carbon isotope ratios (after Barker, 1979. Reproduced by permission of The American Association of Petroleum Geologists)

on the ∂^{34}S values even though sulphur may have been lost during migration. This indicates little or no interaction with the reservoir sulphates under normal circumstances. Water washing and biodegradation alter the ∂^{34}S values, which increase with increasing oil degradation. Hartman & Hammond (1981) have used carbon and sulphur isotopes in tar-source correlations. They used crosspots of sulphur percentage against ∂^{34}S and ∂^{13}C against ∂^{34}S to differentiate different beach tars and to correlate each with a crude oil or natural oil seep.

Summary

When correlating oils it is advisable to use as many parameters as possible in order to increase the likelihood of obtaining a correct correlation. A few positive parameters are of much less value than a few negative indicators. Positive correlations can be due to chance and caused by other effects, e.g. secondary alteration, whereas negative parameters usually result from a definite cause. Hence negative correlations should be especially noted. An example would be two oils with similar gravities, similar alkane distributions, similar CPI, similar nickel/vanadium ratios, etc. Even small differences in these could be explained by one or another of the alteration processes and on such evidence they could be assigned the same source. However, if the stable isotope ratios were different by a significant amount it would mean a different source for the two oils, even if all the other evidence was positive. This is because none of the alteration factors is able to change the stable isotope ratio. The converse is not true and similarities in the isotope ratios do not necessarily mean the same source; but if all the other evidence also indicates the same source it is highly likely that the two oils are related. Thus it is easier to determine that two oils are not related than that they are.

Oil-source correlations

When dealing with oil-to-source rock correlations the problem becomes slightly more difficult because one is comparing an unmigrated bitumen with a migrated hydrocarbon mixture which will have been altered. In addition to this, the quantities that are available are considerably smaller in the case of the extracts, and thus the tests which involve distillation, or any measurement that uses large quantities of material, such as viscosity, gravity, etc., are no longer available. Thus more emphasis is placed upon the analysis of compounds such as the biological markers, as well as carbon isotope studies.

Methods

Welte *et al.* (1975) has listed some of the prime parameters which can be used for oil–source rock correlations. These include investigation of steroid and triterpenoid hydrocarbons of the C_{27+} cyclic molecule type including such compounds as steranes, terpanes, ergastane, cholestane, and hopane, which are determined by glc-ms analysis, ratios of hydrocarbons in the C_{16}–C_{30} *n*-alkane series, isoprenoid compounds, and carbon isotope studies.

One of the major differences between oils and bitumen extracts is that some compounds are present in greater amounts within the source rock. This can be due to a variety of factors, the most prominent of which are differences in the ability of the rock to absorb different compounds. Thus absorption can occur within the source rock, or within the rocks through which the oil migrates. The clays will tend to absorb the more polar compounds and this is shewn by the analysis of shale–sandstone pairs where the reservoired oils (in the sands) contain more paraffins and fewer NSO compounds (heteroatomic compounds containing nitrogen, sulphur and oxygen) than the extracts of their source rocks. In general there is a strong retention of oxygenated compounds in the source rock, whatever its nature, but whereas retention of nitrogen- and sulphur-containing compounds occurs in shales, it is not evident in carbonates. For similar reasons the saturated/aromatic ratios are almost always higher in reservoired oil than in the extracts from their source rock. Baker (1972) reported saturated/aromatic ratios of 3.1 : 1 and 2.1 : 1 for reservoired oils, compared with 1.3 : 1 and 0.8 : 1 for the corresponding source shale extracts.

Thus the relative concentrations of compounds may not be suitable for correlation studies.

In addition to the problems caused by the fact that unmigrated material is being compared with migrated matter, there are other problems such as weathering and subsurface oxidation which can affect the organic material in the source rock. Notes were made about this in Chapter 8 when sampling was discussed; however, it is also of special importance here as a correlation may be missed if the material is oxidized, weathered, or contaminated. Contamination can be a serious problem, giving very negative correlations. However, most contaminants are identifiable by glc-ms analysis.

Examples

Alexander *et al.* (1981) used carbon isotope studies, pristane/phytane ratios, isoprenoid/*n*-alkane ratios, (pristane + nC_{17})/(phytane + nC_{18}) ratios (which they considered independent of generation and migration and more source dependant), CPI, and *n*-alkane distributions for oil–source correlations.

Williams (1974) and Dow (1974) studied the Williston Basin and identified three source rocks, one for each of the three oils. The techniques involved were hydrocarbon type distribution, and carbon isotope ratios (Figure 10.1). Welte *et al.* (1975) have examined twenty potential source rocks and crude oil pairs with particular emphasis on the geochemical fossil/biological marker compounds such as the pristane, phytane, and the tetra- and penta-cyclic steroid and triterpenoid hydrocarbons. However, the hydrocarbon type distributions and the carbon isotope studies were not ignored. They obtained a variety of correlations of varying degrees of reliability but the best correlations were those which resulted from the glc-ms analysis of the C_{27+} cyclics, whereas hydrocarbon type analysis, carbon isotope studies, and the isoprenoid-compound ratios were not successful as oil–source correlation parameters. This is probably due to the fact that the comparisons were between migrated and unmigrated material. Yeh & Epstein (1981) consider that the ∂D and $\partial^{13}C$ values of crude oils appear to be determined by the isotopic composition of the materials from which they originated. Thus the use of isotope compositions could be useful in oil–source correlations.

Philippi (1981) has correlated oils with their sources by the use of high resolution GC analyses of the C_6–C_7 hydrocarbons. The relative amounts of five isomeric hexanes and five isomeric heptanes (each expressed as a ratio to the total non-aromatic C_6 or C_7 hydrocarbons respectively) are summed to give a similarity coefficient. (Ratios greater than unity are converted to their reciprocals.) Good correlations exist if this coefficient is greater than 0.8 (maximum 1.00) and poor correlations are indicated by coefficients of less than 0.73. Stahl (1975) has suggested that by isotope measurements of the saturates, aromatics, polar/asphaltene fractions of crude oils it is possible to predict the carbon isotope ratios of the source kerogen. This may increase the use that can

388

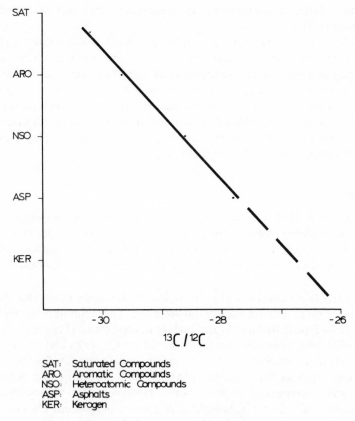

SAT: Saturated Compounds
ARO: Aromatic Compounds
NSO: Heteroatomic Compounds
ASP: Asphalts
KER: Kerogen

Figure 10.4 Source rocks crude correlation by isotope type curve
(after Stahl, 1978. Reproduced by permission of Pergamon Press
Ltd)

be made of isotope studies in oil–source correlations. Stahl constructs carbon
isotope curves using the data from the saturates, aromatics, heteroatomic
compounds and asphaltenes which can be extrapolated to give the kerogen
carbon isotope ratio (Figure 10.4). His predicted values compared well with
measured kerogen $^{13}C/^{12}C$ values and hence this method can be used to
correlate crude oils and the kerogens whence they came.

Leythaeuser *et al.* (1977) looked at an oilfield in Southern Germany which
had four possible sources ranging from Middle Jurassic to Oligocene and by
using compositional data judged the Jurassic to be the source.

The analysis of material from surface seeps follows very much the methods
used for investigating source rock extracts, taking care that the material has not
been oxidized.

Gas correlation

The investigation of gases for correlation purposes can be rather difficult as

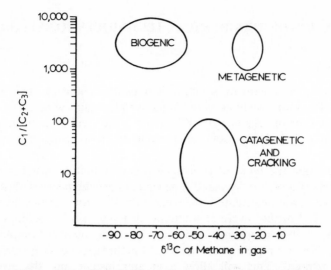

Figure 10.5 Classification of natural gas. (From *Organic Geochemistry for Exploration Geologists*, 1981, Douglas Waples, Burgess Publishing Company, Minneapolis, Minnesota. Reprinted by permission)

they are normally fairly simple mixtures of compounds. However, isotopic composition of the hydrocarbon components is the most commonly used method as it can distinguish biogenic gas from thermogenic gas. Biogenic methane is highly depleted in ^{13}C, having values between -55 to $-90°/oo$ (relative to PDB), whereas thermogenic methane has $\partial^{13}C$ values below $-55°/oo$. However, if one plots biogenic gas, catagenic (wet) gas, and metagenic (dry) gas in relation to their carbon isotope ratios, the ends of their ranges nearly meet, so some care is required. However, the three types are easily separated by using a cross-plot of wetness ($C_1/C_2 + C_3$) against carbon isotope ratio (Bernard *et al.*, 1976; Sackett, 1977; Feux, 1977) (Figure 10.5). Coleman *et al.* (1981) urge some caution in the use of isotope ratios for gas correlations because they have shewn that some methane-oxydizing bacteria can cause the methane to become isotopically heavier with respect to both carbon and hydrogen. However because the change in ∂D is 8–14 times greater than in the $\partial^{13}C$ value it may be possible to identify methane which has been subjected to such alteration.

Bailey *et al.* (1971) suggest that the maximum amount of information should be given when reporting geochemical analyses of crude oils so that other workers can make the maximum use of the data. The information they suggested was vital included API gravity, sulphur content, gross composition by gas–liquid chromatography, the analysis of the paraffins in the saturated fraction, the light gasolene as a percentage of the whole oil, the gas analysis, and USBM (United States Bureau of Mines) correlation curves.

APPLICATION OF GEOCHEMICAL TECHNIQUES TO EXPLORATION

Introduction

In Chapter 9, the means by which four questions could be answered were discussed. The four questions were: 'What is the amount of organic carbon?'; 'What is the type of organic matter?'; 'What is the level of maturation of the organic matter?'; and 'Have any generated hydrocarbons migrated away from the source?'

In this section the types of geochemical survey that are used in different stages of exploration are discussed. The value of geochemistry is illustrated by Pirson (1942) who calculated the success ratios for various exploration methods. His results were : random drilling, 5.8%; geology, 8.2%; geophysics, 14.9%; geochemistry 57.8%. Next the means of employing the answers to those four questions, obtained from that geochemical investigation, are demonstrated. This will allow a greater insight into the area under investigation. In particular, the total oil and gas source potential of the various rock units can be evaluated. The amount of fluid hydrocarbon which has been generated can be assessed and an indication of the potential of the basin under investigation as a mature hydrocarbon province can be obtained.

Reconaissance phase

In the reconaissance period an exploration programme for a relatively unknown area is being considered and is in the first stages of execution. The basin must be studied from a geological point of view to obtain as much information as possible on the sedimentary filling of the basin in order to try and identify potential source and reservoir rock sequences. Often outcrops at the rim of the basin are readily available sources of such information. At the same time the use of sniffer techniques, which were discussed in Chapter 9, can be used to pinpoint areas of higher than normal hydrocarbon concentration. These sniffer methods will find very low concentrations of hydrocarbons which are normally invisible to the human eye, but the search for visible seeps can also take place during this period. These seeps can be related to the geology of the area or with geophysical information in an attempt to identify their underground source. The seepages will require a pathway to the surface and this can be either a porous and permeable sediment, a fault, or the edge of an intrusion. Analysis of the seep material can give indications of the environment and origin of the organic material. Care has to be taken to avoid and identify contamination and to take account of any weathering that has occurred.

At the same time, organic geochemical examination of outcrop and shallow core material should be undertaken. This will involve unweathered surface samples and unweathered core material from shallow boreholes. If the basin is offshore, dredging and shallow coring will obtain equivalent samples. Because these samples will probably originate from the edge of the basin, the most important information that will be obtained from them will be the type and amount of the organic matter present in them. The level of organic maturation of samples obtained from the basin edge will be lower than that in the deeper centre of the basin. A knowledge of the basin geology and local thermal gradients will allow a preliminary assessment of the maturity of the organic matter which is buried more deeply. The analysis of such material will give the first indications of the hydrocarbon potential of the basin.

The methods that are commonly used include the whole range of geochemical techniques which were described in the previous chapter. Correlations between any seeps and the kerogen in the basin edge material will confirm whether or not that kerogen has sourced hydrocarbons at a greater depth. Organic carbon content will be determined on all samples, but account will have to be taken of any recycled dead carbon that may be later be observed. This gives a preliminary screening to check that sufficient organic matter is present. Pyrolysis will also give indications of the generative potential of a source rock and is another useful screening method. The presence of dead carbon will tend to make pyrolysis indicate a lower potential than the sample actually has, because dead carbon increases the carbon content without increasing the pyrolysis products. Thus screening methods are very susceptible to the presence of carbon which has no generative potential, and hence more exhaustive examination will be required. Such detailed analyses will be needed for correlation studies. In the course of these more detailed examinations the samples can be macerated and the elemental analysis of the kerogen (less any dead carbon) obtained. If there is a considerable amount of dead carbon, this method will be more accurate than pyrolysis. The optical examination of the organic matter will allow an assessment of temperature history, and organic matter type. This should lead to a preliminary picture of the source rocks within the basin. If any geological information about depth of burial, extent of the source intervals, presence of migration pathways, suitable reservoir rocks, thermal gradients, early igneous activity, the timing of hydrocarbon generation relative to reservoir and trap formation can be obtained and added to the geochemical data, a more complete picture of the basin will emerge from which decisions about the initial drilling can be made with more confidence.

Initial drilling phase

The second stage is the drilling of the initial wells and during this stage detailed analyses of as much material as possible must be obtained and used to refine the geochemical reconstruction of the history of the basin. This will have to be integrated with the geological history to ensure that the source rock was formed, subjected to sufficent heat for long enough, and that any generated hydrocarbons migrated to a suitable reservoir which was by then available.

The analyses should include the light hydrocarbons, as these give direct information relating to the concentration of hydrocarbons in the sediments. The results of gas yields from mud logging can be plotted both areally and vertically to indicate the principal zone of hydrocarbon accumulation. Cuttings and cores should be collected as frequently as possible, and kept wet in airtight containers until analysis, which should be as soon as possible after sample collection. The analytical methods should include rapid screening analyses by pyrolysis, total organic carbon, and the light C_1–C_7 hydrocarbons. These should be carried out at the wellhead so that a hydrocarbon log can be produced.

Clementz *et al.* (1979) advocate the use of programmed pyrolysis as a well site technique to rapidly identify and evaluate petroleum source beds. The instrument also provides a useful stratigraphic correlation tool. The advantages to be obtained by applying this technique to the well site are: better and more representative samples, instant access of data to well site geologist and the integration of geochemical and conventional mud logging data. In this context the authors have extended the system to offshore locations including drilling ships. However well head pyrolysis suffers from all the disadvantages of such pyrolytic analyses and care is needed to interpret the results. (See Chapter 9). This log can be amplified by the inclusion of the results of the more detailed laboratory investigations which will be undertaken on the samples. (See Chapter 9). The types of organic matter which are present can be established and their level of maturity assessed. The log should also have the temperature, both present and maximum palaeo- included as temperature is very important in the transformation of sedimentary organic matter. These detailed organic geochemical analyses of cores and cuttings allow one to obtain indications of the thresholds of the principle zones of oil and gas generation. Any larger shows of oil, or gas, need to be analysed and correlated with other oils, seeps and the material from possible source rocks.

Basin development

The initial period may cover the drilling of two or three wells and will provide some information as to the history of the organic matter present. This can be used, with other information, to delineate the basin and to predict where other hydrocarbon occurrences can be found. As more and more wells are drilled, the extra information allows greater refinement of the model.

CONSTRUCTION OF GEOCHEMICAL MODELS

The construction of a geochemical model of a hydrocarbon generating basin requires that one can ascertain information about the depositional environment of the organic matter. The shape and dimensions of the basin need to be determined and this is best done by a combination of subsurface mapping based on well data and seismic surveys. The history of the organic matter, particularly with regard to type, temperature, and time, has to be determined. It has already been shown that these are very important factors in the generation of oil.

The next essential is to check whether, at the correct geological period of time, migration routes were available and that suitable reservoir sands had been deposited. The formation of traps must have been completed by the time that the oil was ready to migrate. The last element of the model is a determination of the temperature history of the reservoir to check that any oil reservoired within it will not have been destroyed.

The type of organic matter can be determined, as was shewn in Chapter 9, by a wide range of chemical, physical, and microscopic examinations and this will allow a reconstruction of the depositional environment of the organic matter to be prepared. This model of the depositional environment will have to be in three dimensions because it is known that the types of kerogen can vary widely across a basin. This is due to variations in the type of organic matter which was deposited, and transgressions and regressions will alter the relative amounts of marine and terrestrial matter deposited at one place over periods of geological time. As transgressions occur the water deepens and the source of terrestrial matter becomes more distant, so that the kerogen which is formed will be open marine. As regressions occur the source of terrestrial organic matter becomes closer and greater amounts are incorporated, and thus the kerogen becomes more terrestrial. Likewise, changes in the characteristics of rivers such as speed, course, and contents (both organic and inorganic) will vary the amount of terrestrial material incorporated into a sediment.

More sophisticated kerogen analyses, especially those based on gravity separation methods (Kinghorn & Rahman, 1980) allow a much more detailed picture of the organic matter distribution to be obtained. Two-dimensional representations are not very satisfactory in this context as one wishes to plot at least four kerogen types not only over an area, but also over periods of geological time, i.e. over the thickness of the sediment. Quite often, either the predominant kerogen is mapped or the relative percentages at each interval in a borehole are recorded. The problems increase when extra information such as temperature history, sediment thickness, and potential reservoirs have to be related to the organic matter type. However, the detailed description of the make-up of the kerogen allows a much more accurate prediction of which hydrocarbons are produced from the kerogen.

For the kerogen to be converted to hydrocarbons it is necessary for the organic matter to be subject to the correct heat for a sufficient time. The interchangeability of time and temperature have already been discussed. The method of Lopatin (Lopatin, 1971; Waples, 1980) allows one to plot the history of a sediment on a time–temperature grid (Figure 9.19) in order to ascertain the time the sediment has spent within each temperature zone. The oil and gas windows for that sediment can be determined and thus the products of kerogen maturation at different times can be predicted so long as the kerogen type is known. From geological information as to when migration pathways and suitable reservoirs were available, it is possible to use the Lopatin time–temperature index to predict the products which would have initially been reservoired at each age in the sediment's history from the start of oil generation onwards.

This latter exercise will require modelling of the possible migration pathways and traps. If the Lopatin theory is again used on these, it is also possible to predict whether any oil that did migrate into the reservoir will have survived. All the processes by which petroleum can be altered are related to temperature and thus their effect can be assessed by observing palaeotemperatures. For this, simple parameters such as vitrinite reflectance or spore carbonization may be used. However, it is now recognized that time is as important as temperature and this must also be taken into account by use of Lopatin's method or equivalent methods.

Organic facies and the geochemical model

The building of this regional picture of the organic chemistry of the sediments is often called organic facies analysis. Organic facies may be considered as a mappable subdivision of a designated stratigraphic unit, distinguished from others on the sole basis of the character of its organic matter. The organic facies will depend upon the kerogen type, which in turn depends on its original source (Figure 10.6). If the organic facies are combined with the subsequent temperature history of the sediment, which can be indicated by iso-reflectance or iso-carbonization values, the probable hydrocarbon products of that kerogen and their location can be predicted (Figure 10.7).

In Figure 10.6, four organic facies are identified, A, B, C, and D. Of these, A, B, and C are the facies in which one would expect to find the Tissot & Welte kerogen types I, II, and III respectively. Facies D does not correspond to type IV kerogen or inertinite. This facies is the facies where organic matter is not deposited and/or preserved within an aquatic environment. There are two non-source areas in Figure 10.6. One is where the aquatic environment is

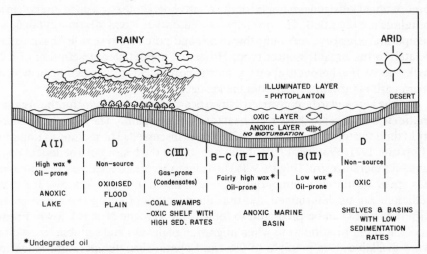

Figure 10.6 Organic facies types (Demaison & Moore, 1980. Reproduced by permission of The American Association of Petroleum Geologists)

always oxic so that the very small amounts of organic matter which come from the neighbouring desert environment will not be preserved. The other non-source area is the oxidized floodplain of the continent where the organic matter is generated but not preserved (Demaison & Moore, 1980).

The organic facies analysis will require a wide range of geochemical analyses. The oxicity of the depositional environment can be inferred from studies of biological markers, e.g. the pristane/phytane ratio or the relative quantities of related steroid molecules. (Nishimura, 1982, Seifert & Moldowan, 1981). Palaeoclimatological and palaeogeographic reconstructions can indicate when conditions were correct for the formation of source and reservoir rocks. Northfield et al. (1981) use isotopic compositions for climatic reconstructions whereas Parrish (1982) shews that one can infer the timing and distribution of upwelling and relate them to sea level change models which together can contribute information on the sites of petroleum source beds.

Palaeoreconstructions are becoming a prerequisite to exploration, as they allow the geochemist to predict whether or not source rocks could have been formed in the area under investigation. If the inferences are favourable the next stage is to look for structures and then to drill. However, if there was never any likelihood of source beds being created it would be a pointless waste of money to continue exploration.

The amount of geochemical information which can be available is immense and the problem is how to make the best use of it and how to relate it all together. Figure 10.7 represents a reconstruction of the deposition of organic matter plus temperature history in terms of vitrinite reflectance. The organic matter which is deposited close to the shore line is basically terrestrial and predominantly type III. The effect of the river can be seen by the fact that the terrestrial organic matter is carried further from the shore opposite the mouth of the river. The iso-reflectance lines on the map indicate the top and bottom of the oil window, i.e. 0.5% and 1.5% R_o respectively. It can be seen that the majority of the type III kerogen and some of the type II sapropelic kerogen is immature, while the majority of the type II kerogen plus a little of the algal type I is producing oil. There is a gas-producing zone in the south-west corner where both types I and II have been at a higher temperature.

The kerogen type boundaries shown between the types are where the predominating type of kerogen changes. However, there will be mixed kerogens on both sides of these boundaries, especially if there have been regressions and transgressions during the deposition of the source rock. An alternative technique is to plot contours showing percentages of the different components of the kerogen based upon detailed separation and examination. These allow a greater precision in the prediction of eventual products.

Use of maps to illustrate source potential

Isopach maps can be plotted shewing the thickness of the source bed, which will allow the quantity of the source rock and a better estimate of the volume of

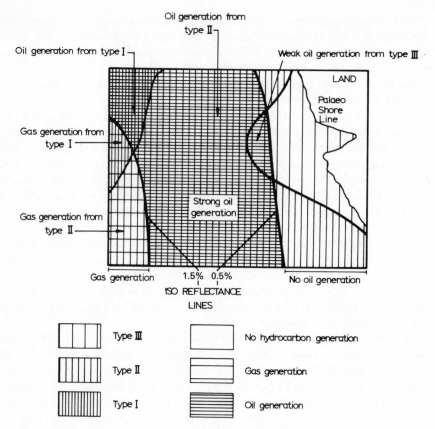

Figure 10.7 Organic facies in an imaginary basin (after Tissot & Welte, 1978. Reproduced by permission of Springer-Verlag, Heidelberg)

final products to be obtained. Likewise, maps can be contoured with values of maximum palaeotemperature or Lopatin's time–temperature index values for any time in the geological past. These latter maps are probably more useful than temperature, iso-reflectance or iso-carbonization value contours as they take into account the quantity of energy available for organic matter alteration.

Various combinations of such maps will indicate the zones of the maximum generation of oil or gas. If this is then combined with information about migration pathways and suitable reservoirs, the model of the basin is nearly complete. It should now be possible to pinpoint the areas which contain source rocks of the correct organic matter and with the correct time–temperature history to have produced oil. The likely migration routes and possible reservoirs can also be highlighted and related to the generation of hydrocarbons. Thus the chance that a reservoir will contain hydrocarbons, and what type of hydrocarbons are likely to be found, can be calculated.

One of the factors which aids migration is the volume increase and therefore the increase in pressure which accompanies the generation of hydrocarbons from kerogen. This pressure increase is available to drive hydrocarbon-phase migration and it will be at its maximum where the generation of hydrocarbons is greatest. This will be not only where the source is thickest, but also where content of the organic matter within the source rock is greatest. At the same time, the temperature has to be high enough for generation to occur and this will mean that any generated oil will have a lower viscosity.

From other geological information it is possible to plot preferred migration routes and also reservoir rocks which were in existence at the time of primary hydrocarbon generation and were able to become traps. From superimposition of such maps on those previously described, it can be possible to predict the destination of any oil that is produced in the area under study. The subsequent history of that reservoir rock will need to be investigated to ensure that any oil which was stored within it has not subsequently been altered or even destroyed. All the agencies of secondary alteration can attack reservoired oil. Again, the time–temperature index type studies can give indications of exposure to thermal attack while evidence of uplifting will show if bacterial attack has been possible.

If there are several possible reservoir rock destinations for the crude oil, the most likely ones are those which are closest to the source of generation, assuming all other factors to be equal.

Estimation of probability of hydrocarbon accumulation

The probability of finding a hydrocarbon accumulation depends upon the probabilities of occurrence of all the independent elements that are required for the accumulation of oil or gas. These elements include the deposition and preservation of a suitable source bed; the correct time–temperature regime to cause maturation; the occurrence of migration; and a suitable reservoir and trap being available at the correct time. Because all these essential conditions are independent of each other, the total probability of the occurrence of a gas or oil pool is the product of the individual probabilities. If one of these probabilities is zero then no oil will occur; for example, if there is no likelihood of a source rock having been formed then no oil can have originated from it. This is a mathematical expression of the conditions necessary for the occurrence of oil which were discussed earlier.

The probability of finding a reservoir and trap can be obtained by non-geochemical studies, e.g. seismic investigations, but the other factors can be investigated by the use of organic geochemistry and especially basin modelling and organic facies analysis. It is therefore possible to improve the chances of finding hydrocarbons by looking in areas where basic conditions are satisfied. The source type and its level of maturation can be investigated by organic geochemical techniques and the results fitted into a conceptual model of the basin. The results can be displayed as a series of maps, one for each parameter: source type, maturity, time–temperature index, and reservoir thickness, which can be lain over one another to shew the areas of maximum probability of hydrocarbon formation and occurrence.

Many workers have been trying to find a way to express the overall source potential of a sediment, either graphically or numerically, based upon various geochemical analyses. Claypool & Reed (1976) described a pyrolysis apparatus where the volatile hydrocarbons which are distilled out of a sample, and the hydrocarbons which are generated by higher-temperature cracking of the kerogen, are indicated by two peaks. The quantity of hydrocarbons produced is proportional to the peak areas. The area of the first peak, which is the distillation of already generated hydrocarbons, was directly related to the C_{15+} solvent-soluble hydrocarbons, whereas the area of the second peak was directly proportional to total organic carbon. These correlations were obtained from the analysis of a large number of samples. However, there must be some doubt as to their application to different kerogen types and to different mineral components. Espitalié et al. (1980) have shown that clay minerals can trap hydrocarbons and thus alter the relative size of the two peaks. The total organic carbon (TOC) is related to the amount of kerogen and solvent–soluble materials in the rock. Because each kerogen has its different convertibility to mobile hydrocarbons, the relationship between peak area and TOC could vary with organic matter type. The percentage of the total organic carbon which corresponds to the solvent-soluble material will depend upon the level of organic maturation. Thus mineral type, organic matter type, and organic maturation could affect these correlations.

Claypool & Reed used their obtained relationships of peak areas with solvent-soluble hydrocarbons and TOC to enable them to plot the two peak areas against each other. This produced a graphical representation of the organic richness and source potential of the rock (Figure 9.15) This is a quicker way of approximating to the quantity of hydrocarbons per gram of organic carbon which can also be used as a maturation indicator (Figure 9.3)

Harwood (1977) plotted the generation of hydrocarbons, as expressed by the solvent-soluble extract, against the original kerogen H/C ratio (Figure 9.10). Using this system with the weight of either liquid or gaseous hydrocarbons and relating these to the richness of the source rock, in terms of percentage of organic carbon, the yield in barrels or MCF (million cubic feet) per acre-foot can be obtained. The genetic potential of a source rock can also be obtained by 'rock-eval' type pyrolysis. These results are normally expressed in kilograms of hydrocarbon per tonne of source rock. All these methods only give information on the organic matter and maturation level. They involve analyses which can be misleading if other tests are not conducted. The effects of inert carbon, kerogen type, and clay minerals have all been discussed earlier and can affect these types of analysis. However, these methods do find quite considerable use, especially for screening and for on-site logging (Clementz et al., 1979)

Waples (1979) considers that the total oil expelled from a source rock is the product of factors related to the quantity of the organic matter, the quality of the organic matter, the percentage generation, and the expulsion efficiency. Waples took a series of commonly used parameters to assess these factors. He considered that the quantity was directly related to the organic carbon content of the source rock. Because the organic carbon content of an 'average' shale is about 1% (Hunt, 1972), Waples normalized the quantity factor (Q_1) for an average shale of 1% TOC as 1.0. Thus the organic content of the sample gives the quantity factor (Q_1) directly. The calibration of TOC vs. Q_1 is on a log–log graph so as to accommodate large values of organic carbon.

The estimation of quality of the organic matter is based upon visual inspection of the kerogen in transmitted light. Waples worked on the simplified assumption that the macerals alginite and exinite are oil generating whereas vitrinite and inertinite do not yield oil, and therefore the relationship between kerogen type and the quality factor is simple and linear. He defined the quality factor as 1.0 for a kerogen of approximately average quality, i.e. 50% alginite and liptinite, as visually determined. As an alternative to this approach he used the measurement of the atomic H/C ratio of the kerogen when immature. He chose a normalized value of 1.0 for the quality factor for an average kerogen of initial immature atomic H/C ratio of 1.0. From either of these scales he was able to obtain a value for the quality factor (Q_2).

The maturity factor (M) is related to vitrinite reflectance (R_o) although the thermal alteration index, TAI (Staplin, 1969) can also be used. This calibration curve, unlike those for quality and quantity, is not a straight line. It is approximately linear over the oil-generation zone but because little generation occurs below $R_o = 0.5\%$ or above $R_o = 1.5\%$, the curve flattens out at the two extremes. Hence by measurement of vitrinite reflectance, Waples obtained a value for percentage generation (M). Expulsion efficiency was the one factor for which a satisfactory correlation with measureable parameters could not be obtained. For the moment he leaves that factor (E) as unity, i.e. 100% expulsion.

Waples' equation for total oil to reach the reservoir becomes:

$$\text{oil} = Q_1 \times Q_2 \times M \times E$$

For a value of the total oil potential of a source rock, the organic matter must be completely mature and the maturation factor M will also be unity. Thus the total oil potential will depend on organic matter quantity and quality, i.e.

$$\text{total oil} = Q_1 \times Q_2$$

Because of the normalization procedures used by Waples an average shale which generates about 80 million barrels of oil per cubic mile of rock will have a total oil value of 1.0. Thus a source rock with a Waples total oil factor of 1.0 will have produced 80×10^6 barrels of bitumen per cubic mile, and *pro rata*. If a value for the maturation factor (M) is included then the value for the oil already generated will be obtained. This is the amount of oil in which one is interested.

Waples considers that this method has potential as a numerical assessment of the hydrocarbon potential of a source rock. Once better assessments of the migration and expulsion efficiency are possible, the Waples method will allow the prediction of the amount of oil that has left the source rock and possibly been trapped, although this method is a fairly coarse assessment of the hydrocarbon potential of a source rock. An examination of all the factors and their method of assessment will shew that they contain pontential errors.

The quantity factor, which is based on total organic carbon, can be erroneously high due to the presence of dead carbon which has no generative potential. The assessment of quality, which is based on the presence of alginite and liptinite, will be approximate because the generation of petroleum is very much more complex than the simplified approach adopted by Waples. The H/C ratio of the kerogen as an assessment of quality may be misleading if the kerogen under investigation is a mixture of kerogen types. The maturation factor is obtained solely from a temperature point of view and it takes no regard for the time that sediment has spent at that temperature (Figure 7.1).

Computer modelling

More sophisticated modelling techniques have recently been developed using computers. Nakayama & Van Siclen (1981) have described a computerized simulation model which synthesizes the processes of petroleum generation, migration, and accumulation under relatively simple conditions. The model considers four geological processes being sequentially performed on a cell. These processes are deposition, compaction, petroleum generation, and migration.

The model is restricted by these conditions. It is two dimensional and represented by geological cross-sections. The sedimentary rocks are all terrigeneous clastics, i.e. shales and sandstones. There is a single petroleum fluid, i.e. either oil and gas, involved in generation and migration. Petroleum migration is considered to be solely the result of buoyancy. Besides these conditions, certain hypotheses are also made and include the following. Compaction is a purely mechanical method of reducing pore size and no account is taken of chemical processes. Porosity is thus a simple function of depth of burial and grain diameter of the sediments. The generative potential is determined by the amount and type of organic matter and maturation is assessed with time taken into account using Connan's equation (Connan, 1974). Hydrocarbon migration is considered to occur only in an upwards direction where there is more oil in the sediment than the residual amount which is found in a sediment after migration.

In spite of these simplifications, meaningful predictions have been obtained, including migration pathways, times for generation, and transformation ratios. The model performed well when applied to real situations so long as there was careful selection of the parameters used.

Welte & Yukler (1981) have produced a three-dimensional deterministic dynamic basin model using geological, geophysical, geochemical, hydrodynamic, and thermodynamic data. Input data consists of heat flux, initial physical and thermal properties of the sediments, palaeobathymetric estimates, sedimentation rates, as well as the amount and type of organic matter. From this the model computes the pressure, temperature, physical and thermal properties of the sediment, the maturity of the organic matter, and the hydrocarbon potential of any source rock as a function of space and time. Hydrocarbon potential maps for any given source rocks in any geological time slice of the basin can be produced as computer printouts.

The construction of the basin model involves the various sediments inputs (type, source, and rate) the depositional environments, palaeobathymetric estimations, changes in mineralogy, and tectonic movements, all of which are obtained from a geological study of the basin. The direction and rate of flow of fluids together with their hydraulic properties can be obtained from hydrodynamic studies, while thermodynamic work will give the rate of flow of heat through both fluids and sediments. The type, amount, and maturity of the organic matter will be determined by organic geochemistry. These are united to form the conceptual model which is compared with the real system and modified as necessary until acceptable, when it is used to produce results.

More sophisticated use is made of the data in this model. Knowledge of palaeopressures and temperatures allow predictions of changes to the sediments. These changes involve the structure and stratigraphy of the sediments as well as alterations in the physical and thermodynamic properties of the fluids. The organic matter is assessed by optical examination of the kerogen and physico-chemical analysis of the kerogen and bitumen. The effect of heat on the alteration of the organic matter and upon the physical properties of the sediments and the fluids was observed. Such parameters as density, viscosity, thermal conductivity, porosity, and permeability are all affected by heat. The thermal conductivity of the sediments and thermal convection of fluids were studied in order to assess their effect on the particular model. Lopatin's method was used to determine the maturity of the organic matter so as to take account of the effects of both time and temperature. Migration was considered to be mainly pressure driven and in a hydrocarbon phase, taking place through available pores or by microfracturing of dense source rocks. Secondary migration and the subsequent formation of oil and gas pools were considered to be the results of buoyancy, capillary pressure, and hydrodynamics.

The model has been successfully applied to real basins to predict the amount and type of oil in place.

EXAMPLES OF GEOCHEMICAL INVESTIGATIONS

Shiboaka *et al.* (1978) have studied the hydrocarbon generation in the Gippsland Basin of Australia. Optical examination shewed that the organic matter contained very large amounts of vitrinite, often as high as 84%, and much smaller quantities of exinite and inertinite. The major products from the thermal maturation of these types of organic matter will be gas from the vitrinite and oil from the exinite, although there will be minor components of oil and gas respectively. Elemental analysis of the kerogen and plotting of the atomic H/C and O/C ratios shewed that the organic matter in the basin is relatively immature and the authors believed that this organic matter was at the stage when both generation and migration were occurring at the present time.

The Windalia oil is a commercial accumulation in a Cretaceous reservoir at Barrow Island, Western Australia. Alexander *et al*. (1981) have made a correlation between this oil and hydrocarbons extracted from 37 cretaceous rock samples from different stratigraphic levels in seven wells on and around Barrow Island. They used seven established and one new correlation parameter to achieve oil–oil and oil–source rock correlations.

Carbon isotope studies were undertaken by converting samples to carbon dioxide and measuring the relative amounts of ^{13}C and ^{12}C with a mass spectrometer with the PDB as standard. Allowance was made during the correlations that within the genetically related series, kerogen–source rock–bitumen and crude oil, the carbon isotope values become increasingly negative (Silverman, 1964). Pristane/phytane ratios, as measured by gas–liquid chromatography, were used by Alexander *et al*. The oxic–anoxic state of the depositional environment is reflected in the pristane/phytane ratio because these compounds are derived from phytol, the former by oxidation and the latter by reduction. (Figure 6.9). In this investigation the pristane/phytane ratios were corrected to take account of differential thermal histories of the oil and extracts before correlations were made.

Isoprenoid and *n*-alkane ratios, especially the pristane/*n*-heptadecane and phytane/*n*-octadecane ratios, were used because these ratios have been shewn to be influenced by the nature of the kerogen and the extent of generation and maturation of petroleum. Alexander *et al*. improved these ratios by taking the pristane plus *n*-heptadecane to phytane plus *n*-octadecane ratio which they considered to be of greater value for correlation, because it is independent of the degree of generation and maturation but dependent upon the source material and its depositional environment.

Other correlation procedures used in this study included hydrocarbon type; ratios of *n*-alkanes to naphthenes representing the nature of the kerogen and the stage of maturation; carbon preference index; distribution of the *n*-alkanes in both extracts and oils. Each of these individual techniques was given a weighting based upon its power to diagnose genetic relationships. The results were summed to give a correlation which was used to relate oils and source rocks.

Magoon & Claypool (1981) have carried out geochemical correlations on oils and source rocks in the Cook Inlet Basin of Alaska. The oils were studied by measuring API gravity and sulphur content, which were related; and by plotting the US Bureau of Mines Hempel distillation analyses in the form of correlation indices. All of the Cook Inlet oils from Tertiary reservoirs have similar correlation curves although they vary in detail. The shallow Trading Bay oil is relatively depleted in paraffins (due to bacterial degradation) whereas the North Cook Inlet high-gravity oil is relatively enriched in paraffins. Detailed analyses of the gasolene-range hydrocarbons were carried out and these shewed that the shallow Trading Bay oil, which has been depleted in *n*-paraffins by the action of hydrocarbon-oxidizing bacteria, is distinctly different from the other oils in respect of these compounds.

Stable isotope studes shewed that all the oils were within $2^o/oo$ of each other (-32 to $-33^o/oo$ relative PDB) and this suggested that the oils in all the major fields originated from a common source which was found by other methods to be Middle Jurassic rocks.

Gas chromatography of the C_{12+} saturates of the soluble bitumen showed that only the Middle Jurassic rocks could have sourced the Cook Inlet oils. Other potential source rocks had dissimilar chromatograms. Carbon isotope studies on the kerogens were not able to identify the source of the Cook Inlet oils. Thermal evolution analysis was used to determine the type, richness, and maturity of the potential source beds. This type of analysis is a programmed temperature pyrolysis which gives peaks relating to already generated and cracked hydrocarbons (see Chapter 9). This pyrolysis technique used the pyrolytic hydrocarbon yield, i.e. peaks I and II, divided by the total organic carbon content expressed as a percentage. This pyrolytic hydrocarbon yield was plotted against depth and is sensitive to the type of organic matter and to the degree of thermal maturity. The results shewed that only the Middle Jurassic rocks were potential sources. These analyses indicate that the oil potential of the Cook Inlet is greatest when Tertiary or Cretaceous reservoir rocks have truncated Middle Jurassic source rocks. The non-marine Tertiary rocks were eliminated as a possible source because their organic matter was thermally immature and of a coaly type which is gas prone when mature. Cretaceous rocks were also eliminated because they were thermally immature. This left only the Middle Jurassic rocks containing adequate amounts of thermally mature oil-prone organic matter with extractable hydrocarbons which resemble those of the Cook Inlet oils.

Magoon & Claypool (1981) have carried out a similar study on the oils from the North Slope of Alaska. A plot of API gravity against sulphur content separated two groups of oil. One group containing less than 0.5% sulphur had API gravity in the range 18 to 32 and the second group whose sulphur content was between 0.8 and 2% and an API gravity range of 13 to 28. Biodegraded oils were recognizable by their higher sulphur content and lower gravity as compared to other oils in their group.

The C_{15} saturated hydrocarbons were analysed by gas chromatography and CPI and pristane/phytane ratios were used to distinguish the two types of oil. Normal oils from South Barrow and Prudhoe Bay had a slight even-to-odd preference (CPI = 0.94–0.99) and pristane/phytane ratios of 0.8 to 1.2, whereas the other oils had a small odd-to-even preference, especially in the C_{20} to C_{28} range (CPI = 1.02–1.05) with pristane/phytane ratios of 1.7 to 2.0. The biodegraded oils had greatly reduced amounts of normal paraffins.

The carbon isotope ratio values were determined for the whole oils as well as the saturated and aromatic fractions. The carbon isotope ratios for the whole oils again differentiated the two groups of oils, one with a $^{13}C/^{12}C$ range of $-29.8^o/oo$. However, carbon isotope ratios for the aromatic and saturated

C_{15+} hydrocarbons appeared to distinguish three groups but the authors consider that one group of oils has two subgroups as analysed in this way. These groups were well differentiated when the $^{13}C/^{12}C$ ratios of the saturated hydrocarbons were plotted against the ratios of the aromatic components. The isotopes ratios of sulphur and nitrogen were also measured and the sulphur isotopes readily separated the two groups of oils whereas the nitrogen isotopes did not.

Chukwemeka & Okoye (1980) have evaluated source beds in the Niger Delta by the use of organic geochemistry. In this study the type of organic matter was characterized by microscopic examination of the kerogen and described by the approximate percentages of the different maceral groups. The concentration of the organic matter was obtained by measurement of the total organic carbon content of each sample. Assessment of the thermal maturity of the sediments involves three criteria which were: the average vitrinite reflectance (Dow, 1977); the odd/even ratio of the C_{28}, C_{29}, and C_{30} alkanes (Philippi, 1965); and the ratios of the concentrations of soluble organic matter to total organic carbon (Cassou *et al.*, 1977). This last indicator was supplemented by variations in the ratios of the total saturated and aromatic or saturated hydrocarbons to total organic carbon (Claypool *et al.*, 1978). All these analyses were carried out on cuttings and sidewall cones.

Brooks & Thusu (1977) studied the oils and Jurassic sediments in the northern North Sea and used geochemical and palynological techniques to determine and define the following source rock characteristics: (a) The depth in the Jurassic sequence at which hydrocarbon generation becomes significant; (b) the lower limits below which liquid hydrocarbons are unlikely to occur; (c) the relationship between types of organic matter, organic maturity, and the generation of hydrocarbons; (d) the thermal history of the Jurassic source rock. The parameters used for this study were organic carbon content and carbon-to-hydrogen compositions; carbonization of structured and dispersed sedimentary organic matter; and the chemical maturity indicators.

The use of these techniques was based on previously published information. The minimum amount of organic carbon required for hydrocarbon generation is 0.5% and in the Jurassic sediments of the northern North Sea the TOC is over eight times this at an average of about 4%, but within the depth range 2,700–2,900 m it is as high as 6.8%. McIver (1967) showed that sedimentary organic matter with relatively high proportions of hydrogen, i.e. greater than 6%, could function as hydrocarbon source material, while Laplante (1974) has shewn that a carbon content of the sedimentary organic matter of between 74 and 76% represents the threshold of significant thermogenic hydrocarbon generation. Carbon ratio was used to determine maturity (Gransch & Eisma, 1966). Microscopic examination of sedimentary organic matter was used for assessment of organic matter type and maturity. Other maturation indices included the ratios of hydrocarbons to non-carbonate carbon (TOC) and the hydrocarbon distribution profiles, as well as the odd/even ratio of high molecular weight *n*-paraffins. In general, the hydrocarbon to organic carbon

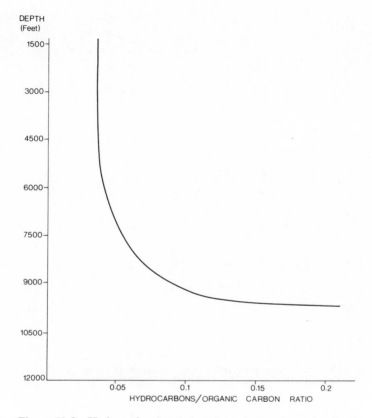

Figure 10.8 Hydrocarbon/total organic carbon ratio versus depth (Brooks & Thusu, 1977. Reproduced by permission of Elsevier Scientific Publishing Co.)

ratio increased with depth (Figure 10.8) while the odd/even ratio decreased with depth until it was nearly unity (Figure 10.9).

In the interval 1,500–2,400 m, the sedimentary organic matter contained mostly sapropelic material with some terrestrial organic matter. The carbon ratio is between 0.33 and 0.68 (see Chapter 9) and the thermal maturation indices indicated a maximum burial temperature of 50–60 °C. The structured organic matter was readily identified as woody and plant debris, while the sapropelic matter changed progressively, with depth, from yellow to light brown. The average carbon content in this section was 3.2% and n-alkane distribution ranged from C_8 to C_{34} with the maxima corresponding to maximum abundance at C_{19} and C_{29}.

In the depth range 2,600–3,200 m, sedimentary organic matter again contained sapropelic material with a terrestrial/woody coaly component. However, the C_R/C_T ratios had increased with increasing depth and were between 0.37 and 0.52. The indicated maximum temperature of the sediment was 100 °C. The colour of the organic material increased from light to medium

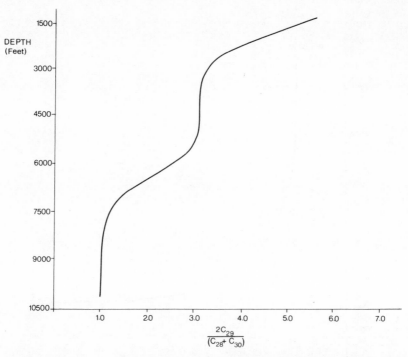

Figure 10.9 Odd–even preference versus depth (Brooks & Thusu, 1977. Reproduced by permission of Elsevier Scientific Publishing Co.)

brown and the average carbon content 5.6%. The n-alkane distribution was between C_8 and C_{34} but with maxima between C_{15} and C_{19}. This section was classified as a good source rock and probably the zone of hydrocarbon generation. The elemental analysis of the kerogen suggested a carbon value of 78% and hydrogen content of 7.6%, both indicative of an oil source in the principal zone of oil generation.

In the depth range 3,200–3,650 m, the C_R/C_T ratio increased to between 0.59 to 0.64 and the maximum temperature of the sediment is indicated as having been 120 to 140 °C. The total organic carbon is 4.9% with the hydrocarbon distribution being between C_8 and C_{24} with maxima between C_{15} and C_{19}. The colour of the sedimentary organic matter is dark brown and the indications are that the organic matter in this zone has passed through the peak of the oil generating stage, but it may still be making a small contribution. Figure 10.9 shews how the carbon number odd/even ratio ($2C_{29}/(C_{28} + C_{30})$) changes with increasing depth. There is a marked odd-number preference in the higher sequence which disappears with increasing depth until the ratio closely approaches unity. The great increase in hydrocarbon to carbon ratio with increasing depth (Figure 10.8) clearly shows the onset of the principal zone of generation of hydrocarbons.

The results indicate that the most productive oil source zone in the northern

North Sea is the Middle Jurassic section of interbedded marine shales and sands which were deposited in a deltaic environment. The Jurassic source rocks at depths between 2,600 and 3,200 m represent the major source of hydrocarbons, but limited additional generation may have occurred in rocks as deep as 3650 m.

Merewether & Claypool (1980) studied shales from Powder River Basin, Wyoming, using total organic carbon, pyrolytic hydrocarbon yield, temperature of maximum pyrolytic yield, and vitrinite reflectance to indicate the amount and character of the organic matter in sampled rocks, as well as depositional environment and depth of burial of strata. The organic carbon content was taken as a direct measure of the amount of organic material preserved with the sediment. The thermal analysis was carried out according to the system of Claypool & Reed (1976). The volatile hydrocarbon content is the proportion of the total pyrolytic hydrocarbon yield that is produced from the rock at a temperature of less than 320 °C, and it is proportional to the content of the organic matter that can be extracted from the rock samples with solvents. Total pyrolytic yield is expressed as a percentage of the dry rock weight, and the temperature during pyrolysis at which the yield of hydrocarbons produced by pyrolysis is a maximum is also recorded. As the total pyrolytic hydrocarbon yield is limited by the hydrogen content of the original organic matter, it reflects both the type of organic matter involved and its degree of thermal maturity. The pyrolytic hydrocarbon to organic carbon ratio indicates the proportion of organic matter in the rock that has been or can be converted to volatile organic compounds by heating. Potential source rocks should have pyrolytic hydrocarbon to organic carbon ratios in excess of 20% at an immature stage of thermal alteration. The ratio of volatile hydrocarbons to total pyrolytic hydrocarbon yield, which is the production index (Barker, 1974), is an indication of the extent of the conversion of solid organic matter to petroleum hydrocarbons, and this ratio should range from 0.05 at the earliest immature stages to a maximum of 0.5 for the post-mature stage of thermal alteration process. Temperatures corresponding to maximum pyrolytic yield can also give information as to the maturity of the organic matter. If the temperature is less than 460 °C this indicates immature organic matter, whereas 500 °C corresponds to the end of oil generation. If samples have a maximum pyrolytic temperature in the range 460–500 °C, it indicates that the organic matter in that sample is in the zone of maximum hydrocarbon generation.

The result of the analysis undertaken by Merewether & Claypool on the shales from Powder River Basin showed that the amount and character of the organic matter varied with depositional environment and thermal maturity of the strata. The material in rocks near the edges of the basin had never been deeply buried, whereas those in the centre of the basin reach down into the oil window. Because the organic material is mostly sapropelic in nature, the rocks in the basin centre could have generated significant amounts of oil and gas.

In a study of samples from the Northern Perth Basin, Australia, Thomas (1979) assessed the organic content by pyrolysis and total organic carbon. The organic matter type was determined by the atomic H/C ratio of the palynologic extracts. Maturity was assessed by the use of vitrinite reflectance and spore discoloration. The organic matter type was shown to be mainly humic and therefore gas prone, so the Northern Perth Basin is mainly a gas province. This assessment agrees with the discovered hydrocarbon accumulations which are mainly gas condensate with secondary occurrences of light paraffinic oils. These oils would be expected from the known degradation products of terrestrial kerogen at a low stage of thermal maturation.

Williams (1974) used a variety of techniques including carbon isotope ratios, gas chromatography, optical rotation and infrared spectroscopy to characterize 184 oils from the Williston Basin (USA) and relate them to their source rocks. These analyses allowed Williams to group the oils into three types and to assign a source rock to each group.

Carbon isotope ratios distinguished types I and II, but could not differentiate type II from type III. Optical rotation data, when combined with carbon isotope ratios, could also separate type I from types II and III and at the same time indicated which oils were the more mature, as their optical rotation decreased with increasing maturity. Gas chromatographic analysis results of the C_4–C_7 hydrocarbons were plotted on a triangular diagram in terms of straight-chain, branched-chain, and alicyclic compounds. This plot separates the three distinct oil types. Likewise, a clear distinction of the three Williston Basin oil types is also provided by the heavy paraffin distributions. The normalized percentage of the straight-chain paraffins from C_{15} to C_{30} are plotted and the distribution pattern for each oil are completely different, thus allowing oil characterization.

Williams obtained oil–source rock correlations by comparing analyses carried out on the oils with those relating to the extracts of bitumen from rock samples. Optical rotation could not be measured in the case of the extracts because of the very small amounts of bitumen. Both carbon isotope data and the heavy-paraffin distribution data provided evidence for relating each of the three oil types to a different source rock.

REFERENCES

Alexander, R., Kagi, R. J., & Woodhouse, G. W. (1981) Geochemical investigation of Windalia oil and extracts of the Winning Group (Cretaceous) potential source rocks, Barrow sub basin, Western Australia. *BAAPG*, **65**(2), 235.

Bailey, N. J. L., Evans, C. R., & Rogers, M. A. (1971) A plea for inclusion of basic sample information when reporting geochemical analyses of crude oils. *Geochim. Cosmochim. Acta*, **35**, 632.

Bailey, N. J. L., Evans, C. R., & Milner, C. W. D. (1974) Applied petroleum geochemistry to the search for oil. *BAAPG*, **58**, 2284.

Baker, C. (1974) Pyrolysis techniques for source rocks. *BAAPG*, **58**, 2349.

Baker, D. R. (1962) Organic geochemistry of the Cherokee group in south eastern Kansas and north eastern Oklahoma. *BAAPG*, **46**(9), 1621.

Bass, N. W. (1963) Composition of crude oils in N. W. Colorado and N. E. Utah. *BAAPG*, **47**, 2039.

Bernard, B. B., Brooks, J. M., & Sackett, W. M. (1976) Natural gas seepage in the Gulf of Mexico. *Earth Planet. Sci. Lett.*, **31**, 48.

Brooks, J., & Thusu, B. (1977) Oil–source rock identification and characterization of the Jurassic sediments in the northern North Sea. *Chem. Geol.*, **20**, 283.

Cassou, A., Connan, J., & Porthauld, B. (1977) Relation between maturation of organic matter and geothermal effect as exemplied in Canada East Coast offshore weels. *Can. Pet. Geol. Bull.*, **25**, 174.

Chukwemeka, W. E., & Okoye, M. D. (1980) Petroleum reservoir–source bed evaluation of Tertiary Niger Delta. *BAAPG*, **64**(8), 1251.

Claypool, G. E., Love, A. H., & Maugham, E. K. (1978) Organic geochemistry incipient metamorphism and oil generation in Black Shale members of the phosphoria formation, Western Interior, USA. *BAAPG*, **62**, 98.

Claypool, G. E., & Reed, P. R. (1976) Thermal analysis techniques for source rock evaluation. *BAAPG*, **60**(4), 608.

Clementz, D. M., Demaison, G. J., & Daly, A. R. (1979) Well site geochemistry by programmed pyrolysis. *Offshore Technology Conference, Houston*. Paper No. OTC 3410.

Coleman, D. D., Risatti, J. B., & Schoell, M. (1981) Fractionation of carbon and hydrogen isotopes by methane oxydizing bacteria. *Geochim. & Cosmochim. Acta*, **45**, 1033.

Connan, J. (1974) Time–temperature relations in oil genesis. *BAAPG*, **58**, 2516.

Demaison, G. J., & Moore, G. T. (1980) Anoxic environments and oil source bed genesis. *BAAPG*, **64**(8), 1179.

Deroo, G., Powell, T. G., Tissot, B., McCrossan, R. G., & Hacquebard, P. A. (1977) The origin and migration of petroleum in the Western Canadian sedimentary basin, Alberta. A geochemical and thermal maturation study. *Geological Survey of Canada Bull. 262*. Ottawa: Geological Survey of Canada.

Dow, W. G. (1974) Application of oil-correlation and source rock data to exploration in the Williston Basin. *BAAPG*, **58**, 1253.

Dow, W. G. (1977) Kerogen studies and geological interpretations. *J. Geochem. Expl.*, **7**, 79.

Erdman, J. G., & Morris, D. A. (1974) Geochemical correlation of Petroleum. *BAAPG*, **58**, 1253.

Espitalié, J., Madec, M., & Tissot, B. (1980) Role of mineral matrix in kerogen pyrolysis. Influence on petroleum generation and migration. *BAAPG*, **64**(1), 59.

Feux, A. N. (1977) The use of stable carbon isotopes in hydrocarbon exploration. *J. Geochem. Expl.*, **7**, 155.

Gransch, J. A., & Eisma, E. (1966) Characterization of insoluble organic matter of sediments by pyrolysis. In: Hobson, G. D., & Spears G. C. (eds), *Advances in Organic Geochemistry*. Pergamon Press, Oxford.

Hartman, B., & Hammond, D. E. (1981) The use of carbon and sulphur isotopes as correlation parameters for the source identification of beach tar in the Southern California borderland. *Geochim. & Cosmochim. Acta*, **45**, 309.

Harwood, R. J., (1977) Oil and gas generation by the laboratory pyrolysis of kerogen. *BAAPG*, **61**, 2082.

Hunt, J. M. (1972) Distribution of carbon in the crust of the earth. *BAAPG*, **56**, 2273.

Kinghorn, R. R. F., & Rahman, M. (1980) The density separation of different maceral groups of organic matter dispersed in rocks. *J. Pet. Geol.*, **2**(4), 499.

Koons, C. B., Bond, J. G., & Peirce, F. L. (1974) Effects of depositional environment and post depositional history on chemical composition of Lower Tuscaloosa oils. *BAAPG*, **58**(7), 1272.

410

Laplante, R. E. (1974) Hydrocarbon generation in Gulf Coast Tertiary sediments. *BAAPG*, **58**, 1281.

Leythaeuser, D., Hollerbach, A., & Hogemann, H. W. (1977) Source rock/crude oil correlations based on the tribution of C_{27+} cyclic hydrocarbons. In R. Campos and J. Goñi (eds), *Advances in Organic Geochemistry*. Madrid, Empresa Nacional Adaso de Investigaciones Mineras, pp. 3–20.

Lopatin, N. V. (1971) Temperature and geological time as factors of carbonization. *Akad. Nauk. SSSR Ivz. Ser. Geol.* No. 3. p. 95.

McIver, R. D. (1967) Composition of kerogen—Clue to its role in the origin of petroleum. *7th World Petroleum Congress Proc., Mexico City.* Vol. 2. London, Elsevier, p. 26.

Magoon, L. B., & Claypool, G. E. (1981) Two types of oil on the North Slope of Alaska—Implications for exploration. *BAAPG*, **65**(4), 629.

Magoon, L. B., & Claypool, G. E. (1981) Petroleum geology of the Cook Inlet Basin—An exploration model. *BAAPG*, **65**(6), 1043.

Merewether, E. A., & Claypool, G. E. (1980) Organic composition of some Upper Cretaceous shale, Powder River Basin, Wyoming. *BAAPG*, **64**(4), 107.

Nakayama, K., & Van Siclen, D. C. (1981) Simulation model for petroleum exploration. *BAAPG*, **65**(7), 1230.

Nishimura, M., (1982) 5 β-isomers of stanols and stanones as potential markers of sedimentary organic quality and depositional palaeoenvironments. *Geochim. & Cosmochim. Acta*, **46**, 423.

Northfield, D. W., DeNiro, M. J., & Epstein, S. (1981) Hydrogen and carbon isotopic ratios of the cellulose nitrate and saponifiable lipid fractions prepared from annual growth rings of a California redwood. *Geochim. & Cosmochim. Acta*, **45**, 1895.

Parrish, J. T. (1982) Upwelling and petroleum source beds, with reference to the Paleozoic. *BAAPG*, **66**, 750.

Philippi, G. T. (1965) On depth time and mechanism of petroleum generation. *Geochim. Cosmochim. Acta*, **29**, 1021.

Philippi, G. T. (1981) Correlation of crude oils with their source formation, using high resolution GLC C_6–C_7 component analyses. *Geochim. & Cosmochim. Acta*, **45**, 1495.

Powell, T. G., & McKirdy, D. M. (1975) The effect of source material, rock type and diagenesis on the *n*-alkane content of sediments. *Geochim. Cosmochim. Acta*, **37**, 623.

Pym, J. G., Ray, J. E., Smith, G. W., & Whitehead, E. C. (1975) Petroleum triterpane fingerprinting of crude oils. *Anal. Chem.*, **47**(9), 1617.

Redding, C. E., Schoell, M., Monin, J. C., & Burand, B. (1979) Hydrogen and carbon isotope composition of coals and kerogens. In: A. G. Douglas and J. R. Maxwell (eds), *Advances in Organic Geochemistry*. Pergamon Press.

Ross, L. M. (1980) Geochemical correlation of San Juan Basin oils — a study. *Oil & Gas J.*, **78**(44), 102.

Rulkötter, J., & Welte, D. H. (1979) Oil–oil and oil–condensate correlation by low cV GC–MS measurements of aromatic hydrocarbons. In: A. G. Douglas and J. R. Maxwell ((eds), *Advances in Organic Geochemistry*. Pergamon Press.

Rulkötter, J., & Wendisch, D., (1982) Microbial alteration of 17 α(H)-hopanes in Madagascar asphalts: removal of C-10 methyl group and ring opening. *Geochim. & Cosmochim. Acta*, **46**, 1545.

Sackett, W. M. (1977) Use of hydrocarbon sniffing in offshore exploration. *J. Geochem. Expl.*, **7**, 243.

Schoell, M., & Milner, C. W. D. (1979) Isotopic crude oil correlation. *9th International Meeting on Organic Geochemistry Abstracts*.

Seifert, W. K. (1977) Source rock–oil correlations by C_{27}–C_{30} biological marker hydrocarbons. In: R. Campos and J. Goñi (eds), *Advances in Organic Geochemistry*. Madrid. Empresa Nacional Adaso de Investigaciones Mineras. pp. 21–44.

Seifert, W. K. (1978) Steranes and terpanes in kerogen pyrolysis for correlation of oils and source rocks. *Geochim. Cosmochim. Acta*, **42**(5), 473.

Seifert, W. K., & Moldowan, J. M. (1981) Palaeoreconstruction by biological markers. *Geochim. & Cosmochim. Acta*, **45**, 783.

Shiboaka, M., Saxby, J. B., & Taylor, G. H. (1978) Hydrocarbons generation in Gippsland Basin, Australia. *BAAPG*, **62**, 1151.

Silverman, S. R. (1964) Investigation of petroleum origin and evolution mechanism by carbon isotope studies. In: H. Craig *et al.* (eds), *Isotopic and Cosmic Chemistry*. Amsterdam, North Holland Publishing Co., pp. 92–102.

Stahl, W. (1974) $^{13}C/^{12}C$ ratio of North German natural gases. *Erdol und Kohle*, **27**, 623.

Stahl, W. (1975a) Carbon isotope ratios of natural gases. *Erdol und Kohle*, **28**, 188.

Stahl, W. (1975b) Source rock–crude oil correlation by isotopic type curves. *Geochim. Cosmochim. Acta*, **42**, 1573.

Staplin, F. L. (1969) Sedimentary organic matter, organic metamorphism, and oil and gas occurrence. *Can. Pet. Geol. Bull.*, **17**(1), 47.

Sutton, C. (1977) Depositional environments and their relation to chemical composition of Java Sea crudes: Paper presented at ASCOPE/CCOP seminar on Generation and Maturation of Hydrocarbons in Sedimentary Basins, Manila.

Thode, H. G. (1981) Sulphur isotope ratios in petroleum research and exploration: Williston Basin. *BAAPG*, **65**(9), 1527.

Thomas, B. M. (1979) Geochemical analyses of hydrocarbon occurrences in the Northern Perth Basin, Australia. *BAAPG*, **63**(7), 1092.

Tissot, B., Durand, B., Espitalié, J., & Combaz, A. (1974). Influence of nature and diagenesis of organic matter in formation of petroleum. *BAAPG*, **58**(3), 499.

Waples, D. W. (1979) Simple method for source rock evaluation. *BAAPG*, 63, 239.

Waples, D. W. (1980) Time and temperature in petroleum formation. Application of Lopatin's method to petroleum generation. *BAAPG*, **64**(6), 916.

Welte, D. H., Hagemann, H. W., Hollerbach, A., Leythaeuser, D., & Stahl, W. (1975). Correlation between petroleum and source rock. *Proc. 9th World Pet. Congr.*, **2**, 179.

Welte, D. H., & Yukler, M. A. (1981) Petroleum origin and accumulation in basin evolution. A quantitative model. *BAAPG*, **65**(8), 1387.

Williams, J. A. (1974) Characterization of oil types in the Williston Basin. *BAAPG*, **58**(7), 1243.

Yeh, H-W., & Epstein, S. (1981) Hydrogen and carbon isotopes of petroleum and related organic matter. *Geochim. & Cosmochim. Acta*, **45**, 753.

Men will forgive a man anything except bad prose.
W. S. Churchill
(Election Speech, Manchester 1906)

INDEX

413